AF774325

Analog Circuits and Signal Processing

Series Editors

Mohammed Ismail, Wayne State University, Detroit, MI, USA

Mohamad Sawan, School of Engineering, Westlake University, Hangzhou, Zhejiang, China

The *Analog Circuits and Signal Processing* book series, formerly known as the *Kluwer International Series in Engineering and Computer Science*, is a high level academic and professional series publishing research on the design and applications of analog integrated circuits and signal processing circuits and systems. Typically per year we publish between 5-15 research monographs, professional books, handbooks, and edited volumes with worldwide distribution to engineers, researchers, educators, and libraries. The book series promotes and expedites the dissemination of new research results and tutorial views in the analog field. There is an exciting and large volume of research activity in the field worldwide. Researchers are striving to bridge the gap between classical analog work and recent advances in very large scale integration (VLSI) technologies with improved analog capabilities. Analog VLSI has been recognized as a major technology for future information processing. Analog work is showing signs of dramatic changes with emphasis on interdisciplinary research efforts combining device/circuit/technology issues. Consequently, new design concepts, strategies and design tools are being unveiled.

Topics of interest include:

Analog Interface Circuits and Systems;
Data converters;
Active-RC, switched-capacitor and continuous-time integrated filters;
Mixed analog/digital VLSI;
Simulation and modeling, mixed-mode simulation;
Analog nonlinear and computational circuits and signal processing;
Analog Artificial Neural Networks/Artificial Intelligence;
Current-mode Signal Processing;
Computer-Aided Design (CAD) tools;
Analog Design in emerging technologies (Scalable CMOS, BiCMOS, GaAs, heterojunction and floating gate technologies, etc.);
Analog Design for Test;
Integrated sensors and actuators;
Analog Design Automation/Knowledge-based Systems;
Analog VLSI cell libraries;
Analog product development;
RF Front ends, Wireless communications and Microwave Circuits;
Analog behavioral modeling, Analog HDL.

Xu Wang · Michael Peter Kennedy

Jitter and Spur Minimization in Fractional-*N* Digital Frequency Synthesizers

Modeling, Simulation, Analysis, and Design Methodologies

Springer

Xu Wang
University College Dublin
Dublin, Ireland

Michael Peter Kennedy
University College Dublin
Dublin, Ireland

ISSN 1872-082X ISSN 2197-1854 (electronic)
Analog Circuits and Signal Processing
ISBN 978-3-032-27185-3 ISBN 978-3-032-27186-0 (eBook)
https://doi.org/10.1007/978-3-032-27186-0

© The Editor(s) (if applicable) and The Author(s), under exclusive license to Springer Nature Switzerland AG 2026

This work is subject to copyright. All rights are solely and exclusively licensed by the Publisher, whether the whole or part of the material is concerned, specifically the rights of translation, reprinting, reuse of illustrations, recitation, broadcasting, reproduction on microfilms or in any other physical way, and transmission or information storage and retrieval, electronic adaptation, computer software, or by similar or dissimilar methodology now known or hereafter developed.
The use of general descriptive names, registered names, trademarks, service marks, etc. in this publication does not imply, even in the absence of a specific statement, that such names are exempt from the relevant protective laws and regulations and therefore free for general use.
The publisher, the authors and the editors are safe to assume that the advice and information in this book are believed to be true and accurate at the date of publication. Neither the publisher nor the authors or the editors give a warranty, expressed or implied, with respect to the material contained herein or for any errors or omissions that may have been made. The publisher remains neutral with regard to jurisdictional claims in published maps and institutional affiliations.

This Springer imprint is published by the registered company Springer Nature Switzerland AG
The registered company address is: Gewerbestrasse 11, 6330 Cham, Switzerland

If disposing of this product, please recycle the paper.

To Han and Rossana

Preface

Fractional-N digital phase-locked loops (DPLLs) have become a cornerstone of modern frequency synthesis in communication, radar, and high-performance clocking systems. As technology scales and integration deepens, spectral purity requirements tighten, while architectures become increasingly nonlinear and digitally intensive. In this context, jitter, phase noise (PN), and spurious tones are no longer secondary performance metrics—they define the feasibility of the overall system.

Despite the maturity of PLL theory, important gaps remain between circuit-level intuition, modeling of nonlinear mechanisms, and system-level linear and nonlinear predictions. In particular, commonly adopted linearized models often rely on implicit assumptions regarding quantization behavior and steady-state signal statistics, assumptions whose validity may change dramatically between integer-N and fractional-N operation. As a result, optimization strategies are sometimes heuristic, and the limits of achievable jitter and spur suppression are not always clearly articulated.

This book develops a statistically consistent and practically verifiable framework for understanding and minimizing jitter and spurs in advanced DPLLs. The analysis emphasizes rigor: linearization is derived from minimum-mean-square-error principles, quantization effects are treated in accordance with the actual probability distribution of steady-state signals, and nonlinear spectral artifacts are traced back to identifiable physical and mathematical origins. The resulting models provide accurate predictions across integer-N and fractional-N modes, with and without digital-to-time converters (DTCs), and establish clear validity limits for simplified representations.

At the same time, the methodology is intentionally universal. Although the discussion centers on fractional-N DPLLs, the principles of statistical linearization, quantization modeling, and nonlinearity-induced spur formation extend naturally to a wide class of digitally intensive mixed-signal systems.

Practicality is equally central. All major theoretical developments are verified using multi-rate discrete-time (MRDT) behavioral simulations. Companion MATLAB source codes are provided to enable readers to construct behavioral models, reproduce analytical results, and perform systematic jitter/PN analysis and minimization of advanced fractional-N DPLLs. These executable tools facilitate reproducible research, parameter exploration, and direct translation of theoretical insight into practical design decisions, making the book both an analytical reference and a hands-on design companion.

The book is organized as follows:

- Chapter 2 introduces an MRDT behavioral modeling methodology for DPLLs and defines consistent simulation-based jitter and PN extraction procedures.
- Chapter 3 establishes a unified and statistically rigorous linearized analysis framework for mid-rise time-to-digital converters (TDCs), valid in both integer-N and fractional-N modes.
- Chapter 4 derives closed-form system-level predictions for jitter and phase noise in DPLLs using the refined TDC models.
- Chapter 5 formulates a jitter-minimization-oriented design methodology for integer-N DPLLs and clarifies the theoretical limits of achievable performance.
- Chapter 6 develops a comprehensive nonideal DTC model capturing both soft and hard nonlinearities and explains their distinct spectral signatures.
- Chapter 7 presents input-dithered quantization techniques that eliminate DTC-quantization-induced spurs at their source with minimal phase noise penalty.
- Chapter 8 integrates DTC enhancement and system-level optimization strategies to approach fractional-spur-free performance with minimum hardware overhead.
- Chapter 9 summarizes the main insights uncovered in this book.

This book is intended for graduate students, researchers, and practicing engineers engaged in frequency synthesis and mixed-signal integrated circuit design. By combining analytical rigor, conceptual clarity, and executable modeling tools, it aims to provide a coherent foundation for the systematic minimization of jitter and spurs in modern digital frequency synthesizers.

Dublin, Ireland
February 2026

Xu Wang
Michael Peter Kennedy

Acknowledgements

We are sincerely grateful to our partners and colleagues from both academia and industry for the inspiring discussions and valuable insights that have helped us strengthen the relevance, rigor, and overall quality of this book. These include, but are not limited to, Prof. Pietro Andreani of Lund University; Prof. Salvatore Levantino and Asst. Prof. Simone M. Dartizio of Politecnico di Milano; Asst. Prof. Teerachot Siriburanon of University College Dublin (UCD); Dr. Francesco Brandonisio of Qualcomm, Cork; and Dr. Dawei Mai of Qualcomm, California.

We also thank Dr. Xu Lu of UCD for carefully reviewing and verifying the correctness and computational efficiency of the source codes provided in this book.

This work was supported by Taighde Éireann—Research Ireland under Grant No. 20/FFP-A/8371, which provided financial support for the research leading to this publication.

Contents

Acronyms

AQE	Accumulated Quantization Error
BBPLL	Bang-Bang Phase-Locked Loop
BPD	Binary (bang-bang) Phase Detector
CDF	Cumulative Distribution Function
CP	Charge Pump
CSC	Constant-Slope Charging
DCD	Digitally Controlled Delay
DCO	Digitally Controlled Oscillator
DDSM	Digital Δ-Σ Modulator
DEM	Dynamic Element Matching
DLF	Digital Loop Filter
DPLL	Digital Phase-Locked Loop
DQI	DTC-Quantization-Induced
DR	Dynamic Range
DTC	Digital-to-Time Converter
EFM	Error-Feedback Modulator
ENOP	Enhanced Nonlinearity-induced nOise Performance
FCW	Frequency Control Word
FS	Full Scale
IBS	Integer Boundary Spur
IDQ	Input-Dithered Quantization
INL	Integral Nonlinearity
LMS	Least-Mean-Square
LSB	Least Significant Bit
LTI	Linear Time-Invariant
MASH	Multi-stAge noise-SHaping
MMD	Multi-modulus Divider
MMDC	Multi-modulus Divider Controller
MMSE	Minimum Mean Square Error
MRDT	Multi-Rate Discrete-Time
NTF	Noise Transfer Function

PD	Phase Detector
PDF	Probability Density Function
PDS	Probability Density-Shaping
PFD	Phase-Frequency Detector
PLL	Phase-Locked Loop
PMF	Probability Mass Function
PMR	Probability Mass Redistribution
PN	Phase Noise
PNN	Periodic Nonlinearity Noise
PSD	Power Spectral Density
PVT	Process-Voltage-Temperature
QE	Quantization Error
RCVS	Reverse-Concavity Variable-Slope
REM	Random Element Mismatch
RMS	Root-Mean-Square
RNG	Random Number Generator
RV	Random Variable
SR	Stochastic Resonance
TDC	Time-to-Digital Converter
VCO	Voltage-Controlled Oscillator

Chapter 1
Introduction

1.1 Frequency Synthesis via Phase-Lock Techniques

Phase-locked loop (PLL)-based frequency synthesizers, such as the traditional charge pump (CP) PLL shown in Fig. 1.1, are used in modern clocking, communication, radar, and instrumentation systems to synthesize high frequencies, e.g., radio frequencies in wireless communications, with reference to a relatively low-frequency source (typically generated by a crystal oscillator) that has a stable oscillation frequency and high spectral purity. This is achieved by controlling the output frequency of a voltage-controlled oscillator (VCO) in a negative feedback manner that locks the phase of the divided output signal, div(t), to the reference signal, ref(t). When the feedback divide ratio, N, is a constant integer, the synthesized output frequency of out(t) becomes an integer multiple of the reference one after a successful locking transient, as in

$$F_{\text{out}} = N \cdot F_{\text{ref}}. \tag{1.1}$$

This is called the "integer-N" mode of operation of the PLL. Note that, due to the integer nature of the divide ratio, the frequency resolution of integer-N frequency synthesis is limited to the reference frequency.

To improve the resolution, PLLs can be operated in the so-called "fractional-N" mode where the synthesized output frequency is a non-integer (fractional) multiple of the reference one. Fractional-N frequency synthesis is typically realized by using a multi-modulus divider (MMD) in the feedback path and using an MMD controller (MMDC) based on a Multi-stAge noise-SHaping (MASH) digital Δ-Σ modulator (DDSM) to modulate the instantaneous divide ratio of the MMD at every reference clock cycle [1]. In this case, the MMDC with modulus M is controlled by the (integer) frequency control word (FCW), X, to produce an integer-valued signal, $y[k]$, whose

© The Author(s), under exclusive license to Springer Nature Switzerland AG 2026
X. Wang and M. P. Kennedy, *Jitter and Spur Minimization in Fractional-N Digital Frequency Synthesizers*, Analog Circuits and Signal Processing,
https://doi.org/10.1007/978-3-032-27186-0_1

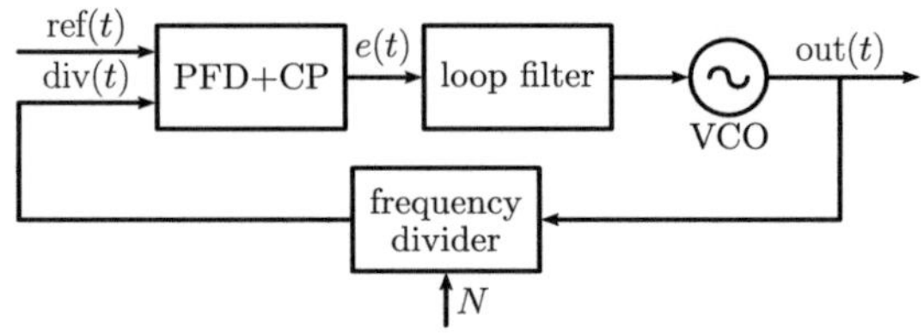

Fig. 1.1 Integer-N analog PLL system model. Phase detection is performed by a phase-frequency detector (PFD)

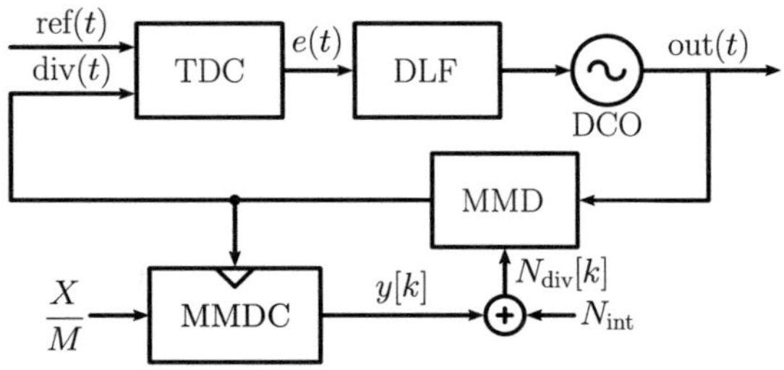

Fig. 1.2 Fractional-N DPLL-system model

average corresponds to the fraction, X/M. The instantaneous divide ratio at the kth reference cycle is thus obtained as

$$N_{\text{div}}[k] = N_{\text{int}} + y[k], \tag{1.2}$$

where N_{int} is the integer part of the nominal divide ratio, N. This fractional-N division is shown in the lower (feedback) branch of Fig. 1.2. Note that the DDSM guarantees that the average value of its output is equal to its input; hence

$$\lim_{k\to\infty} \frac{1}{k} \cdot \sum_{i=1}^{k} y[i] = \frac{X}{M}.$$

When substituted into (1.2), this yields that, on average, the nominal fractional-N divide ratio becomes

$$N = N_{\text{int}} + \frac{X}{M}, \tag{1.3}$$

which implements the desired overall fractional-N frequency synthesis.

Modulation of the divide ratio introduces a quantization error (QE) from the MMDC, which is difference between the instantaneous and nominal divide ratio in (1.2) and (1.3), as in

$$q_{\text{MMDC}}[k] = N_{\text{div}}[k] - N = y[k] - \frac{X}{M}.$$

We will see in the later chapters that this quantization error signal plays a key role in the production of fractional spurs.

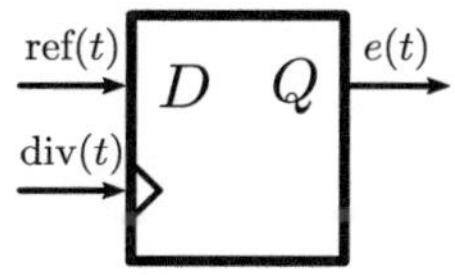

Fig. 1.3 BPD based on a D flip-flop that serves as a 1-bit TDC

Hsu et al. [2] first proposed the architecture of a digital PLL (DPLL) that facilitates $\Delta\Sigma$-based fractional-N operation [3]. The DPLL, as shown in Fig. 1.2, incorporates a multi-bit time-to-digital converter (TDC) rather than a PFD and a CP as the phase detector (PD). This generates a digital phase error signal, $e(t)$, which is passed through a digital loop filter (DLF) and applied to a digitally controlled oscillator (DCO). DPLLs can outperform conventional analog PLLs in Fig. 1.1 that use a PFD followed by a CP and an analog loop filter [4] in terms of design simplicity and consumption of chip power and area, thanks to the full digitalization of the PD and loop filter. Hence, the DPLL has become an attractive design option to serve as a frequency synthesizer that can be integrated into advanced systems-on-chip.

Multi-bit TDCs tend to consume a large area and are power hungry [5, 6]. They also introduce nonlinearity. A major breakthrough for advanced DPLLs has been the adoption of a 1-bit binary (bang-bang) PD (BPD) based on a D flip-flop [7], as shown in Fig. 1.3. This minimizes the complexity and power/area consumption of the TDC and yields an excellent trade-off between the noise performance and the power consumption of the bang-bang PLL (BBPLL) at the system level [8–10].

1.2 Synthesized Phase Noise (PN), Jitter, and Spurs

Ideally, in the steady state, when the loop is successfully locked and the output is stabilized at the desired frequency, one would expect the rising edges of the synthesized high-frequency signal, out(t), to appear at exact integer multiples of T_{out},

$$t_{\text{out}}^{\text{ideal}}[h] = h \cdot T_{\text{out}}, \quad \forall h \in \mathbb{N}, \tag{1.4}$$

where

$$T_{\text{out}} \triangleq \frac{1}{F_{\text{out}}} = \frac{T_{\text{ref}}}{N}$$

represents the nominal synthesized output period by recalling the relationship in Eq. (1.1). However, due to the inevitable noise sources and nonlinearities, this is not satisfied exactly in practice. In the time domain, PLL output noise manifests itself as the random "jittery" excursions in the phases of the synthesized pulses around their desired positions described by (1.4). The key metrics that quantify this phase fluctuation are

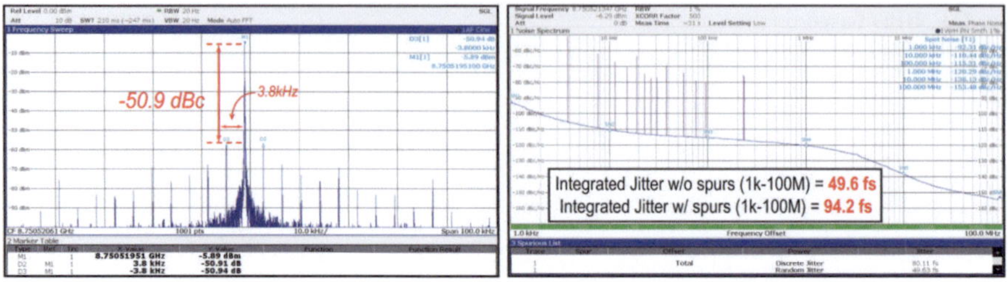

(a) Output spectrum. (b) Output PN spectrum.

Fig. 1.4 Example measured PLL output spectra as shown in [12, Figs. 17a and 19a]. The fractional spurs are at the fundamental and harmonic of 3.8 kHz offset from the 8.75 GHz synthesized center (carrier) frequency

- the root-mean-square (RMS) value of the random phase jitter, which is typically measured in femtoseconds;
- the power spectral density (PSD) of the PN spectrum in dBc/Hz; and
- the undesired spurious tones (or "spurs"), which is also referred to as "deterministic jitter," that are present in the vicinity of the nominal output (carrier) frequency. The powers of the spurs at offset frequencies from the carrier are measured in dBc.

Note that, quantitatively, the jitter power can be calculated as the total power of the spurs added to the PN power and integrated over a specified portion of the spectrum [11]; this is also referred to as the integrated phase noise. Figure 1.4 shows the measured spectrum of the synthesized signal of an example fractional-N DPLL described in [12]. The output spectrum in (a) exhibits spurs at the fundamental and harmonics of 3.8 kHz that are offset from the synthesized carrier frequency at 8.75 GHz; the worst-case fractional spur at 3.8 kHz offset has magnitude equal to –50.9 dBc. These appear also in the output PN spectrum in (b) together with a "smooth" PN level, the in-band component of which is higher than -120 dBc/Hz. The RMS jitter is measured by integrating the area underneath the PN curve from 1 kHz to 100 MHz, which is 49.6 fs and 94.2 fs excluding or including the power associated with the fractional spurs.

In terms of the output PN, DPLLs operating in the integer-N mode have, as their most significant PN contributors, the intrinsic oscillator jitters associated with the reference oscillator and DCO [13, 14]. In addition, the "mathematical" noise generated by the input PD, i.e., the quantization noise of the TDC in DPLLs, also contributes prominently to the system PN [15, 16]. In fractional-N mode, apart from the three noise sources mentioned above, the QE arising from the quantization operation of the MMDC introduced in Sect. 1.1 also gives an important contribution to the overall PN.

In fractional-N PLLs, spurious tones, such as integer-boundary spurs (IBS) [17], typically appear due to the interaction of the MMDC's QE with the inherent nonlinearities associated with the building blocks within the loop [18, 19], e.g., the nonlinearity of the PD.[1]

The PN, jitter, and spur performances of a frequency synthesizer determine the purity of the PLL's synthesized frequency signal. More importantly, these undesired features are also detrimental to the overall performance of the top-level systems that use the PLL. For example, when the synthesized frequency is used as the clocking signal that drives digital signal processing and/or analog-to-digital conversion, the phase jitter associated with the clock edges can cause unwanted metastability and glitches; when it is used as the carrier wave in wireless communications, the PN and spurs can cause adjacent channel leakage and interference to other users [22]. These can cause malfunction of the top-level systems. Excessive PN and spurious responses are strictly prohibited in communications by regulations and communication standards as well.

Various techniques have been proposed in the literature with the aim to understand, predict, and hopefully minimize the synthesized PN and mitigate spurs. In the next section, a brief review of the prior art related to prediction and mitigation strategies will be provided. From there, the objectives and contributions of this book will be outlined.

1.3 Prior Art in Prediction and Mitigation of PN, Jitter, and Spurs

As introduced in Sect. 1.1, the BBPLL which has as its PD a BPD, which is effectively a 1-bit TDC, is characterized by its design simplicity and minimum hardware consumption. However, the nature of the bang-bang detection performed by the BPD, whose output is limited to ± 1, exhibits two fundamental problems, namely, its excess QE power and its easily saturated dynamic range (DR). Prior to this book, much of the literature has sought methods to compensate for the adverse noise and distortion effects induced by these two limitations; the next subsections discuss representative methods in the literature.

1.3.1 Excess Quantization Error (QE) from the Binary (Bang-Bang) Phase Detector (BPD)

In integer-N mode, due to the extremely coarse resolution of the BPD, the BPD contributes more QE than high-resolution TDCs; consequently, this dominates the output PN at low offset frequencies. Integer-N BBPLL design strategies proposed in [10, 16, 23–25] are dedicated to finding the optimized loop design, e.g., the best loop

[1] This interaction also produces excess folded PN at close-in frequencies [20, 21].

bandwidth, in order to rebalance the PN contributions from the BPD and DCO, and thereby to mitigate the excess PN and jitter contributed by the BPD. However, the effectiveness of these methods is limited, because the BPD's QE at source cannot be reduced by changing the loop design *per se*. Hence, the close-in PN of the contribution from the BPD's QE is still dominant.

In contrast to the BBPLL, Seong et al. [26] proposed a new direction to mitigate BPD's PN, that is to use a TDC with one more bit, i.e., a 2-bit mid-rise TDC, in place of the BPD. As the TDC's resolution is improved in this way, the TDC-induced PN can be reduced at source. Despite the novelty, the analysis in [26] falls short in providing a comprehensive linearized analysis framework for such a few-bit TDC that derives its linearized gain and the power of the QE in the steady state, and consequently makes it impossible to derive an accurate linear system model. Therefore, a jitter-minimization-oriented design optimization strategy at the system level still needed to be discovered.

The linearized analysis for a BPD in the integer-N operating mode was previously studied in a wealth of literature, e.g., [7, 10, 24, 25, 27, 28]. However, these studies are all limited to the 1-bit TDC case. Avallone et al. [15] extended the analysis to multi-bit TDCs. In Chap. 3, it will be proven that the linearized estimates in [15] are accurate only in the integer-N mode and cannot be correct in the case of fractional-N analysis. Furthermore, it will be shown in Chap. 5 that the jitter minimization strategy proposed in [15] based on this analysis is flawed and indeed yields suboptimal loop designs. Consequently, we will describe a more accurate approach in the same chapter that leads to an optimal design.

1.3.2 BPD Dynamic Range Saturation

The application of the BBPLL architecture had long been limited to integer-N synthesis. This was because, when using a DDSM-based MMDC to implement fractional-N operation, the injection of the MMDC's QE, which has significantly greater magnitude than the integer-N intrinsic noises, will overwhelm and saturate the DR of the BPD. This in turn drastically affects the loop dynamics and deteriorates the noise performance of the BBPLL. This operating mode, namely, the fractional-N BBPLL without cancellation of the MMDC's QE, yields its worst-case PN/jitter performance limit. The phenomenon of PN/jitter degradation was briefly studied by Xu and Abidi in [28]. However, as will be explained in Chap. 4, the PN/jitter prediction provided in that analysis was only partially correct and lacks accuracy.

Conventionally, in CP-PLLs, e.g., [29, 30], and some DPLLs, e.g., [2], the MMDC's QE can be mitigated by adopting higher-order lowpass loop filters and/or a "digiphase" scheme that matches and subtracts the QE *after* the PD. However, such *a posteriori* approaches are inadequate in fractional-N BBPLLs (and other DPLLs using few-bit TDCs) due to the fact that the output resolution is constrained by the coarse detection of the TDC.

1.3.3 Issues with Digital-to-Time Converter (DTC)-Based QE Cancellation

An *a priori* QE cancellation method for fractional-N DPLLs was first proposed by Zanuso et al. in [31] who introduced a phase interpolator, which is now commonly referred to as a "digital-to-time converter (DTC)," *before* the TDC [3, 32]. As shown in Fig. 1.5, the DTC is digitally controlled by a code representing the MMDC's QE to produce in the time domain a phase delay that compensates for it. In this way, the concerned QE can, in principle, be canceled before its injection into the loop. A DTC-enhanced fractional-N BBPLL was later proposed by Tasca *et al.* in [33]. It will be explained in Chap. 4 why the use of a DTC is necessary in fractional-N DPLLs, such as BBPLLs, that have a TDC with a narrow linear detection region.

Notice that, for advanced digital synthesizers using BPDs and DTCs, the primary nonlinearity concerns are transferred from the TDC to the DTC [34]. Unfortunately, as will be shown in Chap. 6, the DTC's nonidealities: including

1. the *soft (polynomial) integral nonlinearity* (INL) of the DTC's delay line;
2. the inherent *hard quantization nonlinearity* due to the limited resolution of the DTC; and
3. the *random discontinuous nonlinearity* due to element mismatch in the DTC's delay cells;

and their interactions with the MMDC's QE all have detrimental impacts on the system's PN and spur performance.

In terms of the prior art to enhance the DTC's performance, the literature before this book has typically concentrated on analyzing the *soft* INL of the DTC and improving the noise performance related to it. For example, state-of-the-art DTC linearity enhancement techniques, such as adaptive predistortion [32, 35, 36], DTC range reduction [37, 38], DTC second/third-order nonlinearity cancellation [39, 40], and the inverse-constant-slope DTC [41], typically try to compensate for the polynomial INL of the DTC's delay line. Furthermore, "statistical methods" used in

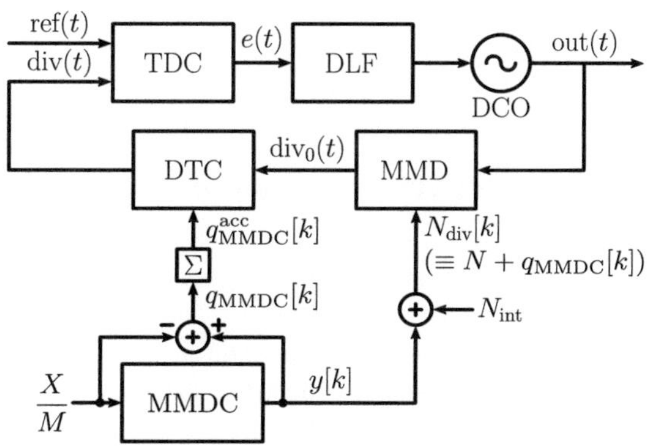

Fig. 1.5 DTC-enhanced fractional-N DPLL-system model

the MMDC, such as probability mass redistribution (PMR) [42], time-invariant-probability modulator [43], probability density-shaping (PDS) $\Delta\Sigma$M [37, 38], and the Enhanced Nonlinearity-induced nOise Performance (ENOP) DDSM [44], improve the performance in relation to fractional spurs caused by the interaction between the MMDC's QE and a soft nonlinearity, e.g., the DTC's approximately polynomial INL. Unfortunately, these methods do not alter the inherent quantization characteristics and random element mismatch associated with the DTC. Hence, they do not help to mitigate the noise/spur responses induced by them.

The literature thus far often overlooked the DTC's hard quantization nonlinearity and its negative impacts. Some works, such as Hu et al. [45], claimed that the DTC's quantization will not produce any spurious tones once a MASH MMDC is used. Others, such as [33, 36], suggest that, if the DTC is sized so that the residual error after DTC-based cancellation is suppressed below the level of the intrinsic noise sources within the loop, random noise will scramble the DTC-related spurious tones so that they will not appear in the DPLL's output spectrum [24, 46]. By contrast, it will be shown in Chap. 6 that, if not eliminated *at source*, the DTC-quantization-induced spurs can be transferred directly to the system's output, which in turn has an impact on the overall spectral purity of the fractional-N DPLL; this invalidates the aforementioned claims.

In conclusion, a better DTC model that takes into account these sources of nonlinearity and that can be used to explain the peculiar PN/spur responses produced by nonideal DTCs is required.

Finally, comprehensive design methodologies at both DTC-block and DPLL-system levels that are oriented to mitigate (and hopefully eliminate) the spurs and minimize the overall system jitter are desired. While approaching the goal of spur/jitter minimization, two implicit requirements for such methodologies must be satisfied; these are

1. when the solution needs to use extra hardware and power in order to achieve the desired performance enhancement, the minimum hardware requirement that guarantees its effectiveness must be found; the proposed solution should not exceed this level—the "good enough" principle;
2. when dithering methods, i.e., proactively applying a noise-like signal to randomize and cancel the spurs, are needed, the PN penalty must be bounded so that it has minimal effect on the overall jitter.

1.4 Objectives and Contributions of the Book

This book offers significant insights and contributions in both theory and practice. In the theoretical aspect, this book has three major objectives.

The first objective is to present a comprehensive linearized analysis framework that can be used to derive accurate estimates of the linearized gain and power of the QE of mid-rise TDCs, including the BPD. The proposed method leads to accurate linearized models for the TDCs operating in both integer-N and fractional-N modes, which take into account the interaction between the TDC's transfer characteristic and different statistical signatures of its steady-state input signals in different modes. The derived estimates conform to the minimum-mean-square-error (MMSE) estimation criterion; their accuracy is verified with behavioral simulations.

The second objective is to provide accurate system PN and jitter predictions for advanced DPLLs incorporating such TDCs when operating in integer-N and fractional-N (with/without DTCs) modes. Based on the system-level models, a jitter-minimization-oriented system optimization strategy is formulated. Using this strategy, the designed DPLL will have its minimum system jitter in theory, while also occupying the minimum hardware. The predicted and optimized system PN and jitter match their behaviorally simulated counterparts.

The third objective is to investigate the behavior of practical DTCs and to present a nonideal DTC model that considers all the nonlinearities associated with the DTC subsystem. Different DTC-related PN/spur responses caused by different sources of nonlinearity are revealed and distinguished. Their individual mitigation techniques, together with a DPLL-system-level design optimization methodology, are proposed; they are applied comprehensively in order to achieve an overall jitter/spur minimization for fractional-N DPLL-based frequency synthesizers.

The original contributions of this book are as follows:

- State-of-the-art TDC linearized analysis frameworks are compared and contrasted. Among them, the method based on MMSE estimation is proven to provide accurate linearized modeling of mid-rise TDCs, including the BPD, in various operation modes.
- Linearized analysis of mid-rise TDCs, taking into account the interaction between the TDC's transfer characteristic and the probability distribution of its input signal, is demonstrated. The steady-state analysis covers integer-N and fractional-N modes of operation.
- A linearized DPLL-system model and comprehensive prediction for the system jitter and PN are proposed. Based on these, a jitter-minimization-oriented design strategy is provided, which optimizes the DPLL's noise performance while using minimum hardware.
- The worst-case PN/jitter performance limits for fractional-N DPLLs are established.
- The soft polynomial and random mismatch integral nonlinearities of a DTC's delay line and the DTC's hard quantization nonlinearity are investigated. A nonideal DTC model is provided, which helps to distinguish the distinct PN and spurious-tone spectral signatures caused by the soft and hard nonlinearities. A comprehensive DTC design methodology is proposed that provides immunity against DTC-related spurs.
- The phenomenon of DTC-quantization-induced (DQI) spurs is explained. A novel DTC enhancement method called input-dithered quantization is presented, which eliminates the DQI spurs at source; this is compared with the DDSM-based DTC requantizer solution reported in the literature.

In terms of practical utility, all the concepts, analysis, and design methods presented in this book are verified by behavioral simulations. Companion MATLAB models and scripts are provide that are designed to facilitate accurate and efficient simulation and analysis using the concepts described in the book.

1.5 Book Outline

This book is arranged as follows.

Chapter 2 introduces a modeling and simulation approach for the DPLL system to capture the multi-rate discrete-time behavior of practical DPLLs. The acquisition methods to record the system's jitter and PN performances are discussed. These methods and the accompanying simulator are used in the book to verify the analysis and proposed novel design methods.

Chapter 3 compares and contrasts state-of-the-art linearized analysis frameworks, while taking into account the interaction between the characteristics of the TDC and the statistical properties of the locked-state input time jitter. The Δ-Σ modulation-augmented fractional-N synthesis is described here. The verified MMSE estimation method is used in linearized analysis of the TDCs in different working scenarios, e.g., integer-N and fractional-N modes.

Chapter 4 uses the TDC linear model that was derived in Chap. 3 to derive accurate estimates of jitters and PN of DPLLs with mid-rise TDCs operating in integer-N and fractional-N modes.

Chapter 5, which is based on the closed-form predictions for the system jitter and PN provided in Chap. 4, proposes a comprehensive jitter minimization strategy for integer-N DPLLs; this helps to mitigate the TDC-induced PN to achieve an output PN spectrum that is dominated by the noises contributed by the reference and DCO only.

Chapter 6 introduces and considers the nonideal behaviors of a practical DTC. DTC macromodels are used to analyze and distinguish the DTC's sources of nonideality and the distinct adverse spectral responses introduced by them. The DTC's inherent quantization is studied in detail. The phenomena of DTC-quantization-induced spurs and random element mismatch-induced spurs that were insufficiently studied in the previous literature are explored. State-of-the-art DTC segmentation methods are also compared and discussed.

Chapter 7 addresses the DTC performance-enhancement techniques related to the inherent quantization operation of the DTC. We describe a family of DQI-spur cancellation methods called input-dithered quantization, which are effective in eliminating DQI spurs at source, with a negligible PN penalty.

Chapter 8 addresses DTC enhancement related to the nonlinear delay line used as the main component within the DTC. A comprehensive design strategy incorporating different classes of techniques is presented; it mitigates the revealed excess PN and spurs incurred by the DTC's nonidealities. The resulting enhanced DTC enables the fractional-N DPLL to approach the fractional-spur-free integer-N PN performance limit with minimum hardware.

Finally, conclusions are presented in Chap. 9.

References

1. B. Miller, R. Conley, A multiple modulator fractional divider. IEEE Trans. Instrum. Meas. **40**(3), 578–583 (1991). (June)
2. C.-M. Hsu, M.Z. Straayer, M.H. Perrott, A low-noise wide-BW 3.6-GHz digital $\Delta\Sigma$ fractional-N frequency synthesizer with a noise-shaping time-to-digital converter and quantization noise cancellation. IEEE J. Solid-State Circuits **43**(12), 2776–2786 (2008). (Dec)
3. S. Levantino, C. Samori, Nonlinearity cancellation in digital PLLs (invited paper), in *Proceedings of the IEEE 2013 Custom Integrated Circuits Conference* (2013), pp. 1–8
4. X. Wang, M.P. Kennedy, A random pulse modulation approach to modeling the flicker and white noise of the charge pump of a fractional-N frequency synthesizer. Int. J. Circuit Theory Appl. **50**(4), 1049–1063 (2022)
5. ...R. Staszewski, K. Muhammad, D. Leipold, C.-M. Hung, Y.-C. Ho, J. Wallberg, C. Fernando, K. Maggio, R. Staszewski, T. Jung, J. Koh, S. John, I.Y. Deng, V. Sarda, O. Moreira-Tamayo, V. Mayega, R. Katz, O. Friedman, O. Eliezer, E. de Obaldia, P. Balsara, All-digital TX frequency synthesizer and discrete-time receiver for bluetooth radio in 130-nm CMOS. IEEE J. Solid-State Circuits **39**(12), 2278–2291 (2004). (Dec)
6. M.Z. Straayer, M.H. Perrott, A multi-path gated ring oscillator TDC with first-order noise shaping. IEEE J. Solid-State Circuits **44**(4), 1089–1098 (2009). (April)
7. N. Da Dalt, Markov chains-based derivation of the phase detector gain in bang-bang PLLs. IEEE Trans. Circuits Syst. II Express Briefs **53**(11), 1195–1199 (2006). (Nov)
8. N. Da Dalt, A design-oriented study of the nonlinear dynamics of digital bang-bang PLLs. IEEE Trans. Circuits Syst. I Regul. Pap. **52**(1), 21–31 (2005). (Jan)
9. N. Da Dalt, Linearized analysis of a digital bang-bang PLL and its validity limits applied to jitter transfer and jitter generation. IEEE Trans. Circuits Syst. I Regul. Pap. **55**(11), 3663–3675 (2008). (Dec)
10. M. Zanuso, D. Tasca, S. Levantino, A. Donadel, C. Samori, A.L. Lacaita, Noise analysis and minimization in bang-bang digital PLLs. IEEE Trans. Circuits Syst. II Express Briefs **56**(11), 835–839 (2009). (Nov)
11. I. Galton, C. Weltin-Wu, Understanding phase error and jitter: definitions, implications, simulations, and measurement. IEEE Trans. Circuits Syst. I Regul. Pap. **66**(1), 1–19 (2019). (Jan)
12. M. Rossoni, S.M. Dartizio, F. Tesolin, G. Castoro, R. Dell'Orto, A.L. Lacaita, S. Levantino, A low-jitter fractional-N digital PLL adopting a reverse-concavity variable-slope DTC. IEEE J. Solid-State Circuits **60**(6), 2122–2133 (2025). (June)
13. M.H. Perrott, M.D. Trott, C.G. Sodini, A modeling approach for Σ-Δ fractional-N frequency synthesizers allowing straightforward noise analysis. IEEE J. Solid-State Circuits **37**(8), 1028–1038 (2002). (Aug)
14. H. Arora, N. Klemmer, J.C. Morizio, P.D. Wolf, Enhanced phase noise modeling of fractional-N frequency synthesizers. IEEE Trans. Circuits Syst. I Regul. Pap. **52**(2), 379–395 (2005). (Feb)
15. L. Avallone, M.P. Kennedy, S. Karman, C. Samori, S. Levantino, Jitter minimization in digital PLLs with mid-rise TDCs. IEEE Trans. Circuits Syst. I Regul. Pap. **67**(3), 743–752 (2020). (March)
16. L. Avallone, M. Mercandelli, A. Santiccioli, M.P. Kennedy, S. Levantino, C. Samori, A comprehensive phase noise analysis of bang-bang digital PLLs. IEEE Trans. Circuits Syst. I Regul. Pap. **68**(7), 2775–2786 (2021). (July)

17. D. Banerjee, *PLL Performance, Simulation and Design* (Dog Ear Publishing, Indianapolis, IN, USA, 2006)
18. Y. Donnelly, M.P. Kennedy, Prediction of phase noise and spurs in a nonlinear fractional-*N* frequency synthesizer. IEEE Trans. Circuits Syst. I Regul. Pap. **66**(11), 4108–4121 (2019). (Nov)
19. M.P. Kennedy, V. Mazzaro, D. Mai, Nonlinearity-induced spurs in fractional-N frequency synthesizers. IEEE Trans. Circuits Syst. II Express Briefs **69**(6), 2617–2622 (2022). (June)
20. S. Levantino, L. Collamati, C. Samori, A.L. Lacaita, Folding of phase noise spectra in charge-pump phase-locked loops induced by frequency division. IEEE Trans. Circuits Syst. II Express Briefs **57**(9), 671–675 (2010)
21. V. Mazzaro, M.P. Kennedy, Folded noise prediction in nonlinear fractional-N frequency synthesizers. IEEE Trans. Circuits Syst. I Regul. Pap. **68**(10), 4038–4048 (2021). (Oct)
22. B. Razavi, *RF Microelectronics*,vol. 2 (Prentice Hall New York, 2012)
23. S. Levantino, G. Marzin, C. Samori, A.L. Lacaita, A wideband fractional-N PLL with suppressed charge-pump noise and automatic loop filter calibration. IEEE J. Solid-State Circuits **48**(10), 2419–2429 (2013). (Oct)
24. G. Marucci, S. Levantino, P. Maffezzoni, C. Samori, Exploiting stochastic resonance to enhance the performance of digital bang-bang PLLs. IEEE Trans. Circuits Syst. II Express Briefs **60**(10), 632–636 (2013). (Oct)
25. T.-K. Kuan, S.-I. Liu, A bang bang phase-locked loop using automatic loop gain control and loop latency reduction techniques. IEEE J. Solid-State Circuits **51**(4), 821–831 (2016). (April)
26. T. Seong, Y. Lee, S. Yoo, J. Choi, A 320-fs RMS jitter and – 75-dBc reference-spur ring-DCO-based digital PLL using an optimal-threshold TDC. IEEE J. Solid-State Circuits **54**(9), 2501–2512 (2019). (Sep.)
27. B. Chun, M.P. Kennedy, Statistical properties of first-order bang-bang PLL with nonzero loop delay. IEEE Trans. Circuits Syst. II Express Briefs **55**(10), 1016–1020 (2008). (Oct)
28. H. Xu, A.A. Abidi, Design methodology for phase-locked loops using binary (bang-bang) phase detectors. IEEE Trans. Circuits Syst. I Regul. Pap. **64**(7), 1637–1650 (2017). (July)
29. S. Pamarti, L. Jansson, I. Galton, A wideband 2.4-GHz delta-sigma fractional-N PLL with 1-Mb/s in-loop modulation. IEEE J. Solid-State Circuits **39**(1), 49–62 (2004). (Jan)
30. A. Swaminathan, K.J. Wang, I. Galton, A wide-bandwidth 2.4 GHz ISM band fractional-*N* PLL with adaptive phase noise cancellation. IEEE J. Solid-State Circuits **42**(12), 2639–2650 (2007). (Dec)
31. M. Zanuso, S. Levantino, C. Samori, A.L. Lacaita, A wideband 3.6 GHz digital $\Delta\Sigma$ fractional-N PLL with phase interpolation divider and digital spur cancellation. IEEE J. Solid-State Circuits **46**(3), 627–638 (2011). (March)
32. S. Levantino, G. Marzin, C. Samori, An adaptive pre-distortion technique to mitigate the DTC nonlinearity in digital PLLs. IEEE J. Solid-State Circuits **49**(8), 1762–1772 (2014). (Aug)
33. D. Tasca, M. Zanuso, G. Marzin, S. Levantino, C. Samori, A.L. Lacaita, A 2.9–4.0-GHz fractional-N digital PLL with bang-bang phase detector and 560-fs_{rms} integrated jitter at 4.5-mW power. IEEE J. Solid-State Circuits **46**(12), 2745–2758 (2011). (Dec)
34. C. Samori, L. Bertulessi, Digital PLLs: the modern timing reference for radar and communication systems, in *ESSCIRC 2021 - IEEE 47th European Solid State Circuits Conference (ESSCIRC)* (2021), pp. 21–27
35. S. Levantino, L. Grimaldi, C. Samori, Analysis of adaptive pre-distortion in DTC-based digital fractional-N PLLs, in *2016 12th Conference on Ph.D. Research in Microelectronics and Electronics (PRIME)* (2016), pp. 1–4
36. A. Mahmoud, P. Andreani, F. Pepe, A 2.8–3.8-GHz low-spur DTC-based DPLL with a class-D DCO in 65-nm CMOS. IEEE Microwave Wirel. Compon. Lett. **27**(11), 1010–1012 (2017). (Nov)
37. W. Wu, C.-W. Yao, C. Guo, P.-Y. Chiang, L. Chen, P.-K. Lau, Z. Bai, S.W. Son, T.B. Cho, A 14-nm ultra-low jitter fractional-N PLL using a DTC range reduction technique and a reconfigurable dual-core VCO. IEEE J. Solid-State Circuits **56**(12), 3756–3767 (2021). (Dec)

38. C. Hwang, H. Park, T. Seong, J. Choi, A 188fs$_{rms}$-jitter and -243dB-FoM$_{jitter}$ 5.2 GHz-ring-DCO-based fractional-N digital PLL with a 1/8 DTC-range-reduction technique using a quadruple-timing-margin phase selector, in *2022 IEEE International Solid- State Circuits Conference (ISSCC)*, vol. 65 (2022), pp. 378–380
39. H. Park, C. Hwang, T. Seong, Y. Lee, J. Choi, 32.1 a 365 fs$_{rms}$-jitter and -63dBc-fractional spur 5.3 GHz-ring-DCO-based fractional-N DPLL using a DTC second/third- order nonlinearity cancelation and a probability-density-shaping $\Delta\Sigma$M, in *2021 IEEE International Solid- State Circuits Conference (ISSCC)*, vol. 64 (2021), pp. 442–444
40. C. Hwang, H. Park, Y. Lee, T. Seong, J. Choi, A low-jitter and low-fractional-spur ring-DCO-based fractional-N digital PLL using a DTC's second-/third-order nonlinearity cancellation and a probability-density-shaping $\Delta\Sigma$M. IEEE J. Solid-State Circuits **57**(9), 2841–2855 (2022). (Sep.)
41. S.M. Dartizio, F. Tesolin, G. Castoro, F. Buccoleri, L. Lanzoni, M. Rossoni, D. Cherniak, L. Bertulessi, C. Samori, A.L. Lacaita, S. Levantino, 4.3 a 76.7 fs-Integrated-jitter and -71.9 dBc in-band fractional-spur bang-bang digital PLL based on an inverse-constant-slope DTC and FCW subtractive dithering, in *2023 IEEE International Solid-State Circuits Conference (ISSCC)* (2023), pp. 3–5
42. Y. Donnelly, M.P. Kennedy, J. Breslin, S. Tulisi, S. Patil, C. Curtin, S. Brookes, B. Shelly, P. Griffin, M. Keaveney, 4.48-GHz fractional- *N* frequency synthesizer with spurious-tone suppression via probability mass redistribution. IEEE Solid-State Circuits Lett. **2**(11), 264–267 (2019). (Nov)
43. T. Seong, Y. Lee, C. Hwang, J. Lee, H. Park, K.J. Lee, J. Choi, A -58dBc-worst-fractional-spur and -234 dB-FoM jitter, 5.5 GHz ring-DCO-based fractional-N DPLL using a time-invariant-probability modulator, generating a nonlinearity-robust DTC-control word, in *2020 IEEE International Solid- State Circuits Conference - (ISSCC)* (2020), pp. 270–272
44. V. Mazzaro, M.P. Kennedy, A family of $\Delta\Sigma$ modulators with high spur immunity and low folded nonlinearity noise when used in fractional-*N* frequency synthesizers. IEEE Trans. Circuits Syst. I Regul. Pap. **69**(11), 4499–4509 (2022). (Nov)
45. A. Hu, Z. Jin, X. Shan, M. Zhang, D. Liu, X. Zou, Fractional spurs reduction technique using probability density shaping sigma-delta modulator and fractional frequency divider. IEEE Trans. Circuits Syst. II Express Briefs **70**(2), 456–460 (2023). (Feb)
46. M.D. McDonnell, Is electrical noise useful? [point of view]. Proc. IEEE **99**(2), 242–246 (2011)

Chapter 2
Modeling Methodology for Digital Phase-Locked Loop (DPLL)

2.1 Introduction

We first introduce methods to perform transient behavioral simulations of closed-loop DPLLs. The key modeling principle is the multi-rate discrete-time (MRDT) (or phase-domain) behavioral modeling for DPLLs. Similar methods have been proposed and used in the literature such as [1–4]; however, these classical methods are typically limited to modeling DPLLs operating in the integer-N mode. In this book, novel customization of the modeling approach is proposed to extend its functionality in order to allow fractional-N simulations (with or without a DTC) that consider various noise sources and nonlinearities within the loop. The modeling approach is explained in Sect. 2.2. Behavioral models of the building blocks of the loop are described in Sect. 2.3.

The behavioral simulation is used to observe in the time domain the locking transient and system jitter of DPLLs, and to acquire in the frequency domain the spectral signatures of the steady-state system PN and spurs. Detailed performance acquisition methods are included in Sect. 2.4.

These MRDT models allow one to simulate nonideal behaviors of practical DPLLs; they help to understand better the PN/jitter performances under specific parameterization, verify the accuracy of the PN predictions made in this work, and assess the effectiveness of various jitter/spur mitigation strategies. Unless otherwise stated, the same simulation methodology is used throughout this book.

2.2 Multi-rate Discrete-Time Behavioral Modeling Principles

To explain the principle of MRDT, we first investigate the integer-N operation of DPLLs. As explained in Sect. 1.1, the integer-N DPLL controls the DCO in order

© The Author(s), under exclusive license to Springer Nature Switzerland AG 2026

X. Wang and M. P. Kennedy, *Jitter and Spur Minimization in Fractional-N Digital Frequency Synthesizers*, Analog Circuits and Signal Processing,
https://doi.org/10.1007/978-3-032-27186-0_2

to synthesize a signal, out(t), whose frequency is an integer multiple of that of the reference signal, ref(t). As a negative feedback loop, the DPLL uses a TDC to quantize the phase (or time) difference between the edges of the reference and the output signal after division by the integer ratio, N. The digital error signal, $e(t)$, that is generated by the TDC is then filtered by the DLF to control the DCO. The DCO synthesizes the desired output frequency. The frequency divider counts N edges of the synthesized signal to produce one edge in div(t), which is fed back and compared with the next edge of the reference signal. The iteration propagates in this way.

There are two key features associated with the behavior of DPLLs, namely, the multi-rate and discrete-time properties.

2.2.1 *Multi-rate Property*

Within the loop, the signals operate at two different rates. In the slow domain, the signals used to control the DCO vary at the (low) reference frequency; these include ref(t), div(t), and the outputs from the TDC and DLF. In the fast domain, the signal generated by the DCO, out(t), has the high output frequency.

2.2.2 *Discrete-Time Property*

The signals ref(t), div(t), and out(t) can be thought of as pulse trains, of which the instantaneous phases associated with their rising edges determine key performance indicators such as the input phase difference and locked-state output PN/jitter/spurs. The phase locking ability of the system can also be examined by inspecting the transient variation of the input phase difference. Therefore, one can represent these signals by exploiting the timestamps of their rising edges (or positive zero-crossing events) in order to improve the speed, while also guaranteeing the accuracy of simulation. Furthermore, in contrast to the continuous-time control over the VCO in CP-PLLs, the control path for the DCO in DPLLs is clocked at the reference rate. Like a Moore machine, the outputs from the TDC and the DLF do not change between two reference clock cycles. Hence, the signals produced by the TDC and DLF can be discrete-time sampled at the reference rate as well.

These observations result in the MRDT model for a DPLL shown in Fig. 2.1. The system is partitioned into two parts according to the corresponding signal rates. On the input side, the signals operating at the reference frequency ($F_{\mathrm{ref}} = 1/T_{\mathrm{ref}}$) are indexed with k; on the output side, the signals operating at the output frequency ($F_{\mathrm{out}} = 1/T_{\mathrm{out}}$) are indexed with h. The DPLL acts as a stable negative feedback system and guarantees that, after a successful locking transient, $F_{\mathrm{out}} = N \cdot F_{\mathrm{ref}}$.

The timestamping notation for the frequency signals is illustrated in Fig. 2.2 using the signals associated with the TDC block as an example. Recall that the TDC takes as input the reference clock and divided output signals, denoted ref(t) and div(t),

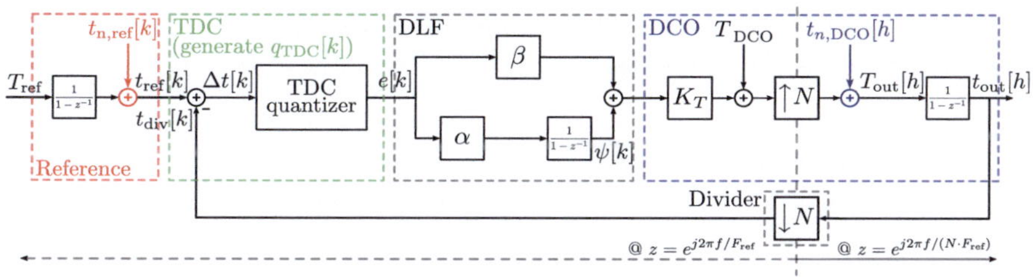

Fig. 2.1 MRDT system model of an *integer-N* DPLL

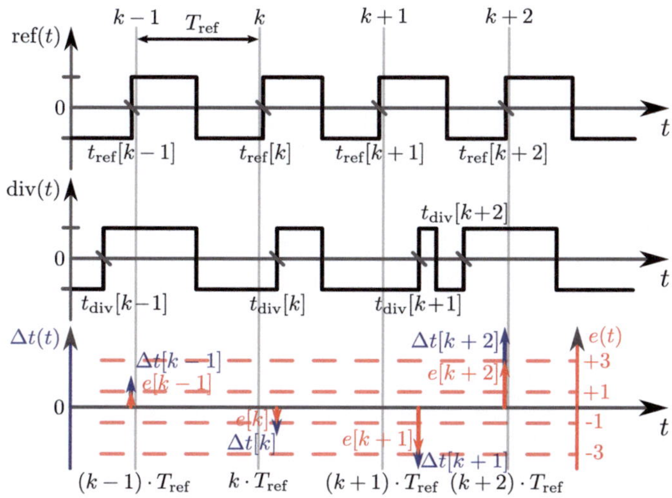

Fig. 2.2 Example timing diagram illustrating ref(t), div(t), $t_{\mathrm{ref}}[k]$, $t_{\mathrm{div}}[k]$, $\Delta t[k]$, and $e[k]$, where $t_{\mathrm{ref}}[k]$, $t_{\mathrm{div}}[k]$ are the "timestamps" taken for ref(t) and div(t), whose difference $\Delta t[k]$ is recorded for each kth sample, which is quantized to $e[k]$ using a four-level quantizer

respectively, that are pulse sequences as functions of t, the "wall time." At the kth reference clock cycle, the TDC detects the time difference, $\Delta t[k]$, between the kth rising edges of ref(t)[1] and div(t). Their timestamps are represented by $t_{\mathrm{ref}}[k]$ and $t_{\mathrm{div}}[k]$. The TDC then quantizes $\Delta t[k]$ to produce the error signal $e[k]$ at the kth reference cycle. Note that, in the discrete-time domain, $\Delta t[k]$ and $e[k]$ record the signed magnitudes of the kth samples of the impulse sequences $\Delta t(t)$ and $e(t)$. The timestamp $t_{\mathrm{out}}[h]$ at the hth output cycle is taken to represent the output signal, out(t), in the same manner. Note that $t_{\mathrm{ref}}[k]$, $t_{\mathrm{div}}[k]$, and $t_{\mathrm{out}}[h]$ are modeled in Fig. 2.1 to represent ref(t), div(t), and out(t) in Fig. 1.1.

[1] Notice that the rising edges of ref(t) do not necessarily appear at exact integer multiples of T_{ref}, $t = k \cdot T_{\mathrm{ref}}$, $k \in \mathbb{N}$, because of the inevitable reference clock jitter.

MATLAB Top File for MRDT Simulation of Fractional-*N* DPLL and Post-Simulation Result Analysis

Listing 2.1 shows the top-level MATLAB script that is used to parameterize the DPLL system, perform behavioral simulation of the MRDT system model, and perform post-simulation analysis in the time and frequency domains.

Listing 2.1 MATLAB top file to performa MRDT simulation and post-simulation analysis.

```
clc; close all; clear all;
%%%%%%%%%%%%%%%%%%%%%%%%%%%%%%%%%%%%%%%%%%%%%%%%%%%%%%%
%                    PARAMETERIZATION STAGE                   %
%%%%%%%%%%%%%%%%%%%%%%%%%%%%%%%%%%%%%%%%%%%%%%%%%%%%%%%
ncycles              = 2^20; % Number of simualted (reference) cycles
%% Reference
F_ref                = 100e6; % Reference frequency
%% MMDC
MASH_order           = 2; % 0=Integer-N; fractional-N: 1=EFM1, 2=MASH 1-1, 3=MASH 1-1-1
Nb_MMDC              = 20; % Number of bits of the MMDC
N_int                = 24; % Integer part of the nominal divide ratio
X                    = 8193; % FCW
M                    = 2^Nb_MMDC; % Modulus
N                    = N_int + X/M * (MASH_order~=0); % Nominal divide ratio; int-N: =N_int, frac-N: = N_int+X/M
%% TDC
Nb_TDC               = 1; % 1=Bang-bang, 2=2-bit, 3=3-bit
t_res                = 1; % TDC input resolution; =1 (very big) for the bang-bang case
%% DCO
F_DCO                = N * F_ref; % DCO free-running frequency; = N*F_ref for simulation simplicity
KF                   = 833; % DCO frequency gain
KT                   = KF/(F_DCO)^2; % DCO period gain
%% DLF
beta                 = 47.7729; % DLF proportional path gain
R                    = 2^-8; % beta->alpha ratio
alpha                = R * beta; % DLF integral path gain
%% DTC
DTC_ON               = 1; % 0=DTC OFF, 1=DTC ON
DTC_quant_ON         = 1; % 0=Nonquantized DTC, 1=Quantized DTC
DTC_nonlin_ON        = 1; % 0=Linear DCD, 1=Nonlinear DCD
res_DTC              = 2^10; % DTC input resolution
% DTC DCD nonlinearity coefficients:
alpha0               = +4.10;
alpha1               = +0.998;
alpha2               = -3.2e-6;
alpha3               = +4.9e-10;
% Sanity checks:
DTC_ON               = (MASH_order ~= 0) && DTC_ON;
DTC_quant_ON         = DTC_ON && DTC_quant_ON;
DTC_nonlin_ON        = DTC_ON && DTC_nonlin_ON;
%% Noises
S_phi_ref            = -155; % reference noise PSD in dB
S_phi_DCO_Deltaf     = -108; % VCO noise PSD in dB; 10log(S_phi_dco(Deltaf))
Deltaf               = 10^6; % Deltaf

%%%%%%%%%%%%%%%%%%%%%%%%%%%%%%%%%%%%%%%%%%%%%%%%%%%%%%%
%                    INITIALIZATION STAGE                     %
%%%%%%%%%%%%%%%%%%%%%%%%%%%%%%%%%%%%%%%%%%%%%%%%%%%%%%%

```

```
tic; % Start the time count
%% Reference + synchronous jitter
[t_ref,sigma_t_n_ref] = ref_clk_gen(F_ref,S_phi_ref,ncycles);
%% DCO + accumulating jitter
[T_DCO,sigma_t_n_DCO] = DCO_clk_gen(F_DCO,S_phi_DCO_Deltaf,Deltaf,
    N,ncycles);
%% MMDC
[N_div,q_acc]         = MMDC(MASH_order,X,M,N_int,ncycles);
%% DTC
t_DTC                 = DTC_ON * DTC(q_acc,1/F_DCO,M,res_DTC,
    DTC_quant_ON,DTC_nonlin_ON,alpha0,alpha1,alpha2,alpha3);
%% Preallocation
% TDC
t_div        = zeros(1,ncycles);
t_div(1:2)   = t_ref(1:2);
Deltat       = zeros(1,ncycles);
e            = zeros(1,ncycles);
% DLF
psi          = zeros(1,ncycles);
c_DLF        = zeros(1,ncycles);
% DCO (high frequency)
t_out        = zeros(1,ceil(N)*ncycles);
h            = N_int; % This creates an initial input time
    differnece of T_ref.
t_out(1:h) = cumsum(1./F_DCO*ones(1,h));

%%%%%%%%%%%%%%%%%%%%%%%%%%%%%%%%%%%%%%%%%%%%%%%%%%%%%%%%%%%%%%%
%                       ITERATION STAGE                       %
%%%%%%%%%%%%%%%%%%%%%%%%%%%%%%%%%%%%%%%%%%%%%%%%%%%%%%%%%%%%%%%

fprintf(1,'\nIteration starts ...\n');

for k=2:ncycles
    %% Bang-bang TDC
    [Deltat(k),e(k)] = TDC(t_ref(k),t_div(k),t_res,Nb_TDC);
    %% LF
    [c_DLF(k),psi(k)] = LF(e(k),psi(k-1),beta,alpha);
    %% DCO
    t_out(h+1:h+N_div(k)) = DCO(t_out(h),T_DCO(h+1:h+N_div(k)),
        c_DLF(k),KT);
    %% Divider path
    t_div(k+1) = t_out(h+N_div(k)) - t_DTC(k);
    %% Update
    h = h + N_div(k);
end

fprintf(1,'\nDone (simulation time = %g s)\n',toc);

%%%%%%%%%%%%%%%%%%%%%%%%%%%%%%%%%%%%%%%%%%%%%%%%%%%%%%%%%%%%%%%
%                       ANALYSIS STAGE                        %
%%%%%%%%%%%%%%%%%%%%%%%%%%%%%%%%%%%%%%%%%%%%%%%%%%%%%%%%%%%%%%%

analysis_time_domain;
analysis_locked_state;
analysis_DTC;
```

The top script is comprised of four stages that are executed in sequence, namely:

1. **Parameterization stage**: The loop and noise parameters are set here. The definition of the parameters will be introduced in the following section. The loop

operates in the integer-N mode when the variable `MASH_order` $= 0$ in Line 9, and otherwise operates in the fractional-N mode.

2. **Initialization stage**: Here, the timestamps of the reference clock and the free-running DCO clock are generated. A sequence of instantaneous divide ratio $N_{\text{div}}[k]$ is also generated by invoking the `MMDC` module. The `DTC` function is called that is controlled by the accumulated QE (AQE) of the MMDC to produce corrective time delays for its compensation.[2] Preallocation of memory space for the signals that will be updated cycle by cycle during the iteration stage is also done here; this increases the simulation speed.
3. **Iteration stage**: A big `for` loop iterates through the user-defined number of reference cycles to behaviorally simulate the MRDT DPLL model. Within each cycle, the behavioral model of each building block is executed sequentially; their outputs are recorded for post-simulation analysis.
4. **Analysis stage**: Finally, in this stage, the simulated results are analyzed in the time domain (to check the locking transient, statistical distribution, etc.) and in the frequency domain (to calculate the simulated jitter, plot the simulated PN, compare against their predicted counterparts, etc.). For the conciseness of code presentation, this stage will invoke different scripts for different analyses, which will be discussed in detail in the following discussion.

In the next section, we will introduce the behavioral model of each of the building blocks in Fig. 2.1, together with its MATLAB model.

2.3 Models of Building Blocks

In this book, the inevitable jitters produced by the oscillators within the loop, i.e., the reference and DCO, are taken into account. The coarse quantization of the TDC yields QE as well. These noise sources are marked in different colors in the integer-N system model (Fig. 2.1). In the fractional-N mode, the MMDC's QE and DTC's nonlinearities also contribute to the system jitter and PN. The methods to model the jitters of the reference and DCO will be described in this section.

Note that the analysis and design methodology related to TDC, MMDC, and DTC are key contributions of this book; they require a considerable volume to be discussed fully. Therefore, their behavioral models and MATLAB implementation are presented in the chapters where they are studied in detail, namely, the TDC model is discussed in Sect. 3.2, the MMDC model is in Sect. 3.5, and the DTC model is discussed in Chap. 6.

---

[2] Here, we assume the gain of the DTC has been calibrated to match the original injection of the MMDC's AQE. Methods to model the autocalibration scheme are discussed in Chap. 6.

2.3.1 Reference with Synchronous Jitter

In DPLL jitter analysis, the intrinsic oscillator jitters can be classified by two canonical types, namely, the reference synchronous jitter ($t_{\mathrm{n,ref}}[k]$) and the DCO random-walk jitter ($t_{\mathrm{n,DCO}}[h]$). Their injection points are marked in red and blue, respectively, in Fig. 2.1. Here, the reference signal affected by the reference jitter is described first.

PLL frequency synthesizers typically use a crystal oscillator to provide the reference frequency, whose PN profile is mostly white: namely, it has a $1/f$ component at close-in frequencies and a flat spectrum above about one kilohertz offset [5–7]. The white PN approximation conforms to the characteristic of a clock source that generates *synchronous* jitter which is directly associated with each timestamp of the transition events and does not move the center frequency of the clock [8, 9]. The timestamp of the kth cycle of the reference clock can be expressed as

$$t_{\mathrm{ref}}[k] = k \cdot T_{\mathrm{ref}} + t_{n,\mathrm{ref}}[k], \tag{2.1}$$

whose jitter term, $t_{n,\mathrm{ref}}[k]$, is synchronous with the current cycle and has no memory of previous jitter samples. This relationship is graphed in the "Reference" (red) block in Fig. 2.1.

The variance of the jitter is expressed as $\sigma^2_{t_{n,\mathrm{ref}}}$. From (2.1), the kth period of the reference can be obtained as

$$T_{\mathrm{ref}}[k] = t_{\mathrm{ref}}[k] - t_{\mathrm{ref}}[k-1] = T_{\mathrm{ref}} + t_{n,\mathrm{ref}}[k] - t_{n,\mathrm{ref}}[k-1], \tag{2.2}$$

which is implicitly modeled in [3].

MATLAB Code to Generate Reference Clock with Synchronous Jitter

Listing 2.2 shows the MATLAB model that is implemented to generate the timestamps of the reference clock signal with synchronous jitter. This function is called in Line 51 of the main file, Listing 2.1.

Listing 2.2 MATLAB behavioral model generating the reference clock used in MRDT simulation.

```
function [t_ref,sigma_t_n_ref] = ref_clk_gen(F_ref,S_phi_ref,
    ncycles)
    %% Calculate the RMS value of reference jitter
    sigma_t_n_ref = sqrt(10^(S_phi_ref/10)/F_ref)/(2*pi);
    %% Generate ncycles random jitters following Gaussian
        distribution
    t_n_ref         = sigma_t_n_ref * randn(1,ncycles);
    %% Associate the jitters with reference timestamps
    t_ref           = 1/F_ref *  (1:ncycles) + t_n_ref;
end
```

During the parameterization stage of simulation, namely in Line 41 of Listing 2.1, it is convenient to specify the intensity of the white reference jitter by its PSD in the phase domain, namely $S_{\phi_{\mathrm{ref}}}$. This is related to the RMS reference jitter in the time domain via

$$S_{\phi_{\rm ref}} = \left(\frac{2\pi}{T_{\rm ref}}\right)^2 \cdot \frac{\sigma^2_{t_{n,\rm ref}}}{F_{\rm ref}}, \tag{2.3}$$

where $2\pi/T_{\rm ref}$ transfers the signal from the time to the phase domain, and $\sigma^2_{t_{n,\rm ref}}/F_{\rm ref}$ represents the PSD of the reference jitter, which is uniformly distributed in the noise bandwidth, $F_{\rm ref}$.

Line 3 calculates $\sigma_{t_{n,\rm ref}}$ according to (2.3); this value is passed as an output of the function, because it is needed in the analysis stage of the top file to predict the system jitter and PN, which is discussed in Chap. 4; this variable can be kept local if such analysis is not needed. The sequence of $t_{n,\rm ref}$ is generated in Line 5, and Line 7 yields the desired sequence of timestamps $t_{\rm ref}$ as per (2.1).

2.3.2 *Digital Loop Filter*

After $\Delta t[k]$ is measured and quantized by the TDC, the resulting error signal $e[k]$ is lowpass filtered by the DLF, which, in a type-II DPLL, is usually a proportional-integral controller that has a proportional path with the coefficient, β, and an integral path with the coefficient, α [2, 10]. Hence, its output can be represented as

$$c_{\rm DLF}[k] = \beta \cdot e[k] + \psi[k], \tag{2.4}$$

where $\psi[k] = \psi[k-1] + \alpha \cdot e[k]$ is the output of the integral path of the DLF.

The DLF is shown in the "DLF" (gray) block in Fig. 2.1. Due to the purely digital nature, the DLF *per se* is assumed to produce no additional noise/jitter when passing $e[k]$ to control the DCO.

Listing 2.3 shows the MATLAB model of the DLF defined by (2.4). This function is called in Line 82 within the iteration stage of Listing 2.1.

Listing 2.3 MATLAB model of the DLF.

```
function [c_DLF,psi] = LF(e,psi_1,beta,alpha)
    %% Integral path
    psi    = psi_1  + alpha * e;
    %% Integral+proportional paths
    c_DLF  = psi    + beta  * e;
end
```

2.3.3 Digitally Controlled Oscillator (DCO) with Accumulating Jitter

A free-running DCO generates an output transition referenced to its previous transition with a jitter representing the uncertainty in the time difference between two consecutive transition events. Thus, the timestamp of a particular transition contains an accumulation of all the jitters of previous cycles, in which case the jitter is termed *accumulating* [9], random-walk [3, 11, 12], or period [13] jitter. For a DCO with accumulating jitter, the clock *period* of the hth cycle is represented as

$$T_{\mathrm{DCO}}[h] = T_{\mathrm{DCO}} + t_{n,\mathrm{DCO}}[h]. \tag{2.5}$$

The variance of the period jitter ($t_{n,\mathrm{DCO}}[h]$) is expressed as $\sigma^2_{t_{n,\mathrm{DCO}}}$. Based on (2.5), the timestamp of the hth cycle is

$$t_{\mathrm{DCO}}[h] = \sum_{i=1}^{h} T_{\mathrm{DCO}}[i] = h \cdot T_{\mathrm{DCO}} + \sum_{i=1}^{h} t_{n,\mathrm{DCO}}[i]. \tag{2.6}$$

MATLAB Code to Generate Free-Running DCO Clock Periods with Period Jitter

As will be introduced shortly, during the iteration stage of simulation, the DCO's oscillation *period* is controlled cycle to cycle via the output of the DLF. Therefore, it is convenient to generate the periods—rather than timestamps—of the free-running DCO within the initialization stage, i.e., Line 53 of Listing 2.1. Listing 2.4 shows the model of the free-running DCO.

Listing 2.4 MATLAB code generating the free-running DCO periods used in MRDT simulation.

```
function [T_DCO,sigma_t_n_DCO] = DCO_clk_gen(F_DCO,S_phi_DCO,
    Deltaf,N,ncycles)
    %% Calculate the RMS value of DCO jitter by using S_phi_DCO(
        Deltaf)
    sigma_t_n_DCO = sqrt(Deltaf^2*10^(S_phi_DCO/10)/F_DCO^3);
    %% Generate ceil(N)*ncycles random jitters following Gaussian
        distribution
    t_n_DCO = sigma_t_n_DCO * randn(1,ceil(N)*ncycles);
    %% Associate the jitters with DCO periods
    T_DCO = 1/F_DCO*ones(1,ceil(N)*ncycles) + t_n_DCO;
end
```

Similar to the convenient specification of the intensity of a noisy reference clock, it is commonplace to represent the intensity of the DCO's jitter by the PSD of its output-referred PN; namely, the phase-domain PSD of $t_{\mathrm{n,DCO}}[h]$ (marked in blue in Fig. 2.1) transferred to the output via the digital integrator, $1/\left(1 - z^{-1}\right)$, as the DCO's period jitter gets accumulated to affect the PN associated with its timestamps (see (2.6)).

Similar to (2.3), the PSD of the PN generated by the DCO is related to the RMS value of the DCO's period jitter in the time domain via

$$S_{\phi_{\mathrm{DCO}}}(f) = \left(\frac{2\pi}{T_{\mathrm{DCO}}}\right)^2 \cdot \frac{\sigma^2_{t_{n,\mathrm{DCO}}}}{F_{\mathrm{DCO}}} \cdot \frac{1}{\left|1 - z^{-1}\right|^2}, \tag{2.7}$$

where $z = e^{j2\pi f/F_{\mathrm{DCO}}}$, as discussed in Fig. 2.1.

Because, in PN analysis, we typically care about PN contributions of the components in the spectrum, where $f \ll F_{\mathrm{DCO}}$, Eq. (2.7) can be simplified by adopting the "narrowband approximation," which yields $z^{-1} \approx 1 - j2\pi f/F_{\mathrm{DCO}}$; this becomes

$$S_{\phi_{\mathrm{DCO}}}(f) \approx \sigma^2_{t_{n,\mathrm{DCO}}} \cdot \frac{F^3_{\mathrm{DCO}}}{f^2},$$

which conforms to the PN profile of a typical DCO that has a −20 dB/dec PSD.

For a representative DCO, the PSD of its PN specified at a certain offset frequency is typically reported in the form:

$$S_{\phi_{\mathrm{DCO}}}(\Delta f) = \frac{F^3_{\mathrm{DCO}}}{\Delta f^2} \cdot \sigma^2_{t_{n,\mathrm{DCO}}}, \tag{2.8}$$

which is implicitly used in [3].

In Listing 2.4, Line 3 calculates $\sigma_{t_{n,\mathrm{DCO}}}$ according to (2.8); similar to the reference generator, this value is passed as an output of the function to be used in system jitter and PN prediction, which is discussed in Chap. 4; this variable can be kept local if such analysis is not needed. The sequence of $t_{n,\mathrm{DCO}}$ is generated in Line 5, and Line 7 yields the desired sequence of free-running DCO's periods T_{DCO}, following (2.5).

On the other hand, when the DCO is controlled via the DLF, the control code $c_{\mathrm{DLF}}[k]$ in (2.4) is scaled by the DCO's period gain, K_T in s/LSB, where "LSB" is short for least significant bit, and added to the DCO's free-running period in (2.5). Hence, the hth output period in the kth reference clock cycle can be expressed as

$$T_{\mathrm{out}}[h] = K_T \cdot c_{\mathrm{DLF}}[k] + T_{\mathrm{DCO}} + t_{n,\mathrm{DCO}}[h],$$
$$\forall h \in \{N \cdot k + 1, N \cdot k + 2, \ldots, N \cdot (k+1)\}. \tag{2.9}$$

Notice that, in one reference clock cycle (indexed by k), the DLF's output is held and the DCO is controlled to yield N output periods (indexed by h). In integer-N operation, N is an unchanged integer constant in all the reference clock cycles; in fractional-N operation, it is instead the instantaneous divide ratio at the kth clock cycle, $N_{\mathrm{div}}[k]$, yielded by the MMDC; this is described in more detail in Sect. 3.5. The "1-to-N" upsampling operation is modeled in behavioral simulation and is denoted by the "$\uparrow N$" box in the "DCO" (blue) block in Fig. 2.1.

Finally, at the high sampling rate, the hth timestamp of the output signal is obtained by adding the hth output period in (2.9) to the previous timestamp, which is equivalent to the accumulation of all the previous output periods, as in

$$t_{\text{out}}[h] = t_{\text{out}}[h-1] + T_{\text{out}}[h] = \sum_{i=1}^{h} T_{\text{out}}[i]. \tag{2.10}$$

MATLAB Behavioral Model of DLF-Controlled DCO

Listing 2.5 shows the MATLAB model of the DCO to implement (2.9) and (2.10). This function is executed in every iteration in Line 84 of Listing 2.1.

Listing 2.5 MATLAB model of the DCO controlled by the DLF's output.

```
1 function t_out = DCO(t_out_1,T_DCO,c_DLF,KT)
2     %% Output (Controlled DCO oscillation) periods
3     T_out = T_DCO + KT * c_DLF;
4     %% Output timestamps
5     t_out = t_out_1 + cumsum(T_out);
6 end
```

Note that, when characterizing a voltage or digitally controlled oscillator, its control signal variation mapped to the change in the oscillation frequency is commonly described by the *frequency gain* K_F in Hz/V or Hz/LSB. Since our MRDT behavioral model operates in the time domain, it is easier to map the control to the oscillation period via the *period gain* K_T as in (2.9) and Line 3 of Listing 2.5.

The relationship between K_F and K_T can be derived by recalling Eq. (2.9) that describes an arbitrary hth oscillation period of the controlled DCO, which is

$$T_{\text{out}}[h] = T_{\text{DCO}} + K_T \cdot c_{\text{DLF}}[h], \tag{2.11}$$

where we omit the jitter term whose amplitude is insignificant relative to the remaining terms; this is equivalent to

$$T_{\text{out}}[h] = \frac{1}{\frac{1}{T_{\text{DCO}}} - K_F \cdot c_{\text{DLF}}[h]}, \tag{2.12}$$

which represents the free-running oscillation period by the reciprocal of the oscillation frequency ($1/T_{\text{DCO}}$) that is changed by the control code via the frequency gain, $K_F \cdot c_{\text{DLF}}[h]$. Note that an increase in oscillation period corresponds to a decrease in frequency, and vice versa, hence the minus sign in the denominator. Equating (2.11) and (2.12) and rearranging yield

$$T_{\text{DCO}} \cdot K_F - \frac{1}{T_{\text{DCO}}} \cdot K_T + K_T \cdot K_F \cdot c_{\text{DLF}}[h] = 0.$$

The third term on the left-hand side ($K_T \cdot K_F \cdot c_{\mathrm{DLF}}[h]$) can be neglected, because when comparing it with the first term ($T_{\mathrm{DCO}} \cdot K_F$), omitting the common factor K_F, $K_T \cdot c_{\mathrm{DLF}}[h]$ is negligible compared to T_{DCO}. Recalling (2.11), the change in the oscillation period is much smaller than the free-running period *per se*. Similarly, the third term is also negligible compared to the second term ($-K_T/T_{\mathrm{DCO}}$) by utilizing the relationship in (2.12).

Finally, the relationship simplifies to

$$K_T = T_{\mathrm{DCO}}^2 \cdot K_F,$$

which is calculated in Line 21 in the main parameterization of Listing 2.1.

2.3.4 Frequency Divider

In integer-N DPLLs, the frequency divider in the feedback path counts N (a fixed integer) output edges of $\mathrm{out}(t)$ to produce one edge of the divided signal, whose timestamp is $t_{\mathrm{div}}[k+1]$; this is used as the input divided timestamp in the $(k+1)$th iteration. This "N-to-1" downsampling operation is modeled in behavioral simulation and is denoted by the "$\downarrow N$" box in the "Divider" (gray) block in Fig. 2.1. Therefore, the timestamp yielded by the frequency divider is obtained by calculating

$$t_{\mathrm{div}}[k+1] = t_{\mathrm{out}}\left[N \cdot (k+1)\right]. \tag{2.13}$$

Using a retimer at the output of the frequency divider [14] can suppress the divider's jitter to be negligible compared to the other intrinsic noise sources; therefore, ideal (jitter-free) frequency dividers are assumed in this book. Consequently, the transfer function in (2.13) does not yield additional divider-related noise.

In the fractional-N mode, the nominal divide ratio, N, is not an integer. As introduced in Sect. 1.1, fractional-N synthesis is typically realized by using an MMD in place of the integer divider; the instantaneous divide ratio, $N_{\mathrm{div}}[k]$, is modulated by the DDSM-based MMDC. The system model of the DDSM-based fractional-N DPLL is shown in Fig. 2.3. In this case, the function of the MMD is modeled as

$$t_{\mathrm{div}}[k+1] = t_{\mathrm{out}}\left[N_{\mathrm{div}}[k] \cdot (k+1)\right]. \tag{2.14}$$

The method using the MMDC to generate $N_{\mathrm{div}}[k]$ is introduced in the next chapter.

The MATLAB model of the MMD is shown in Line 86 in the top iteration of Listing 2.1.

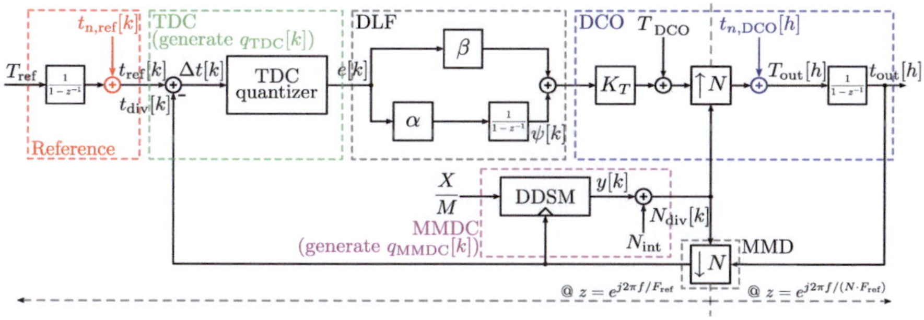

Fig. 2.3 MRDT system model of a *fractional-N* DPLL (without DTC)

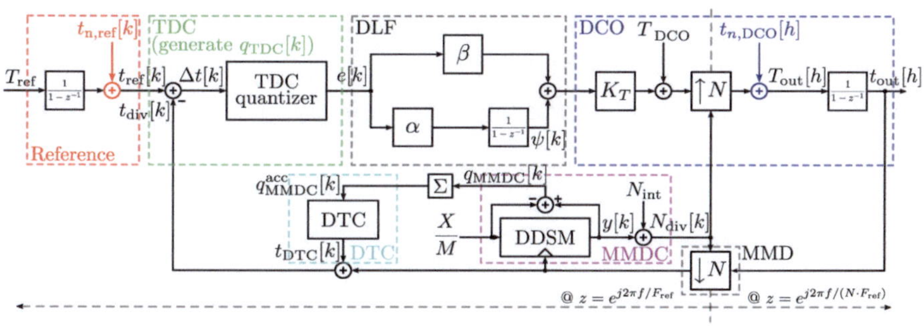

Fig. 2.4 MRDT system model of a *DTC-enhanced fractional-N* DPLL

2.3.5 DTC

Because the MMDC's AQE injected into the loop can drastically deteriorate the system jitter and PN, advanced PLLs use DTCs in the feedback (and/or reference) path of the loop to mirror and cancel this error. The DTC is denoted by the (cyan) "DTC" block in Fig. 2.4. The core component of the DTC is a tunable delay line that is digitally controlled by the MMDC's AQE, $q^{\mathrm{acc}}_{\mathrm{MMDC}}[k]$, to yield a corrective time (or phase) delay, $t_{\mathrm{DTC}}[k+1]$, that is added to the divided signal in (2.14) in order to cancel the QE.

A real DTC features various nonlinearities. For example, a quantization operation must take place if the time-domain output that can be produced by the DTC has coarser resolution than its digital-domain control code; the DTC's delay line itself typically manifests an integral nonlinearity and random element mismatch. In this book, the key contributions are the modeling approach that takes into account the hard quantization, soft integral, and element mismatch nonlinearities associated with the DTC; analyses of the detrimental spectral responses caused by them; and mitigation methods. Detailed DTC models are discussed in Chaps. 6, 7, and 8.

2.4 Determination of Simulated Performance Metrics

This section introduces methods to determine quantitatively the system jitters and PN from behavioral simulation of the MRDT DPLL introduced in previous sections. Three physical phenomena are of particular interest in this book, namely, the output jitter, the output PN spectrum, and the input jitter. Together, these quantify the noise performance of the DPLL system.

MATLAB Script for Post-Simulation Analysis in Time Domain

Listing 2.6 shows the MATLAB script that is executed in Line 97 in Listing 2.1 after the simulation is finished. This script can be used to perform time-domain analysis to check the locking transient (whether the loop has locked as expected) and to calculate the statistical distributions of key signals.

Listing 2.6 MATLAB script for post-simulation time-domain analysis.

```
%%%%%%%%%%%%%%%%%%%%%%%%%%%%%%%%%%%%%%%%%%%%%%%%%%%%%%%%
%                TIME-DOMAIN ANALYSIS                  %
%%%%%%%%%%%%%%%%%%%%%%%%%%%%%%%%%%%%%%%%%%%%%%%%%%%%%%%%

%% Entire simulation
figure('Position',[300 300 1000 300])
yyaxis left
stairs(t_ref(3:end),'linewidth',2)
hold on
stairs(t_div(3:end),'linewidth',2)
xlabel('Samples')
ylabel('Timestamp [s]')
yyaxis right
stairs(Deltat(3:end),'linewidth',2)
ylabel('Input Time Difference [s]')
legend('t_{ref}','t_{div}','{\Delta}t','Location','southeast');
title('All Samples of Simulation')
grid on
grid minor
xlim([-inf inf]);
ax=gca;
ax.LineWidth=3;
ax.FontSize=16;

%% Locking transient
figure('Position',[300 300 1000 300])
yyaxis left
stairs(t_ref(3:100000),'linewidth',2)
hold on
stairs(t_div(3:100000),'linewidth',2)
xlabel('Samples')
ylabel('Timestamp [s]')
yyaxis right
stairs(Deltat(3:100000),'linewidth',2)
ylabel('Input Time Difference [s]')
legend('t_{ref}','t_{div}','{\Delta}t','Location','southeast');
title('Beginning of Simulation')
grid on
grid minor
xlim([-inf inf]);
ax=gca;
```

```
ax.LineWidth=3;
ax.FontSize=16;

%% Distribution of AQE
if MASH_order ~= 0
    figure('Position',[0 0 900 700]+100)
    histogram(q_acc/M, 100,'Normalization', 'pdf', 'FaceColor',
        [0.5 0.5 0.5])
    ax=gca;
    ax.LineWidth=3;
    ax.FontSize=22;
    xlabel(ax,'AQE [LSB/M]');
    ylabel(ax,'PDF [LSB^{-1}]');
    hold on
    grid on
    title('Distribution of Simulated AQE')
end
```

The first two plots show the entire simulated waveforms of the timestamps, t_{ref} and t_{div}, and the input time difference, Δt and the locking process at the beginning.

When the loop operates in the fractional-N mode (`MASH_order` $\neq 0$), the statistical distributions (histograms) of the simulated AQE of the MMDC is also plotted.

2.4.1 *Output Jitter*

The transient simulation starts with the initial value of the input time difference ($\Delta t[k]$ in Fig. 2.1) set to $\Delta t[0] = T_{ref}$; this mimics the worst-case phase difference between the ref(t) and div(t) in theory. If the DPLL functions correctly, it will progressively close the input time difference as the iteration goes on, i.e., as k grows. After a successful locking transient, the locked-state time jitter of the synthesized output signal is measured from the array (with Ω data samples, where Ω is an integer) of the simulated output timestamps, namely, $t_{out}[h]$ in (2.10). The algorithm is explained as follows.

First, after discarding the transient data to characterize the locked state, the simulated output periods are calculated via

$$T_{out}[h] = t_{out}[h] - t_{out}[h-1], \qquad \forall h \in \{1, 2, 3, \ldots, \Omega\},$$

of which the mean value of the output period is calculated using

$$T_{out}^{\mu} = \frac{1}{\Omega} \cdot \sum_{i=1}^{\Omega} T_{out}[i].$$

The period jitter associated with each output period is hence

$$T_{n,\text{out}}[h] = T_{\text{out}}[h] - T_{\text{out}}^{\mu}, \qquad \forall h \in \{1, 2, 3, \ldots, \Omega\}.$$

The time jitter is the accumulation of the period jitter as

$$t_{n,\text{out}}[h] = \sum_{i=1}^{h} T_{n,\text{out}}[i], \qquad \forall h \in \{1, 2, 3, \ldots, \Omega\},$$

which can be centered to zero mean by removing its DC component,

$$t_{n,\text{out}}^{\mu} = \frac{1}{\Omega} \cdot \sum_{i=1}^{\Omega} t_{n,\text{out}}[i],$$

if there is any, this guarantees that the plotted power spectrum of the output time jitter does not exhibit peaking at DC. The zero-mean output time jitter is finally calculated as

$$t_{n,\text{out}}^{\text{AC}}[h] = t_{n,\text{out}}[h] - t_{n,\text{out}}^{\mu}, \qquad \forall h \in \{1, 2, 3, \ldots, \Omega\}. \tag{2.15}$$

The simulated RMS output jitter, $\sigma_{t_{\text{n,out}}}$, whose closed-form prediction will be derived in Chap. 4, can be measured by taking the standard deviation of the samples of $t_{n,\text{out}}^{\text{AC}}[h]$ in (2.15).

2.4.2 *Output PN*

To plot the power spectrum of the simulated output PN, the output time jitter in (2.15) is first transferred to the phase domain by calculating

$$\phi_{n,\text{out}}^{\text{AC}}[h] = \frac{2\pi}{T_{\text{out}}^{\mu}} \cdot t_{n,\text{out}}^{\text{AC}}[h], \qquad \forall h \in \{1, 2, 3, \ldots, \Omega\}. \tag{2.16}$$

The simulated output PN's PSD is then plotted using Welch's PSD estimation, i.e., "pwelch" function in MATLAB.

The spectrum of the total output PN can be broken down into the contributions from the noise sources within the loop. Theoretical predictions for each of the individual contributions and their aggregated total PN, $S_{\phi_{\text{out}}}(f)$, are provided in Chap. 4.

MATLAB Function to Characterize Steady-State Timestamp Sequence

Listing 2.7 shows the MATLAB function that is used for determining the key characteristics of the simulated sequence of an oscillator's timestamps.

Listing 2.7 MATLAB function to characterize the sequence of timestamps of a simulated clock signal.

```
function [F_mean,t_n_rms,phi_n_rms,f_PSD,PSD_phi] = acquisition(t)
%% Ensure column vector
[~,ncol]=size(t);
if ncol ~= 1
t = t';
end
%% Periods
T           = diff(t);
%% Average period
T_mean      = mean(T);
F_mean      = 1/T_mean;
%% Period jitter
T_n         = T - T_mean;
%% Zero-mean accumulated time jitter
t_n         = cumsum(T_n);
t_n         = t_n - mean(t_n);
%% RMS time jitter
t_n_rms     = std(t_n);
%% Phase error
phi_n       = 2*pi/T_mean * t_n;
phi_n_rms = std(phi_n);
%% PSD of phase error
npsd = 2^(nextpow2(length(phi_n)/8)-1);
Fpsd = 1/(npsd*T_mean);
f_PSD = 0:Fpsd:Fpsd*(floor(npsd/2)-1);
[PSD_phi,~] = pwelch(phi_n,window(@hann,npsd),npsd/2,npsd,1/T_mean
    ,'Two-sided');
PSD_phi = PSD_phi(1:floor(npsd/2));
```

The function takes as its input the timestamp sequence; it performs the calculation following the above procedure, and returns the average frequency (`F_mean`), RMS time jitter (`t_n_rms`), RMS PN (`phi_n_rms`), and the PSD of the PN (`PSD_phi`) evaluated at the frequencies (`f_PSD`) of the sequence.

The function is called in Line 25 of Listing 4.1, which performs the locked-state analysis in Line 98 of Listing 2.1.

2.4.3 Input Jitter

Apart from the output jitter and PN, the input time difference, $\Delta t[k]$, in the locked state, which is also referred to as the "input time jitter," is of great interest. This is primarily because the statistical distribution and RMS value of $\Delta t[k]$ in the locked state determine the linear gain and power of the QE of the TDC, which in turn affect the linearized analysis of the DPLL at the system level. This is explained in detail in Chap. 3.

In terms of the acquisition of simulated results related to the input time jitter, the distribution of the sampled locked-state $\Delta t[k]$ can be plotted and compared with the theoretically predicted input distribution. The simulated RMS input jitter can be obtained by measuring the standard deviation of the locked-state $\Delta t[k]$ samples; this can be compared with its theoretically predicted counterpart, i.e., $\sigma_{\Delta t}$, as elaborated in Chap. 4.

As a final remark, all the theoretically predicted values of the quantities mentioned in this section will be shown to match their simulated results later in this book. Theory versus simulation comparisons will be provided alongside the analytical discussions to verify and confirm the accuracy of analysis and predictions in this work.

2.5 Summary

This chapter contains three components:

1. In Sect. 2.2, the MRDT behavioral modeling methodology used in this book is introduced; this is used to model and simulate accurately practical DPLLs that have MRDT behaviors.
2. In Sect. 2.3, the models of the building blocks in DPLLs are introduced in order; namely, reference source (with jitter), DLF, DCO (with jitter), and frequency divider. The behavioral models of the TDC, MASH MMDC, and DTC are introduced in later chapters.
3. In Sect. 2.4, the calculation methods for the simulated output PN, input and output jitters are introduced; these are used in the following chapters to demonstrate the performance of the DPLLs that we model in the book.

References

1. N. Da Dalt, A design-oriented study of the nonlinear dynamics of digital bang-bang PLLs. IEEE Trans. Circuits Syst. I Regul. Pap. **52**(1), 21–31 (2005). (Jan)
2. R.B. Staszewski, P.T. Balsara, Phase-domain all-digital phase-locked loop. IEEE Trans. Circuits Syst. II Express Briefs **52**(3), 159–163 (2005). (March)
3. L. Avallone, M. Mercandelli, A. Santiccioli, M.P. Kennedy, S. Levantino, C. Samori, A comprehensive phase noise analysis of bang-bang digital PLLs. IEEE Trans. Circuits Syst. I Regul. Pap. **68**(7), 2775–2786 (2021). (July)
4. Y. Hu, T. Siriburanon, R.B. Staszewski, Multirate timestamp modeling for ultralow-jitter frequency synthesis: a tutorial. IEEE Trans. Circuits Syst. II Express Briefs **69**(7), 3030–3036 (2022). (July)
5. B. Razavi, *RF Microelectronics*, vol. 2 (Prentice Hall New York, 2012)
6. T. Feng, H. Xianhe, W. Wei, F. Wei, Analysis of phase noise and timing jitter in crystal oscillator, in *2007 International Conference on Communications, Circuits and Systems* (2007), pp. 1103–1106

7. X. Huang, J. Jiao, F. Sun, W. Fu, The prediction, simulation and verification of the phase noise in low-phase-noise crystal oscillator, in *2015 Joint Conference of the IEEE International Frequency Control Symposium & the European Frequency and Time Forum* (2015), pp. 129–132
8. K. Kundert, *Modeling and Simulation of Jitter in Phase-Locked Loops* (Springer, Boston, MA, US, 1997), pp. 359–379. https://doi.org/10.1007/978-1-4757-2602-2_16
9. K. Kundert, Modeling jitter in PLL-based frequency synthesizers (2006). www.designers-guide.org
10. R.B. Staszewski, P.T. Balsara, *All-digital Frequency Synthesizer in Deep-submicron CMOS* (Wiley, 2006)
11. M. Zanuso, D. Tasca, S. Levantino, A. Donadel, C. Samori, A.L. Lacaita, Noise analysis and minimization in bang-bang digital PLLs. IEEE Trans. Circuits Syst. II Express Briefs **56**(11), 835–839 (2009). (Nov)
12. L. Avallone, M.P. Kennedy, S. Karman, C. Samori, S. Levantino, Jitter minimization in digital PLLs with mid-rise TDCs. IEEE Trans. Circuits Syst. I Regul. Pap. **67**(3), 743–752 (2020). (March)
13. I. Galton, C. Weltin-Wu, Understanding phase error and jitter: definitions, implications, simulations, and measurement. IEEE Trans. Circuits Syst. I Regul. Pap. **66**(1), 1–19 (2019). (Jan)
14. L. Romano, S. Levantino, S. Pellerano, C. Samori, A. Lacaita, Low jitter design of a 0.35 μm-CMOS frequency divider operating up to 3GHz, in *Proceedings of the 28th European Solid-State Circuits Conference* (2002), pp. 611–614

Chapter 3
Enhanced Linearized Analysis of Mid-Rise Time-to-Digital Converter (TDC)

3.1 Introduction

TDCs, as introduced in Chap. 2, are implemented at the input of a DPLL to serve as the PD that measures and quantizes the cycle-wise (analog) time differences between the edges of the reference clock and the divided output signal. The quantized (digital) time difference signal is lowpass filtered by the DLF to control the DCO that generates the high-frequency output signal, e.g., a radio-frequency signal for wireless communication purposes. The DPLL acts as a negative feedback system to guarantee that, after a successful locking transient, the frequency of the output signal in the locked state is N times that of the reference signal, where N is the nominal divide ratio of the divider.

The TDC's hard quantization nonlinearity makes it nontrivial to estimate its linearized gain and the power of its QE in the locked state, especially for TDCs that have a limited number of bits, such as a 1-bit BPD and 2-bit/3-bit TDCs. Such TDCs with a mid-rise transfer characteristic are often preferred because of their design simplicity, high linearized gain (as will be derived in this book), and the resulting ultra-low consumption of area and power in the TDC and DLF. It is crucial to perform accurate linearized analysis of these in order to facilitate linearized analysis at the system level.

The literature [1–9] has revealed the dependency of the linearized gain of the TDC on the statistical distribution of the steady-state input jitter. A DPLL operating in different modes exhibits different jitter distributions at the input to the TDC. For example, in the integer-N mode of operation, the input jitter of a frequency synthesizer is dominated by the contributions from the reference clock jitter, TDC's QE, and DCO period jitter, which, when combined, can be approximated by a zero-mean Gaussian distribution. Such a Gaussian approximation was assumed in [2, 4, 5, 7] and verified in [3, 8, 9]. On the other hand, in the fractional-N mode of operation, the instantaneous divide ratio of the MMD is cycle-to-cycle modulated by the MMDC among several integer levels. We will show in the next chapter that, if the

© The Author(s), under exclusive license to Springer Nature Switzerland AG 2026
X. Wang and M. P. Kennedy, *Jitter and Spur Minimization in Fractional-N Digital Frequency Synthesizers*, Analog Circuits and Signal Processing,
https://doi.org/10.1007/978-3-032-27186-0_3

MMDC's QE is not canceled, the loop can effectively operate in an open-loop-like mode. In this case, the input jitter is dominated by the AQE of the MMDC and its probability distribution, which is non-Gaussian.

Different methods for linearizing mid-rise TDCs have been reported. For example, to linearize the BPD in an integer-N BBPLL, a linearized analysis framework that conforms to the MMSE estimation principle[1] has been used in [4, 7, 9]; this can be regarded as a special application of Bussang's theorem [10]. On the other hand, Avallone et al. [8] used a linearization method that involves calculating the gradient of the expected output at the locking point for a multi-bit TDC to describe the linearized gain. We would expect that, in general, these two methods should result in different algebraic expressions for estimating the TDC's linearized gain and the variance of the QE. Both [7, 8] assume a strictly Gaussian input jitter distribution to approximate the random jitter in the locked state due to the well-characterized contributions from the intrinsic jitter sources when operating in the integer-N mode. Interestingly, the two methods give the *same* estimates in the case of a BPD. This motivates us to analyze other types of input distributions to check whether the two methods still yield the same estimates in those cases. Where the approaches do not agree, this chapter will show which method works better, in the sense that the estimates match simulated results.

This chapter is arranged as follows. In Sect. 3.2, we first formulate the behavioral model of the mid-rise TDCs of interest. The MATLAB functions for these are provided.

In Sect. 3.3, another canonical distribution is first considered for the TDC input jitter, namely, an assumed uniform distribution,[2] and we use it to compare and contrast the estimates made using these two methodologies. The linearized analysis of a 2-bit TDC and a BPD is demonstrated as examples. Behavioral simulation results confirm that the "MMSE estimator" provides estimates that are more accurate for the linearized gain and the variance of the TDC QE, regardless of the nature of the input distribution, as opposed to the method based on the "small-signal approximation." This suggests that the "MMSE estimator" method is the better analysis framework to use in order to linearize quantizers involving TDCs.

In Sect. 3.4, we use the MMSE estimator and extend the linearized analysis to a generic mid-rise TDC with arbitrary N_b bits (including a BPD) operating in integer-N mode.

Prior to the fractional-N analysis, we introduce in Sect. 3.5 the working principles for fractional-N synthesis driven by MASH DDSMs. MATLAB models for these are also provided. Section 3.6 provides linearized analysis for the TDC in a fractional-N DPLL by extension from the integer-N case. Accurate analysis will be demonstrated for the fractional-N case based on the MMSE method, for which input distributions

[1] An MMSE estimator aims to ascribe a linear gain to the quantizer that minimizes the variance of the estimation error.

[2] In reality, the steady-state input jitter presented to the TDC used as the PD of a PLL does not have a uniform distribution. The assumption that the input jitter has a uniform distribution is deliberately chosen here in order to distinguish the accuracies of the discussed linearization frameworks.

dominated by the QE produced by MASH 1-1 and 1-1-1 MMDCs are investigated as examples. Behavioral simulation results confirm the accuracy of the analysis.

The accuracy of the linearized models of the TDC operating in the integer-N and fractional-N modes derived in these two sections is verified in Sect. 3.7 using behavioral simulations.

Finally, in Sect. 3.8, we derive and verify validity limits for pseudo-linear operation of the optimized mid-rise TDCs. We explain why the hard TDC quantizer can be approximated as a linear component in most of the pseudo-linear region, outside of which the TDC will apply nonlinear compression-type distortion to its input signal. An optimized multi-bit TDC exhibits a wider pseudo-linear region than a BPD, and consequently motivates the use of a few-bit TDC in place of a BPD.

3.2 TDC Behavioral Model

The TDC's behavioral model is depicted by the "TDC" (green) block in Fig. 2.1. Note that the quantization operation of the "TDC quantizer" produces the TDC's QE, which is nontrivial to estimate when the TDC has only a few bits. The linearized analysis of the TDC and its QE minimization will be discussed in greater detail in this and later chapters.

In the discrete-time domain, recalling the "timestamping" notation introduced in Chap. 2 and illustrated in Fig. 2.2, during the kth reference clock cycle, the TDC first detects the time difference between the timestamps of the reference and divided signals, as in

$$\Delta t[k] = t_{\text{ref}}[k] - t_{\text{div}}[k]. \tag{3.1}$$

Thereafter, the TDC behaves like an N_b-bit analog-to-digital converter that quantizes the analog input time difference $\Delta t[k]$ to produce a digital "error" signal, denoted $e[k]$. The behavioral model of the TDC is shown in Fig. 3.1a.

In this chapter, we consider ideal mid-rise TDCs with *equidistant* quantization thresholds having a time resolution, t_{res}, as examples; linearized analysis extended to TDCs with arbitrary thresholds will be presented in Chap. 5.

When the loop is close to lock, $-T_{\text{ref}}/2 < \Delta t < T_{\text{ref}}/2$, where T_{ref} is the period of the reference signal, and cycle slipping does not occur. The quantization characteristic

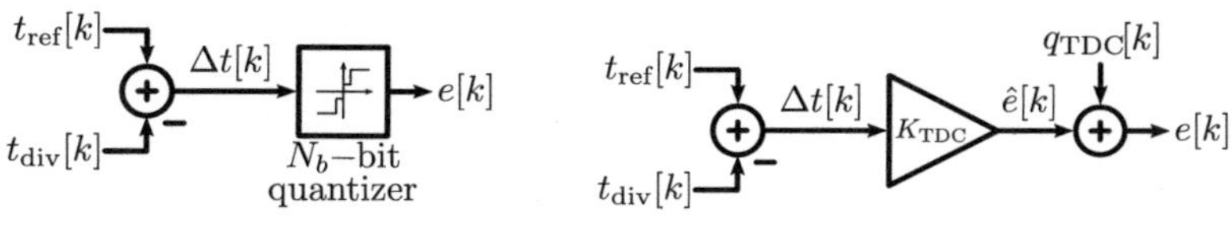

Fig. 3.1 Discrete-time TDC models

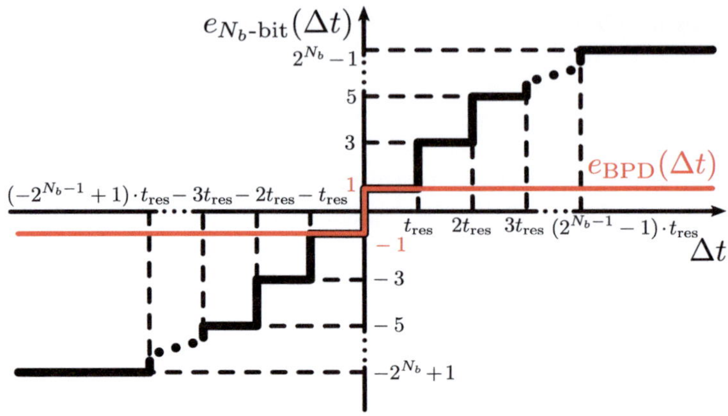

Fig. 3.2 Input-output characteristics of (i) black: a generic N_b-bit quantizer of a mid-rise TDC with equidistant quantization thresholds expressed in (3.2) and (ii) red: a BPD expressed in (3.4)

of a generic N_b-bit TDC quantizer with mid-rise equidistant quantization levels is described by[3]

$$e_{N_b\text{-bit}}(\Delta t) = \begin{cases} -2^{N_b}+1 & \text{for } \Delta t < \left(-2^{N_b-1}+1\right)\cdot t_{\mathrm{res}}, \\ & \vdots \\ -3 & \text{for } -2t_{\mathrm{res}} \le \Delta t < -t_{\mathrm{res}}, \\ -1 & \text{for } -t_{\mathrm{res}} \le \Delta t < 0, \\ +1 & \text{for } 0 \le \Delta t < t_{\mathrm{res}}, \\ +3 & \text{for } t_{\mathrm{res}} \le \Delta t < 2t_{\mathrm{res}}, \\ & \vdots \\ 2^{N_b}-1 & \text{for } \Delta t \ge \left(2^{N_b-1}-1\right)\cdot t_{\mathrm{res}}. \end{cases} \tag{3.2}$$

Note that, in principle, arbitrary quantization and mismatch nonlinearities can be included in the TDC's model; this can be done by adjusting the quantization thresholds in (3.2). Chap. 5 models representative nonlinearities of the TDC in this manner to demonstrate the robustness of the proposed jitter-minimized system design. The transfer characteristic in (3.2) is shown in black in Fig. 3.2.

The transfer characteristic of the quantizer of a 2-bit TDC is described by

[3] The quantized outputs with a spacing of 2, i.e., $\pm1, \pm3, \pm5, \ldots, \pm2^{N_b}-1$, are set to facilitate a fair comparison with a BPD that outputs ±1s.

$$e_{2\text{-bit}}(\Delta t) = \begin{cases} -3 & \text{for } \Delta t < -t_{\text{res}}, \\ -1 & \text{for } -t_{\text{res}} \leq \Delta t < 0, \\ +1 & \text{for } 0 \leq \Delta t < t_{\text{res}}, \\ +3 & \text{for } \Delta t \geq t_{\text{res}}. \end{cases} \quad (3.3)$$

The BPD is a special case of mid-rise TDC. It does not have an explicit input resolution, and its transfer characteristic is expressed as

$$e_{\text{BPD}}(\Delta t) = \begin{cases} -1 & \text{for } \Delta t < 0, \\ +1 & \text{for } \Delta t \geq 0, \end{cases} \quad (3.4)$$

as shown in red in Fig. 3.2.

MATLAB Model of Mid-Rise TDCs

Listing 3.1 shows the MATLAB model of a mid-rise TDC with equidistant quantization thresholds. Line 3 calculates the input time difference (Δt) between the reference and divided signal edges as per (3.1); this variable is passed as an output of this function so that its statistics, waveform, and spectrum, etc. can be analyzed after simulation, which will be shown in later discussion; Δt can be kept local if such analyses are not needed.

Listing 3.1 MATLAB behavioral model of mid-rise TDCs with programmable number of bits equal to 1, 2, and 3.

```
1  function [Deltat,e] = TDC(t_ref,t_div,t_res,Nb_TDC)
2      %% Input time difference
3      Deltat = t_ref - t_div;
4      if Nb_TDC == 1
5          %% Bang-bang TDC
6              e = sign(Deltat);
7      elseif Nb_TDC == 2
8          %% 2-bit TDC
9          if      Deltat <  t_res
10             e = -3;
11         elseif (Deltat >= t_res) && (Deltat < 0)
12             e = -1;
13         elseif (Deltat >= 0)     && (Deltat < t_res)
14             e = 1;
15         else
16             e = 3;
17         end
18     elseif Nb_TDC == 3
19         %% 3-bit TDC
20         if      Deltat <  -3*t_res
21             e = -7;
22         elseif (Deltat >= -3*t_res) && (Deltat < -2*t_res)
23             e = -5;
24         elseif (Deltat >= -2*t_res) && (Deltat < -1*t_res)
25             e = -3;
26         elseif (Deltat >= -1*t_res) && (Deltat < 0)
27             e = -1;
28         elseif (Deltat >= 0)        && (Deltat < 1*t_res)
29             e = 1;
```

```
30            elseif (Deltat >= 1*t_res)  && (Deltat < 2*t_res)
31                e = 3;
32            elseif (Deltat >= 2*t_res)  && (Deltat < 3*t_res)
33                e = 5;
34            else
35                e = 7;
36            end
37        end
38    end
```

If the TDC is modeled as a BPD, the bang-bang quantization in (3.4) can be easily modeled using the MATLAB `sign(·)` function, as shown in Line 6. For TDCs with a number of bits ≥ 2, `if... elseif... end` statements are used to make the Δt-based decision for quantized e, as shown in Lines 9–17, according to (3.3). Notice that, during the iteration stage of simulation, the function of TDC detection (as in Line 80 of Listing 2.1) is performed once per simulated reference cycle. Such decision statements being executed repetitively can degrade the simulation speed.

Alternatively, we can simplify the model for the generic mid-rise equidistant quantization by using [11]

$$e_{N_b\text{-bit}}(\Delta t) = 2 \cdot \left\lfloor \frac{\Delta t}{t_{\text{res}}} \right\rfloor + 1,$$

where $\lfloor \cdot \rfloor$ denotes the floor function. Listing 3.2 shows the simplified TDC model.

Listing 3.2 Simplified MATLAB model of a generic mid-rise TDC.

```
function [Deltat,e] = TDC_simple(t_ref,t_div,t_res)
    %% Input time difference
    Deltat = t_ref - t_div;
    %% Generic mid-rise TDC
    e       = 2 * floor(Deltat/t_res) + 1;
end
```

Note that, although the simplified TDC model in Listing 3.2 accelerates the simulation compared to the previous model in Listing 3.1, it is less accurate in the sense that the simplified model essentially describes a mid-rise TDC with infinite dynamic range without saturation; whereas, in the other one, the TDC's output will be clipped to the maximum quantized values, if $|\Delta t|$ goes beyond the edge quantization thresholds (corresponding to the top and bottom cases in, e.g., (3.2) and (3.3)). This inaccuracy can affect the transient and locked-state simulations.

3.3 Comparative Study of TDC Linearized Analysis Frameworks

The generic TDC depicted in Fig. 3.2 exhibits a hard quantization nonlinearity and can be modeled by a linear gain element, to whose output is added the QE, as shown in Fig. 3.1b, as in

$$e[k] = K_{\text{TDC}} \cdot \Delta t[k] + q_{\text{TDC}}[k]. \tag{3.5}$$

Linearized analysis involves estimating the effective gain (K_{TDC}) and predicting the power of the QE (q_{TDC}). The next subsections compare two frameworks for TDC linearized analysis: the first is an MMSE estimator, while the second [8] is extended from [1], which conceptually originates from the idea of conventional linear analysis based on a small-signal approximation (linearization) at an operating point.

In the context of linearized analysis, the input, output, and QE of the TDC quantizer, $\Delta t[k]$, $e[k]$, and $q_{\text{TDC}}[k]$, are effectively stochastic processes when the loop operates in the locked state. We assign the random variables (RV's), Δt, E, and Q_{TDC} to represent their ensembles in order to facilitate statistical analysis hereinafter.

3.3.1 *Minimum-Mean-Square-Error Estimator*

When the loop is locked, the QE of a mid-rise TDC has zero mean. By rearranging (3.5), the variance (power) of the QE can be expressed as

$$\mathbb{E}\left(Q_{\text{TDC}}^2\right) \triangleq \mathbb{E}\left((E - K_{\text{TDC}} \cdot \Delta t)^2\right) = \mathbb{E}\left(E^2\right) + K_{\text{TDC}}^2 \cdot \mathbb{E}\left(\Delta t^2\right) - 2 \cdot K_{\text{TDC}} \cdot \mathbb{E}\left(\Delta t \cdot E\right),$$

where $\mathbb{E}(\cdot)$ denotes the expectation operator.

According to the MMSE principle [10], the best prediction for the effective gain is the one that minimizes the mean-square (quantization) error between the estimated (linearized) and actual (quantized) outputs. This can be found when

$$\frac{\partial \mathbb{E}\left(Q_{\text{TDC}}^2\right)}{\partial K_{\text{TDC}}} = 2 \cdot K_{\text{TDC}} \cdot \mathbb{E}\left(\Delta t^2\right) - 2 \cdot \mathbb{E}\left(\Delta t \cdot E\right) = 0.$$

Hence, the best estimate of the linearized gain becomes

$$\hat{K}_{\text{TDC}} = \frac{\mathbb{E}\left(\Delta t \cdot E\right)}{\mathbb{E}\left(\Delta t^2\right)}, \tag{3.6}$$

and $\partial^2 \mathbb{E}\left(Q_{\text{TDC}}^2\right) / \partial K_{\text{TDC}}^2 = 2 \cdot \mathbb{E}\left(\Delta t^2\right) > 0$ confirms that the estimate in (3.6) indeed minimizes the estimation error. In other words, the MMSE estimator calculates the TDC's effective gain as the ratio of the cross-correlation between the output (E) and input (Δt) to the autocorrelation of the input.

Similarly, recalling the relationship in (3.5), the variance of the output can be calculated as

$$\mathbb{E}\left(E^2\right) \triangleq K_{\mathrm{TDC}}^2 \cdot \mathbb{E}\left(\Delta t^2\right) + \mathbb{E}\left(Q_{\mathrm{TDC}}^2\right) + 2 \cdot K_{\mathrm{TDC}} \cdot \mathbb{E}\left(\Delta t \cdot Q_{\mathrm{TDC}}\right). \tag{3.7}$$

In the last term, $\mathbb{E}\left(\Delta t \cdot Q_{\mathrm{TDC}}\right) \triangleq \mathbb{E}\left(\Delta t \cdot E - K_{\mathrm{TDC}} \cdot \Delta t^2\right) = \mathbb{E}\left(\Delta t \cdot E\right) - K_{\mathrm{TDC}} \cdot \mathbb{E}\left(\Delta t^2\right)$, substituting the MMSE estimate in (3.6) yields that

$$\mathbb{E}\left(\Delta t \cdot Q_{\mathrm{TDC}}\right) = 0. \tag{3.8}$$

This indicates that the MMSE estimate for the TDC's linearized gain guarantees that the inputs (Δt) and QE (Q_{TDC}) are uncorrelated, i.e., their cross-correlation equals zero. It can be shown that, in this case, the best estimated output ($\hat{K}_{\mathrm{TDC}} \cdot \Delta t$) is also uncorrelated with the estimation error.

Substituting (3.8) into (3.7) gives the power of the QE estimated by the MMSE estimator, that is:

$$\mathbb{E}\left(Q_{\mathrm{TDC}}^2\right) = \mathbb{E}\left(E^2\right) - \hat{K}_{\mathrm{TDC}}^2 \cdot \mathbb{E}\left(\Delta t^2\right). \tag{3.9}$$

Note that this relationship holds only if the relationship (3.8) is satisfied, i.e., the estimation error and the input are uncorrelated; this is inherently guaranteed by the MMSE estimator. By contrast, other estimation methods do not necessarily guarantee this lack of correlation; using (3.9) in this case can lead to incorrect estimation results. An example will be shown in Sect. 3.3.3, where, considering a non-Gaussian distribution for Δt, the "small-signal approximation" method which estimates the linearized gain with an expression that is different from (3.6) and the QE power with (3.9), yields linearized estimates that do not match their simulated counterparts for such mid-rise TDCs.

As the distribution of Δt in the locked state has zero mean, and Q_{TDC} is also zero mean, we use $\sigma_{\Delta t}^2 = \mathbb{E}\left(\Delta t^2\right)$ and $\sigma_{Q_{\mathrm{TDC}}}^2 = \mathbb{E}\left(Q_{\mathrm{TDC}}^2\right)$ to simplify (3.6) and (3.9) as

$$\hat{K}_{\mathrm{TDC}} = \frac{\mathbb{E}\left(\Delta t \cdot E\right)}{\sigma_{\Delta t}^2}, \tag{3.10}$$

and

$$\sigma_{Q_{\mathrm{TDC}}}^2 = \mathbb{E}\left(E^2\right) - \hat{K}_{\mathrm{TDC}}^2 \cdot \sigma_{\Delta t}^2. \tag{3.11}$$

The cross-correlation $\mathbb{E}\left(\Delta t \cdot E\right)$ in (3.10) and the output autocorrelation $\mathbb{E}\left(E^2\right)$ in (3.11) can be calculated explicitly by analyzing the probability distribution of the output that is deterministically mapped from the input RV. This will be achieved in the following subsections by considering the DPLL's input jitter distributions in the integer-N and fractional-N modes of operation. The mean-square input time jitter, $\sigma_{\Delta t}^2$, is assumed to be known; it can be predicted as described in the next chapter, simulated, or measured.

3.3.2 "Small-Signal Approximation at the Locking Point"

By contrast with the MMSE estimator approach, the method for TDC linearization introduced in [8], using the same terminology as above, is

$$K_{\mathrm{TDC}} = \left. \frac{\partial \mathbb{E}\left(E(x + \Delta t)\right)}{\partial x} \right|_{x=0}, \tag{3.12}$$

where the output E is defined as a function of $x + \Delta t$ and x represents a deviation from the origin. At the point x, the statistical quantities are evaluated. This method calculates the gradient of the expected TDC output and evaluates it at the "locking point," $x = 0$. Afterward, the method to calculate the variance of the TDC QE is the same as (3.11) for the MMSE estimator. This linearization method, by taking the first-order derivative of the output about the operating point, conforms to the idea of conventional linear analysis based on "small-signal approximation." However, the difference between (3.10) and (3.12) means that they do not necessarily provide the same estimation results. More importantly, the linear gain derived from (3.12) cannot guarantee the minimization of the QE, which in turn cannot guarantee that the QE and the estimated output are uncorrelated. Using (3.11) after (3.12) can lead to an unreasonable estimation of the variance of the TDC QE, as will be shown in the following subsection.

3.3.3 Linear Model with (Hypothetical) Uniform Input Distribution

This subsection derives explicit estimates for the TDC's effective gain and QE variance. A 2-bit TDC or BPD interacting with an input jitter that has a uniform distribution is considered, and we show the different results produced by the two methods. The input probability density function (PDF) is expressed as

$$f_{\Delta t}(x) = \begin{cases} \frac{1}{2b} & \text{for } -b < x < b, \\ 0 & \text{otherwise,} \end{cases} \tag{3.13}$$

which feeds a 2-bit TDC quantizer with the characteristic described in (3.3). We assume that $t_{\mathrm{res}} < b$ to make full use of the quantization levels.

Estimation with MMSE Estimator

The cross-correlation in (3.10) is calculated as

$$\mathbb{E}(E \cdot \Delta t) = -3 \cdot \int_{-b}^{-t_{\text{res}}} x \cdot f_{\Delta t}(x)\mathrm{d}x - 1 \cdot \int_{-t_{\text{res}}}^{0} x \cdot f_{\Delta t}(x)\mathrm{d}x$$
$$+1 \cdot \int_{0}^{t_{\text{res}}} x \cdot f_{\Delta t}(x)\mathrm{d}x + 3 \cdot \int_{t_{\text{res}}}^{b} x \cdot f_{\Delta t}(x)\mathrm{d}x,$$

where $f_{\Delta t}(x) = \frac{1}{2b}$. Noting that, $\int_p^q x\mathrm{d}x = \frac{1}{2}(q^2 - p^2)$, the expression above can be simplified as

$$\mathbb{E}(E \cdot \Delta t) = \frac{3}{2}b - \frac{t_{\text{res}}^2}{b}.$$

The denominator of (3.10) represents the variance of the zero-mean uniform input distribution, which is

$$\sigma_{\Delta t}^2 = \frac{b^2}{3}. \tag{3.14}$$

Together, the linearized gain for the TDC is expressed as

$$K_{\text{TDC}} = \frac{\left(\frac{3}{2}b - \frac{t_{\text{res}}^2}{b}\right)}{b^2/3} = \frac{9}{2b} - \frac{3 \cdot t_{\text{res}}^2}{b^3}, \tag{3.15}$$

which is negatively correlated with the quantization resolution t_{res}. Thus, for a given boundary of b for the input span, the linearized gain K_{TDC} (or the slope of the output-input plot) decreases when t_{res} increases, and vice versa.

The autocorrelation term in (3.11) is evaluated as

$$\mathbb{E}\left(E^2\right) = (-3)^2 \cdot \int_{-b}^{-t_{\text{res}}} f_{\Delta t}(x)\mathrm{d}x + (-1)^2 \cdot \int_{-t_{\text{res}}}^{0} f_{\Delta t}(x)\mathrm{d}x$$
$$+1^2 \cdot \int_{0}^{t_{\text{res}}} f_{\Delta t}(x)\mathrm{d}x + 3^2 \cdot \int_{t_{\text{res}}}^{b} f_{\Delta t}(x)\mathrm{d}x,$$

which, recalling $f_{\Delta t}(x) = \frac{1}{2b}$ and $\int_p^q \mathrm{d}x = q - p$, yields

$$\mathbb{E}\left(E^2\right) = 9 - 8 \cdot \frac{t_{\text{res}}}{b}. \tag{3.16}$$

Substituting (3.14), (3.15), and (3.16) into (3.11) yields the variance of the TDC's QE as

$$\sigma_{Q_{\text{TDC}}}^2 = \frac{9}{4} - \frac{8 \cdot t_{\text{res}}}{b} + \frac{9 \cdot t_{\text{res}}^2}{b^2} - \frac{3 \cdot t_{\text{res}}^4}{b^4}. \tag{3.17}$$

The estimation for a BPD can be obtained by assuming $t_{\text{res}} = b$ for (3.15) and (3.17), giving $K_{\text{BPD}} = 3/2b$ and $\sigma^2_{Q_{\text{BPD}}} = 1/4$.

Estimation with the Small-Signal Method

For this case, we have that

$$\begin{aligned}\mathbb{E}\left(E(x+\Delta t)\right) =& -3\cdot\int_{-b}^{-t_{\text{res}}-x} f_{\Delta t}(a)\mathrm{d}a - 1\cdot\int_{-t_{\text{res}}-x}^{-x} f_{\Delta t}(a)\mathrm{d}a \\ &+1\cdot\int_{-x}^{t_{\text{res}}-x} f_{\Delta t}(a)\mathrm{d}a + 3\cdot\int_{t_{\text{res}}-x}^{b} f_{\Delta t}(a)\mathrm{d}a \\ =&3-2\cdot F_{\Delta t}(-t_{\text{res}}-x)-2\cdot F_{\Delta t}(-x)-2\cdot F_{\Delta t}(t_{\text{res}}-x),\end{aligned}$$

where a is a dummy variable in place of the RV (Δt) for the expectation calculation, and $F_{\Delta t}(x) = \int_{-\infty}^{x} f_{\Delta t}(a)\mathrm{d}a$ is the cumulative distribution function (CDF) of Δt. Substituting into (3.12) yields

$$K_{\text{TDC}} = 2\cdot f_{\Delta t}(-t_{\text{res}}) + 2\cdot f_{\Delta t}(0) + 2\cdot f_{\Delta t}(t_{\text{res}}),$$

which, with $t_{\text{res}} < b$ assumed and, following (3.13), becomes

$$K_{\text{TDC}} = \frac{3}{b}. \tag{3.18}$$

Substituting (3.14), (3.16), and (3.18) into (3.11) yields the variance of the TDC QE as

$$\sigma^2_{Q_{\text{TDC}}} = 6 - 8\cdot\frac{t_{\text{res}}}{b}. \tag{3.19}$$

Similar analysis for a BPD yields that, using the small-signal method, $K_{\text{BPD}} = 1/b$ and $\sigma^2_{Q_{\text{BPD}}} = 2/3$.

Notice that, for both the 2-bit TDC and BPD cases, the linear estimates which have been derived using these methods are different. Counter-intuitively, unlike (3.15), the linearized effective gain in (3.18) derived for a 2-bit TDC using the small-signal method has a form that is independent of t_{res}, i.e., the quantization characteristics of the TDC *per se*. To check this result, simulations can be performed to find the linear gain of the TDC that minimizes the variance of the TDC QE, the estimation criterion introduced previously, while varying the t_{res} setting. For a fixed value of b, the estimation by the small-signal method is valid if and only if the best estimated K_{TDC} is constant at $3/b$ and independent of t_{res}.

In the next subsection, the dependency of K_{TDC} on t_{res} will be demonstrated by simulation. The results invalidate the small-signal approximation method [8], while matching the estimates produced using the MMSE estimator detailed in this subsection.

3.3.4 Verification

To compare the accuracy of the two methods, a test bench is set up as shown in Fig. 3.3. In the behavioral simulation, a sequence of 2^{18} random data samples $\Delta t[k]$ is generated following a uniform distribution with $\Delta t \sim U(-1, 1)$, i.e., $b = 1$. The jitter samples are quantized to produce $e[k]$ and scaled by a linear gain factor, K_{TDC}, to produce $\hat{e}[k]$. The variance of the QE, $\sigma^2_{Q_{\mathrm{TDC}}}$, is measured from the sequence of QEs, $q_{\mathrm{TDC}}[k] = e[k] - \hat{e}[k]$. As per the estimation criterion introduced in Sect. 3.3, the best estimate for K_{TDC} should be the one that minimizes the variance, $\sigma^2_{Q_{\mathrm{TDC}}}$. Therefore, in the simulation, one thousand K_{TDC} settings are swept through to find the minima of the $\sigma^2_{Q_{\mathrm{TDC}}}$ versus K_{TDC} plots, the coordinates of which should match the best estimated result.

For the BPD case, the MMSE estimator gives $K_{\mathrm{BPD}} = 3/2$ and $\sigma^2_{Q_{\mathrm{BPD}}} = 1/4$, while the small-signal method returns $K_{\mathrm{BPD}} = 1$ and $\sigma^2_{Q_{\mathrm{BPD}}} = 2/3$, which are marked using "star" and "triangle" symbols, respectively, in Fig. 3.4a. As the simulation results show, the linearized gain that minimizes the QE variance matches the prediction by the MMSE estimator, while that predicted by the small-signal approach does not.

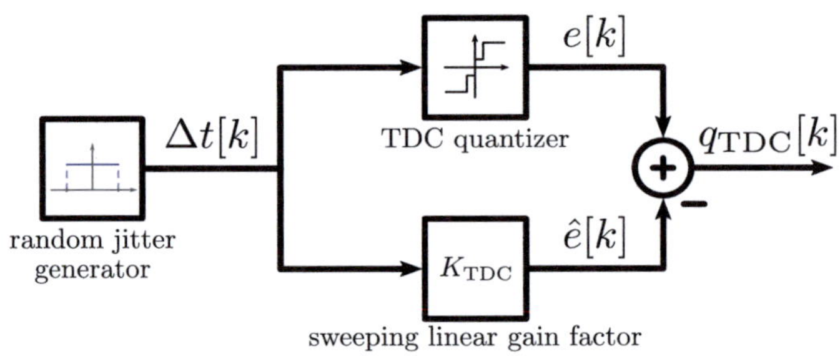

Fig. 3.3 Test bench to verify the estimation of the TDC's effective gain

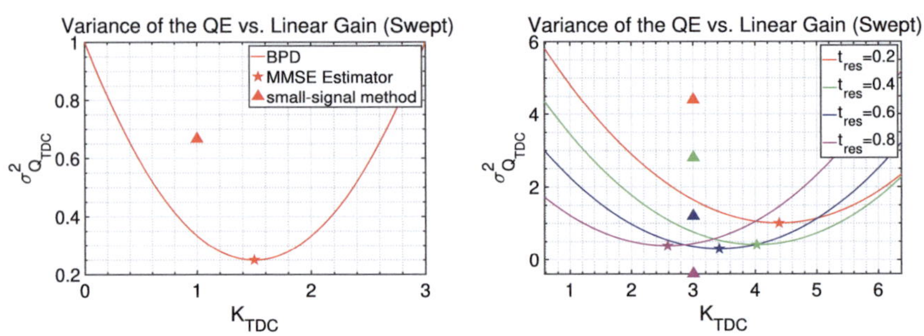

(a) Uniform Δt, BPD.

(b) Uniform Δt, 2-bit.

Fig. 3.4 Verification results showing the simulated $\sigma^2_{Q_{\mathrm{TDC}}}$ versus K_{TDC} plots. The estimates given by the small-signal approximation method and the MMSE estimator are marked with triangles and stars, respectively

Table 3.1 Estimated versus simulated linear gain and QE variance of the 2-bit TDC

	MMSE by (3.15), (3.17)		"Small-signal" by (3.18), (3.19)		Simulation	
t_{res}	K_{TDC}	$\sigma^2_{Q_{TDC}}$	K_{TDC}	$\sigma^2_{Q_{TDC}}$	K_{TDC}	$\sigma^2_{Q_{TDC}}$
0.2	4.38	1.005	3	4.4	4.385	1.004
0.4	4.02	0.413		2.8	4.025	0.415
0.6	3.42	0.301		1.2	3.410	0.303
0.8	2.58	0.381		–0.4	2.581	0.379

For the 2-bit TDC case, four TDC resolutions are considered, and the estimates using both methods arc tabulated in Table 3.1. The simulated results are extracted in Fig. 3.4b, where the estimates are marked using the same symbols as before. Simulations show that estimation using the MMSE estimator matches the simulation exactly, whereas the small-signal method calculates a value of K_{TDC} that is independent of t_{res}, as mentioned in the last subsection, and even gives an aphysical negative variance when $t_{res} = 0.8$. Altogether, the MMSE is demonstrated to outperform the small-signal method and to give the best estimates of the TDC's linear gain and QE variance.

3.3.5 *Conclusion*

In this section, two methods have been compared and contrasted for calculating the effective gain and the QE variance of a 2-bit TDC and a BPD considering its excitation with a uniformly distributed input signal, namely, the MMSE estimator method and the small-signal approximation method. Simulations confirm that the MMSE estimator always provides estimates that conform to the estimation criterion, making it the preferred method to use in linearized analysis. In the next sections, we will use the MMSE estimator to perform linearized analysis in the case of a generic N_b-bit TDC when the loop operates in the integer-N and fractional-N modes.

3.4 TDC Linear Model in Integer-N Mode

For an integer-N DPLL designed to have a moderate bandwidth that balances the noise contributions from the TDC and the DCO, the loop will operate in a so-called "stochastic resonance" regime when locked [2, 3, 12]. In this regime, the limit-cycle quantization of the TDC will be randomized by the physical noise sources. Consequently, the loop typically exhibits an input time jitter that has a zero-mean Gaussian distribution, $\Delta t \sim \mathcal{N}(0, \sigma^2_{\Delta t})$, the PDF of which can be expressed as

$$f_{\Delta t}(x) = \frac{1}{\sqrt{2\pi} \cdot \sigma_{\Delta t}} \cdot e^{-\frac{x^2}{2\sigma_{\Delta t}^2}}. \tag{3.20}$$

In this section, we first consider a generic N_b-bit TDC with equidistant thresholds, as presented in (3.2) and in black in Fig. 3.2. To simplify the notation, we normalize the TDC's input threshold to the RMS input jitter, as in

$$\gamma = \frac{t_{\text{res}}}{\sigma_{\Delta t}}. \tag{3.21}$$

For this case, the cross-correlation in (3.10) is expressed as

$$\mathbb{E}\left(\Delta t \cdot E\right) = \int_{-\infty}^{\infty} e_{N_b\text{-bit}}(x) \cdot x \cdot f_{\Delta t}(x) \mathrm{d}x. \tag{3.22}$$

$e_{N_b\text{-bit}}(x)$, as defined in (3.2), is an odd function of x, x is odd *per se*, and the zero-mean Gaussian PDF, $f_{\Delta t}(x)$, is an even function. Hence, the integrand in (3.22) is an even function of x, which simplifies the integral as

$$\mathbb{E}\left(\Delta t \cdot E\right) = 2 \cdot \int_{0}^{\infty} e_{N_b\text{-bit}}(x) \cdot x \cdot f_{\Delta t}(x) \mathrm{d}x.$$

Substituting the quantization characteristic in (3.2) yields

$$\mathbb{E}\left(\Delta t \cdot E\right) = 2\Bigg[\int_{0}^{t_{\text{res}}} x \cdot f_{\Delta t}(x)\mathrm{d}x + 3\int_{t_{\text{res}}}^{2t_{\text{res}}} x \cdot f_{\Delta t}(x)\mathrm{d}x$$
$$+ \cdots + \left(2^{N_b} - 1\right)\int_{\left[2^{(N_b-1)}-1\right]t_{\text{res}}}^{\infty} x \cdot f_{\Delta t}(x)\mathrm{d}x\Bigg]. \tag{3.23}$$

Note that definite integrals in the form of $\int_a^b x \cdot f_{\Delta t}(x)\mathrm{d}x$ appear in a sequence in the above equation; these represent the "segmental expectation" of a distribution, i.e., the expected value of a random variable for its outcomes confined within the segment from a to b. For a zero-mean Gaussian distribution, whose PDF is defined by (3.20), the generic segmental expectation can be calculated as

$$\int_a^b x \cdot f_{\Delta t}(x)\mathrm{d}x = \frac{\sigma_{\Delta t}}{\sqrt{2\pi}} \cdot \left(e^{-\frac{a^2}{2\sigma_{\Delta t}^2}} - e^{-\frac{b^2}{2\sigma_{\Delta t}^2}}\right).$$

A detailed derivation procedure for this can be found in [13]. Therefore, recalling (3.21), Eq. (3.23) becomes

$$\mathbb{E}\left(\Delta t \cdot E\right) = \sqrt{\frac{2}{\pi}} \cdot \sigma_{\Delta t} \cdot \left(1 + 2 \cdot \sum_{k=1}^{2^{(N_b-1)}-1} e^{-\frac{k^2\gamma^2}{2}}\right),$$

which, substituted into (3.10), yields

$$\hat{K}_{\mathrm{TDC}} = \sqrt{\frac{2}{\pi}} \cdot \frac{1}{\sigma_{\Delta t}} \cdot \left(1 + 2 \cdot \sum_{k=1}^{2^{(N_b-1)}-1} e^{-\frac{k^2\gamma^2}{2}}\right), \tag{3.24}$$

in LSB/s.

The output autocorrelation in (3.11) is

$$\mathbb{E}\left(E^2\right) = \int_{-\infty}^{\infty} e^2_{N_b\text{-bit}}(x) \cdot f_{\Delta t}(x)\mathrm{d}x = 2 \cdot \int_{0}^{\infty} e^2_{N_b\text{-bit}}(x) \cdot f_{\Delta t}(x)\mathrm{d}x, \tag{3.25}$$

because $e^2_{N_b\text{-bit}}(x)$ and $f_{\Delta t}(x)$ are both even functions of x. Expanding (3.25) yields

$$\begin{aligned}\mathbb{E}\left(E^2\right) = 2 \cdot 1 \cdot \int_0^{t_{\mathrm{res}}} f_{\Delta t}(x)\mathrm{d}x + 2 \cdot 3^2 \cdot \int_{t_{\mathrm{res}}}^{2t_{\mathrm{res}}} f_{\Delta t}(x)\mathrm{d}x + \ldots \\ +2 \cdot (2^{N_b} - 1)^2 \cdot \int_{[2^{(N_b-1)}-1]t_{\mathrm{res}}}^{\infty} f_{\Delta t}(x)\mathrm{d}x,\end{aligned}$$

which, exploiting the Gaussian error function,[4]

$$\mathrm{erf}\left(\frac{x}{\sqrt{2}\sigma_{\Delta t}}\right) = 2 \cdot F_{\Delta t}(x) - 1 = \sqrt{\frac{2}{\pi}} \cdot \frac{1}{\sigma_{\Delta t}} \int_0^x e^{-\frac{a^2}{2\sigma^2_{\Delta t}}}\mathrm{d}a,^4 \tag{3.26}$$

and recalling (3.21), can be simplified as

$$\mathbb{E}\left(E^2\right) = (2^{N_b} - 1)^2 - 8 \cdot \sum_{k=1}^{2^{(N_b-1)}-1} k \cdot \mathrm{erf}\left(\frac{k \cdot \gamma}{\sqrt{2}}\right). \tag{3.27}$$

Finally, substituting (3.24) and (3.27) into (3.11) yields the variance of the TDC's QE as

$$\sigma^2_{Q_{\mathrm{TDC}}} = \left[(2^{N_b} - 1)^2 - 8 \cdot \sum_{k=1}^{2^{(N_b-1)}-1} k \cdot \mathrm{erf}\left(\frac{k \cdot \gamma}{\sqrt{2}}\right)\right] - \frac{2}{\pi} \cdot \left(1 + 2 \cdot \sum_{k=1}^{2^{(N_b-1)}-1} e^{-\frac{k^2\gamma^2}{2}}\right)^2, \tag{3.28}$$

in LSB^2.

Note that the predictions for the linear gain and the variance of the QE of an N_b-bit TDC given in (3.24) and (3.28) are the same as those derived in [8] using the "small-signal" method. This can be regarded as a coincidental result that arises from the special statistical properties of the Gaussian distribution.

---

[4] Here, $F_{\Delta t}(x) = \int_{-\infty}^{x} f_{\Delta t}(a)\mathrm{d}a$ is the CDF of the RV, Δt.

Note that, in principle, when γ is large, i.e., $t_{\mathrm{res}} \gg \sigma_{\Delta t}$, most of the input jitter will be accommodated by the two middlemost quantization levels; in this case, the outputs will be mostly ± 1s, which approximates to the behavior of a BPD. In the limit, substituting $\gamma \to \infty$ and $N_b = 1$ into (3.24) and (3.28) yields the MMSE estimates for the effective gain and the power of the QE of a BPD in (3.4) as

$$\hat{K}_{\mathrm{BPD}} = \sqrt{\frac{2}{\pi}} \cdot \frac{1}{\sigma_{\Delta t}},$$
$$\sigma^2_{Q_{\mathrm{BPD}}} = \frac{\pi - 2}{\pi}, \tag{3.29}$$

which are precisely the well-known results in the literature for integer-N BBPLLs, e.g., [2, 6, 7, 9].

Note that DPLLs operating in integer-N mode can exhibit their best possible PN and jitter performances. The analysis in this section that assumes Gaussian input jitter is also valid for advanced fractional-N DPLLs which use a DTC to cancel the QE of the MMDC. When the DTC functions successfully in such systems, the output PN and jitter approach their integer-N levels. Hence, the input jitter also has an approximate Gaussian distribution, similar to the integer-N case. This is verified in Chap. 7, where the simulated input jitter of a fractional-N DPLL using a finite-resolution DTC is shown to exhibit a zero-mean Gaussian distribution in the locked state.

3.5 Introduction to Fractional-*N* Operation Enabled by a Multi-modulus Divider Controller (MMDC)

Fractional-N operation is achieved via cycle-to-cycle modulation of the instantaneous feedback divide ratio among several integer values, typically using a DDSM as the MMDC. The MMDC, denoted by the (magenta) "MMDC" block in Fig. 2.3, is used to generate $N_{\mathrm{div}}[k]$ to set the kth divide ratio of the MMD.

3.5.1 Digital-Accumulator-Based First-Order Error Feedback Modulator (EFM1)

In its digital implementation, a first-order DDSM can be realized using a modulo-M accumulator, as shown in Fig. 3.5 [14], whose outputs can be expressed as

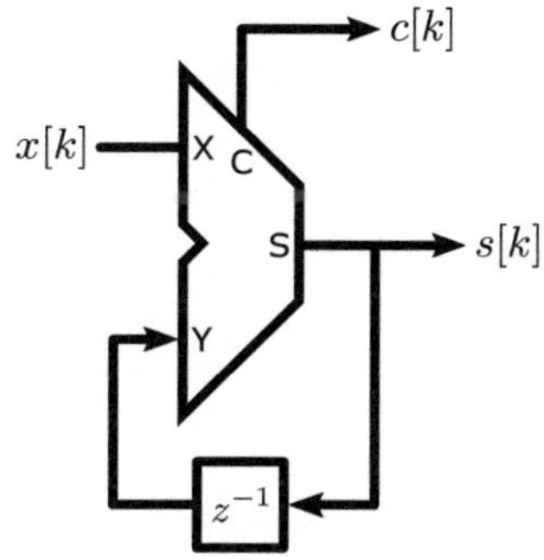

Fig. 3.5 A first-order error feedback modulator (EFM1) implementation based on a modulo-M accumulator

$$c[k] = \begin{cases} 1, & \text{if } x[k] + s[k-1] \geq M, \\ 0, & \text{otherwise,} \end{cases} \tag{3.30a}$$

$$s[k] = x[k] + s[k-1] - M \cdot c[k], \tag{3.30b}$$

where $x[k] \in \{0, 1, \ldots, M-1\}$ is the input, $c[k]$ is the output "carry" bit, and $s[k]$ is the "sum" signal. Note that, as will be discussed in the following subsection, a higher-order DDSM is typically implemented by cascading multiple such stages; in some recent innovations, such as the ENOP DDSM, the constituent stage can accommodate multiple full-scale inputs and allow a multi-bit carry signal. Therefore, to generalize the analysis, Eq. (3.30a) can be updated as

$$c[k] = \left\lfloor \frac{x[k] + s[k-1]}{M} \right\rfloor, \tag{3.31}$$

where $\lfloor \cdot \rfloor$ denotes the floor (largest integer less than) function. Equation (3.30b) is unchanged in this case.

Listing 3.3 shows the MATLAB model of the modulo-M adder depicted in Fig. 3.5 as per (3.31) and (3.30b).

Listing 3.3 MATLAB model of the modolo-M adder comprising a first-order DDSM.

```
function [C,S] = adder_modulo_M(X, Y, M)
    %% Carry (the quotient after division by M)
    C = floor((X + Y)/M);
    %% Sum (the remainder after modulo-M operation)
    S = X + Y - M * C;
end
```

Rearranging (3.30b) yields the output as

$$c[k] = \frac{x[k]}{M} + \frac{s[k-1] - s[k]}{M}, \tag{3.32}$$

which, when transformed to the z domain and rearranged into the form of a generic first-order DDSM, i.e., "$Y(z) = X(z) + \left(1 - z^{-1}\right) \cdot Q(z)$," becomes

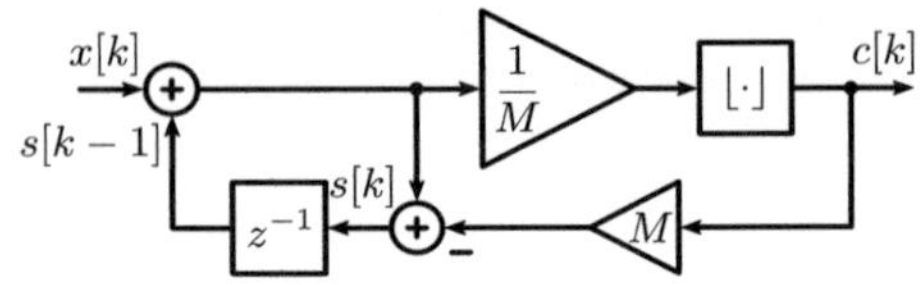

Fig. 3.6 Behavioral model of an EFM1 with generic modulus equal to M

$$C(z) = \frac{X(z)}{M} + \left(1 - z^{-1}\right) \cdot \frac{-S(z)}{M}, \tag{3.33}$$

which applies a linear signal transfer function to the signal term $(X(z)/M)$ and a first-order highpass-shaping noise transfer function (NTF) to the QE term $(-S(z)/M)$. One can notice from (3.33) that the original QE at source is

$$Q(z) = -\frac{S(z)}{M}, \tag{3.34}$$

which is directly related to $s[k]$.

As shown in Fig. 3.5, since the error term is fed back to be added to the input, the first-order DDSM is also called a first-order error feedback modulator. Figure 3.6 depicts the behavioral model of the EFM1 as per the above discussion, which is adopted in the literature, e.g., [15–18].

In fractional-N PLLs, in order to implement a fractional divide ratio, $N = N_{\text{int}} + X/M$ in (1.3). A constant integer FCW, X, can be applied in place of $x[k]$ in (3.32). The output from the MMDC, $c[k]$, can be added to the integer part, N_{int}, to yield the instantaneous divide ratio, $N_{\text{div}}[k]$, as in

$$N_{\text{div}}[k] = N_{\text{int}} + c[k] = N_{\text{int}} + \frac{X}{M} + \frac{s[k-1] - s[k]}{M}. \tag{3.35}$$

By definition, the QE of a quantizer is calculated as the quantized output subtracting the desired input. Substituting (1.3) into (3.35), the QE of an EFM1 MMDC becomes

$$q_{\text{EFM1}}[k] \triangleq N_{\text{div}}[k] - N = \frac{s[k-1] - s[k]}{M}. \tag{3.36}$$

As will be shown in Chap. 4, the MMDC's accumulated QE is injected into the loop and affects the system jitter; this is defined by

$$q_{\text{MMDC}}^{\text{acc}}[k] = \sum_{i=0}^{k} q_{\text{MMDC}}[i]. \tag{3.37}$$

Substituting (3.36) into (3.37) and considering the MMDC as being initialized to $s[k] = 0, \forall k \leq 0$ yields the EFM1's AQE as

$$q_{\text{EFM1}}^{\text{acc}}[k] = \frac{-s[k]}{M},$$

which is directly related to the sum signal of the accumulator. In the case of a constant X, $s[k]$ is a periodic sawtooth waveform with wrapping at overflow and hence its frequency spectrum is rich in tones. If an EFM1 is used as the MMDC, these spurious tones can degrade the output PN significantly. Therefore, MASH DDSMs based on combining the carry outputs of cascaded EFM1 stages are preferred to be used as the MMDC; these help to improve the randomness of the AQE as well as implementing higher-order noise-shaping.

3.5.2 *Cascaded-EFM1-Based Multi-stage Noise-Shaping (MASH) MMDC*

MASH MMDCs are implemented by cascading multiple EFM1 stages, where the input stage is driven by X and the following stages are driven by the sum signal of the preceding stage. The carry bits of the stages are combined in an error cancellation network to produce $y[k]$ which is added to N_{int} to yield $N_{\text{div}}[k]$, as shown in (1.2). For example, second- and third-order MASH (denoted MASH 1-1 and MASH 1-1-1, respectively) MMDCs are shown in Fig. 3.7.

Recalling (3.32), the carry signal from the nth accumulator, where $n \geq 2$, is

$$c_n[k] = \frac{s_{n-1}[k]}{M} + \frac{s_n[k-1] - s_n[k]}{M}, \tag{3.38}$$

where a subscript indicates the stage with which a signal is associated. The Z-transform of (3.38) is

$$C_n(z) = \frac{S_{n-1}(z)}{M} + \left(1 - z^{-1}\right) \cdot \frac{-S_n(z)}{M}. \tag{3.39}$$

Recalling the definition in (3.34), Eq. (3.39) can be rewritten as

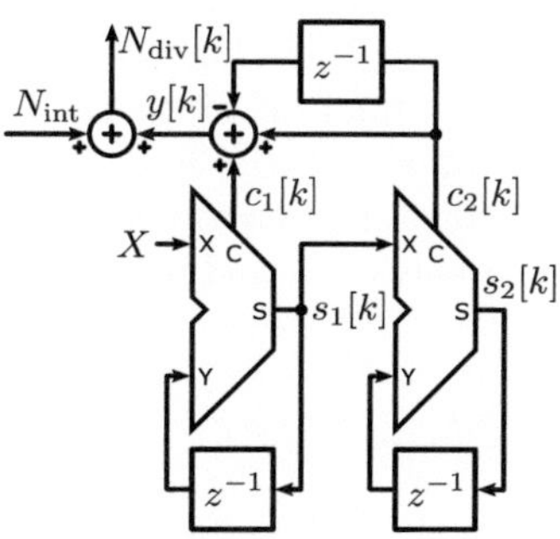

(a) MASH 1-1.

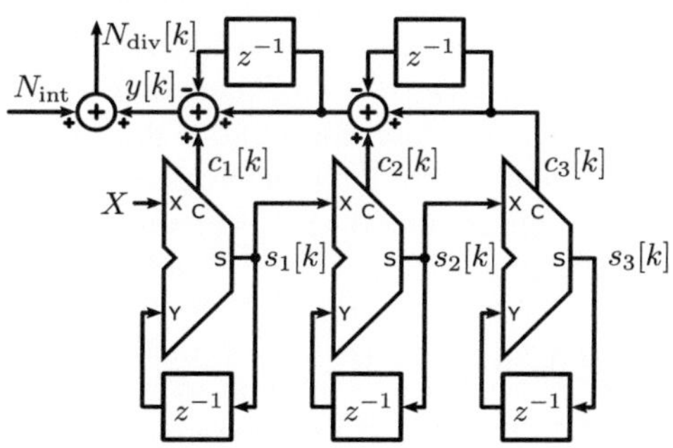

(b) MASH 1-1-1.

Fig. 3.7 Implementations of MASH MMDCs

$$C_n(z) = -Q_{n-1}(z) + \left(1 - z^{-1}\right) \cdot Q_n(z), \tag{3.40}$$

where $Q_n(z)$ stands for the QE injected into the nth stage.

Inspecting Fig. 3.7, the output of the MASH DDSM, $y[k]$, is obtained (represented in the z domain) as

$$Y(z) = \sum_{n=1}^{l} \left(1 - z^{-1}\right)^{n-1} \cdot C_n(z), \tag{3.41}$$

where l is the order (number of EFM1 stages) of the MASH. Substituting (3.40) into (3.41), recalling that the input to the first stage is X and assuming that all of the stages have the same modulus, yields

$$Y(z) = \frac{X}{M} + \left(1 - z^{-1}\right)^{l} \cdot Q_l(z), \tag{3.42}$$

where $Q_l(z) = -S_l(z)/M$ is the QE injected into the lth (final) stage of the MASH. Equation (3.42) shows that an lth-order MASH DDSM cancels all the QEs arising from the stages prior to the final stage. It also implements lth-order highpass noise-shaping of the QE injected into the final stage. The overall MMDC's QE is hence

$$Q_{\mathrm{MMDC}}(z) \triangleq Y(z) - \frac{X}{M} = \left(1 - z^{-1}\right)^{l} \cdot Q_l(z). \tag{3.43}$$

Finally, integrating (3.43) yields the MMDC's AQE as

$$Q_{\mathrm{MMDC}}^{\mathrm{acc}}(z) \triangleq \frac{Q_{\mathrm{MMDC}}(z)}{1 - z^{-1}} = \left(1 - z^{-1}\right)^{l-1} \cdot Q_l(z), \tag{3.44}$$

which indicates that, when using an lth-order MASH DDSM, if the output is sufficiently randomized, the MASH provides $(l-1)$th-order highpass shaping of the white QE. In other words, the PSD of the AQE of an lth-order MMDC has a $+20 \cdot (l-1)$ dB/dec slope at low frequencies.

Figure 3.8 shows the simulated power spectra of the AQE of MASH 1-1 and 1-1-1 DDSMs matching the prediction given by (3.44). In this example, the DDSM operates at $F_{\mathrm{ref}} = 100$ MHz. The simulated AQE is normalized to the LSB of the DDSM's output, $y[k]$; therefore, the PSD of the lth-stage QE equals $1/(12 \cdot F_{\mathrm{ref}})$, i.e., the power of the QE of a 1-bit quantizer uniformly distributed in the noise bandwidth, F_{ref}. As predicted, the MASH 1-1 and 1-1-1 DDSMs impose first- and second-order noise-shaping to their AQE.

MATLAB Model of MASH MMDCs with Programmable Order

Listing 3.4 shows the MATLAB model of the MMDCs that generate the sequence of instantaneous divide ratio $N_{\mathrm{div}}[k]$ in the fractional-N mode. The order of the MMDC is programmable, namely when `MASH_order` $= 0, 1, 2, 3$, the MMDC is switched

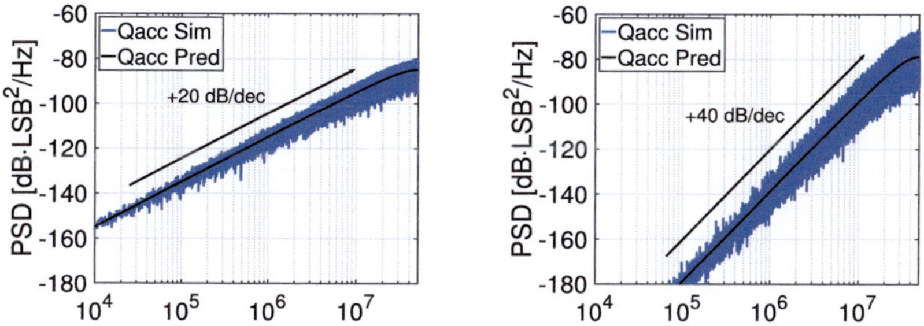

(a) MASH 1-1. (b) MASH 1-1-1.

Fig. 3.8 Simulated PSDs of the AQE of the second- and third-order MASH DDSMs versus their theoretical prediction by (3.44)

to work as an integer-N, EFM1, MASH 1-1, and 1-1-1 controller. In the integer-N case (`MASH_order` $= 0$), $N_{\mathrm{div}}[k] = N_{\mathrm{int}},\ \forall k \in \mathbb{N}^{+}$, namely the divide ratio is always equal to the nominal integer ratio; in the fractional-N cases (`MASH_order` $= 1, 2, 3$), the average value of $N_{\mathrm{div}}[k]$ equals the nominal divide ratio, $N = N_{\mathrm{int}} + X/M$.

Listing 3.4 MATLAB model of MASH MMDCs with different order.

```
function [N_div,q_acc] = MMDC(MASH_order,X,M,N_int,ncycles)
    %% Local preallocation
    if MASH_order >= 1
        c1 = zeros(1,ncycles);
        s1 = zeros(1,ncycles);
    end
    if MASH_order >= 2
        c2 = zeros(1,ncycles);
        s2 = zeros(1,ncycles);
    end
    if MASH_order == 3
        c3 = zeros(1,ncycles);
        s3 = zeros(1,ncycles);
    end
    q_acc    = zeros(1,ncycles);
    %% Integer-N mode generation
    N_div   = N_int * ones(1,ncycles);
    %% EFM1 generation
    if      MASH_order == 1
        for k=2:ncycles
            [c1(k),s1(k)] = adder_modulo_M(X,s1(k-1),M);
            N_div(k)      = N_int + c1(k);
            q_acc(k)      = - s1(k);
        end
    %% MASH 1-1 generation
    elseif MASH_order == 2
        for k=2:ncycles
            [c1(k),s1(k)] = adder_modulo_M(X,     s1(k-1),M);
            [c2(k),s2(k)] = adder_modulo_M(s1(k),s2(k-1),M);
            N_div(k)      = N_int + c1(k) + c2(k)-c2(k-1);
            q_acc(k)      = - s2(k) + s2(k-1) ;
        end
    %% MASH 1-1-1 generation
    elseif MASH_order == 3
```

```
35          for k=3:ncycles
36              [c1(k),s1(k)] = adder_modulo_M(X,      s1(k-1),M);
37              [c2(k),s2(k)] = adder_modulo_M(s1(k),s2(k-1),M);
38              [c3(k),s3(k)] = adder_modulo_M(s2(k),s3(k-1),M);
39              N_div(k)      = N_int + c1(k) + c2(k)-c2(k-1) + c3(k)
                   -2*c3(k-1)+c3(k-2);
40              q_acc(k)      = - s3(k) + 2*s3(k-1) - s3(k-2);
41          end
42      end
43  end
```

Note that local preallocation of memory spaces is programmed in Lines 3–14; this helps to speed up the simulation. The `if...elseif...end` decision statement to control the order of the MMDC is always executed outside of the `for` loop to generate $N_{\text{div}}[k]$s cycle by cycle; this also helps to facilitate fast simulation.

This function is called in Line 55 in the initialization stage of the main file, to generate the cycle-wise divide ratios that are ready to be used in the MMDC in the main simulation iterations.

Note that, in simulation and also hardware implementation, the MMDC's AQE signal can be extracted via the inverse Z-transform of (3.44). This is modeled in Lines 23, 31, and 40 of Listing 3.4.

3.6 TDC Linear Model in Fractional-*N* Mode

In practice, it is preferable to use an MMDC of order greater than one, e.g., MASH 1-1, 1-1-1, etc., to randomize its output. If the MMDC's QE is not canceled prior to its injection into a mid-rise TDC, the input jitter will be overwhelmed by the AQE arising from the MMDC. As we will derive in this section, this direct injection will attenuate the linear gain of the TDC and make it produce much more input-referred QE. The degraded TDC gain in turn shrinks the loop bandwidth drastically and drives the DPLL into an effectively "open-loop" mode of operation with degraded PN and jitter performance. This phenomenon has been reported in [7, 19] and will be analyzed in the next chapter.

In this case, as shown in Fig. 3.9, the pattern and distribution of the input jitter, Δt, are dominated by those of the AQE of the MMDC, whose probability distribution, in turn, depends on the architecture of the MMDC being used. This distribution is bounded and is typically non-Gaussian. For example, the PDFs of the AQEs of different-order MASH MMDCs are provided in [17]; those of ENOP MMDCs are provided in [18]. Closed-form expressions for the PDFs can be derived explicitly using the procedure presented in [20].

In this section, we present a linearized estimation method for the TDC operating in the fractional-N mode without cancellation of the MMDC's QE. As examples, we choose fractional-N operation facilitated by commonly used MASH 1-1 and MASH 1-1-1 MMDCs.

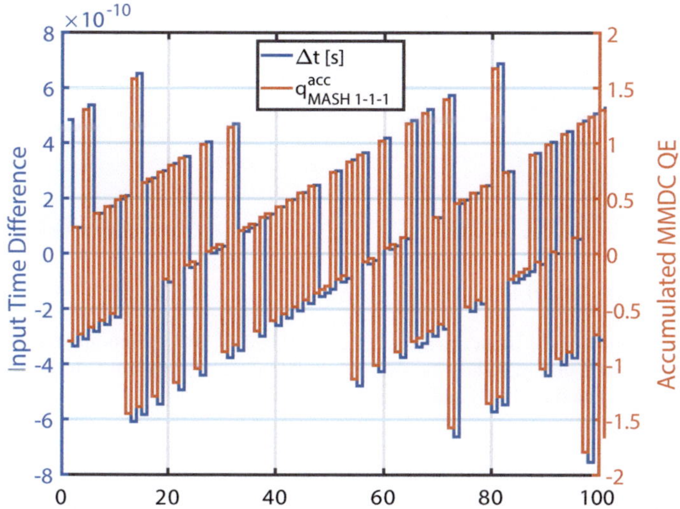

Fig. 3.9 Simulated locked-state TDC input time difference (Δt) of a MASH 1-1-1-enabled fractional-N BBPLL without DTC-based cancellation; the pattern of Δt is dominated by the MMDC's AQE ($q^{\mathrm{acc}}_{\mathrm{MASH\ 1\text{-}1\text{-}1}} \in (-2, 2)$) transferred to the time domain

3.6.1 With a MASH 1-1 MMDC—Triangular Δt Distribution

The AQE, $Q^{\mathrm{acc}}_{\mathrm{MMDC}}$, from a sufficiently randomized MASH 1-1 MMDC has a zero-mean triangular distribution within the range of $(-1, 1)$, whose PDF is expressed as [17]

$$f_{Q^{\mathrm{acc}}_{\mathrm{MASH\ 1\text{-}1}}}(x) = \begin{cases} x+1 & \text{for } -1 \le x < 0, \\ -x+1 & \text{for } 0 \le x \le 1. \end{cases}$$

In the time domain, the MMDC's AQE referred to the TDC input is scaled by the period of the DCO, T_{DCO}, [19, 21]. As the magnitude of the scaled QE is much greater than the integer-N input jitter, it dominates the input jitter, as in $\Delta t \approx T_{\mathrm{DCO}} \cdot Q^{\mathrm{acc}}_{\mathrm{MMDC}}$. Therefore, after PDF transformation to the time domain [22], the PDF of the input jitter in this case can be expressed as

$$f_{\Delta t}(x) = \begin{cases} \frac{x}{T^2_{\mathrm{DCO}}} + \frac{1}{T_{\mathrm{DCO}}} & \text{for } -T_{\mathrm{DCO}} \le x < 0, \\ \frac{-x}{T^2_{\mathrm{DCO}}} + \frac{1}{T_{\mathrm{DCO}}} & \text{for } 0 \le x \le T_{\mathrm{DCO}}, \end{cases} \tag{3.45}$$

which is shown in Fig. 3.10a.

Using a procedure similar to that used in Sect. 3.4, one can derive the TDC's linear gain and the variance of the QE as

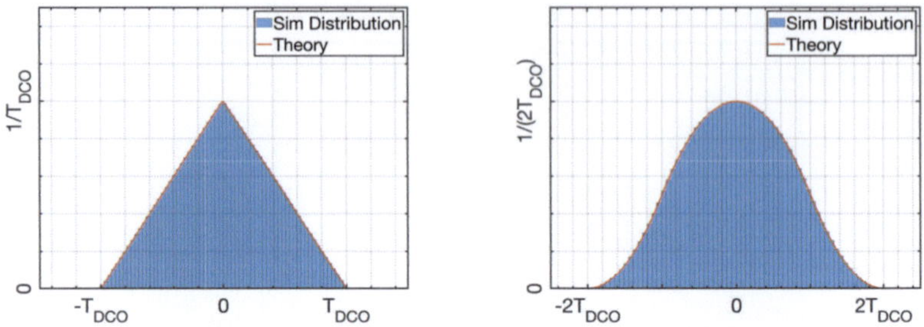

(a) MASH 1-1. (b) MASH 1-1-1.

Fig. 3.10 Distribution of the input time jitter of a fractional-N DPLL without DTC-based cancellation; the input jitter is dominated by the AQE of the MMDC. **a** MASH 1-1; **b** MASH 1-1-1. The distributions are predicted in (3.45) and (3.46), respectively

$$\hat{K}_{\text{TDC}} = \frac{2\cdot\left(2^{N_b}-1\right)}{T_{\text{DCO}}} - 4\cdot\sum_{k=1}^{2^{(N_b-1)}-1}\frac{3\cdot k^2 t_{\text{res}}^2}{T_{\text{DCO}}^3} - \frac{2\cdot k^3 t_{\text{res}}^3}{T_{\text{DCO}}^4},$$

$$\sigma^2_{Q_{\text{TDC}}} = (2^{N_b}-1)^2 - 16\cdot\sum_{k=1}^{2^{(N_b-1)}-1}\frac{-k^3 t_{\text{res}}^2}{2\cdot T_{\text{DCO}}^2} + \frac{k^2 t_{\text{res}}}{T_{\text{DCO}}}$$

$$-\frac{T_{\text{DCO}}^2}{6}\cdot\left[\frac{2\left(2^{N_b}-1\right)}{T_{\text{DCO}}} - 4\cdot\sum_{k=1}^{2^{(N_b-1)}-1}\frac{3k^2 t_{\text{res}}^2}{T_{\text{DCO}}^3} - \frac{2k^3 t_{\text{res}}^3}{T_{\text{DCO}}^4}\right]^2.$$

Substituting $N_b = 1$ into the above expressions yields the estimates for a BPD as

$$\hat{K}_{\text{BPD}} = \frac{2}{T_{\text{DCO}}} \text{ and } \sigma^2_{Q_{\text{BPD}}} = \frac{1}{3}.$$

3.6.2 With a MASH 1-1-1 MMDC—Bell-Shaped Δt Distribution

A MASH 1-1-1 MMDC causes a zero-mean "bell-shaped" non-Gaussian-distributed AQE, whose PDF is expressed as [17]

$$f_{Q^{\text{acc}}_{\text{MASH 1-1-1}}}(x) = \begin{cases} \frac{1}{4}\cdot(x+2)^2 & \text{for } -2 \le x < -1, \\ -\frac{1}{4}\cdot x^2 + \frac{1}{2} & \text{for } -1 \le x < 1, \\ \frac{1}{4}\cdot(x-2)^2 & \text{for } 1 \le x \le 2. \end{cases}$$

In this case, the PDF of the dominant input jitter after PDF transformation can be expressed as

$$f_{\Delta t}(x) = \begin{cases} \frac{1}{4T_{\mathrm{DCO}}}\left(\frac{x}{T_{\mathrm{DCO}}}+2\right)^2 & \text{for } -2T_{\mathrm{DCO}} \le x < -T_{\mathrm{DCO}}, \\ \frac{1}{2T_{\mathrm{DCO}}} - \frac{x^2}{4T_{\mathrm{DCO}}^3} & \text{for } -T_{\mathrm{DCO}} \le x < T_{\mathrm{DCO}}, \\ \frac{1}{4T_{\mathrm{DCO}}}\left(\frac{x}{T_{\mathrm{DCO}}}-2\right)^2 & \text{for } T_{\mathrm{DCO}} \le x \le 2T_{\mathrm{DCO}}, \end{cases} \tag{3.46}$$

as shown in Fig. 3.10b.

As will be derived in Chap. 5, the optimized quantization levels of the mid-rise TDC in a DPLL should be in the vicinity of the zero crossing and span a range equal to a few σs of the integer-N input jitter, which is much smaller than the DCO period ($\sigma_{\Delta t} \ll T_{\mathrm{DCO}}$). Take, for example, radio-frequency synthesizers that synthesize, e.g., a few GHz output signals; in these cases, the RMS integer-N jitter, $\sigma_{\Delta t}$, (in the order of hundreds of femtoseconds) is roughly a factor of 10^{-3} times that of the T_{DCO} [19]. Therefore, all the quantization levels should be allocated within the input range of $(-T_{\mathrm{DCO}}, T_{\mathrm{DCO}})$. Using a similar procedure to the previous cases, the linear gain and the variance of the TDC's QE can be derived as

$$\hat{K}_{\mathrm{TDC}} = \frac{7 \cdot (2^{N_b}-1)}{6 \cdot T_{\mathrm{DCO}}} - 2 \cdot \sum_{k=1}^{2^{(N_b-1)}-1} \frac{k^2 t_{\mathrm{res}}^2}{T_{\mathrm{DCO}}^3} - \frac{k^4 t_{\mathrm{res}}^4}{4 \cdot T_{\mathrm{DCO}}^5},$$

$$\sigma_{Q_{\mathrm{TDC}}}^2 = (2^{N_b}-1)^2 - 4 \cdot \sum_{k=1}^{2^{(N_b-1)}-1} \frac{k^2 t_{\mathrm{res}}}{T_{\mathrm{DCO}}} - \frac{k^4 t_{\mathrm{res}}^3}{6 \cdot T_{\mathrm{DCO}}^3}$$
$$- \frac{T_{\mathrm{DCO}}^2}{2} \cdot \left[\frac{7\,(2^{N_b}-1)}{6 \cdot T_{\mathrm{DCO}}} - 2 \cdot \sum_{k=1}^{2^{(N_b-1)}-1} \frac{k^2 t_{\mathrm{res}}^2}{T_{\mathrm{DCO}}^3} - \frac{k^3 t_{\mathrm{res}}^3}{4T_{\mathrm{DCO}}^4}\right]^2.$$

Substituting $N_b = 1$ into the above expressions yields the estimates for a BPD as

$$\hat{K}_{\mathrm{BPD}} = \frac{7}{6 \cdot T_{\mathrm{DCO}}} \text{ and } \sigma_{Q_{\mathrm{BPD}}}^2 = \frac{23}{72}.$$

3.7 Verification

To verify the accuracy of the linearized analysis above, we perform behavioral simulations that sweep through a sequence of linear gains and find the one that minimizes the power of the estimation error as per the MMSE estimation criterion introduced in Sect. 3.3. In the behavioral verification, we simulate a BPD and a *hypothetical* 10-bit TDC as examples. Three cases of input distributions investigated in this section are checked, namely, Gaussian, triangular, and bell-shaped distributions that correspond to the input jitters of the cases of integer-N and fractional-N enabled by MASH 1-1 and MASH 1-1-1 MMDCs, respectively. For the Gaussian case, we set the RMS input jitter $\sigma_{\Delta t} = 400$ fs, whereas, for the fractional-N cases, the DCO period is set to be

Table 3.2 Predicted versus simulated linear gain and variance of the QE (taking three significant figures)

			Predicted		Simulated	
Δt	N_b	t_{res}	$\hat{K}_{TDC}$	$\sigma^2_{Q_{TDC}}$	K_{TDC}	$\sigma^2_{Q_{TDC}}$
Gaussian (int-N)	BPD	N/A	$2.00 \cdot 10^{12}$	0.363	$2.00 \cdot 10^{12}$	0.363
	10-bit	1 fs	$1.60 \cdot 10^{15}$	$2.48 \cdot 10^4$	$1.60 \cdot 10^{15}$	$2.48 \cdot 10^4$
Triangular (MASH 1-1)	BPD	N/A	$5 \cdot 10^9$	0.333	$5 \cdot 10^9$	0.333
	10-bit	1 fs	$5.12 \cdot 10^{12}$	$3.47 \cdot 10^5$	$5.12 \cdot 10^{12}$	$3.47 \cdot 10^5$
Bell-shaped (MASH 1-1-1)	BPD	N/A	$2.92 \cdot 10^9$	0.319	$2.91 \cdot 10^9$	0.321
	10-bit	1 fs	$2.98 \cdot 10^{12}$	$3.34 \cdot 10^5$	$2.98 \cdot 10^{12}$	$3.36 \cdot 10^5$

1000 times the integer-N input jitter, as discussed in Sect. 3.6.2, that is $T_{DCO} = 400$ ps. The BPD has only ± 1 outputs, and hence no explicit time resolution, while a resolution of $t_{res} = 1$ fs is selected for the 10-bit case; this mitigates the integer-N TDC's QE.

The equations derived in this section are used to estimate the linearized gain and the variance of the QE of the TDC. The predicted and simulated results are tabulated in Table 3.2, which demonstrates the match between the analytically predicted and simulated results.

Table 3.2 also shows that, when switching from the integer-N to fractional-N modes, the effective gain of the TDC is reduced by about 10^3 times in all the cases analyzed. This confirms the reason for PN/jitter degradation in fractional-N DPLLs with such mid-rise TDCs, when the unmitigated MMDC's QE is injected into the loop.

3.8 Validity Limits of Linearized Modeling

In this section, we study the validity limits of linear modeling for mid-rise TDCs that can be optimized following the QE-minimization-oriented design strategy that will be presented in Chap. 5. We will show here that, considering a Gaussian-distributed input jitter, e.g., in integer-N and DTC-enhanced fractional-N modes, such TDCs feature a limited pseudo-linear dynamic range in the proximity of zero crossing that can be described explicitly. Within this region, the TDC can be regarded as a linear gain in the signal chain; outside the region, the TDC forces compression distortion of the input signal.

This is important because, in the DTC-cancellation-based fractional-N mode of operation mentioned above, the DTC is typically designed to ensure that, after inexact cancellation by the practical DTC that exhibits inevitable nonlinearity, the residual error is below the integer-N RMS input jitter in order to approach the integer-N PN/jitter [19]. While the spurs associated with the residual error *per se* can be mitigated as much as possible, if the range of the residual error extends beyond the TDC's pseudo-linear region, spurs can regrow due to its clipping nonlinearity, which in turn deteriorates the spectral purity of the synthesized signal. This entails a further consideration for practical DTC designs in DPLLs with mid-rise TDCs, especially for BBPLLs which have a relatively small linear input range, as we will show.

3.8.1 Pseudo-Linear Operation Region

To derive the effective transfer characteristics of the mid-rise TDCs in the locked state, we refer to the linearized analysis method in [8, Sect. II-A].

The method can be interpreted as follows. Although the N_b-bit TDC in open-loop operation features a hard quantization nonlinearity, it can be regarded in closed-loop operation as a quantizer that is dithered by the Gaussian random input jitter, Δt. The dithered quantization operation for an arbitrary input, x, is hence $e_{N_b\text{-bit}}(x + \Delta t)$. Note that x here can be interpreted as the residual error after the DTC's inexact cancellation. Since the output from a dithered quantizer is also stochastic, it is necessary to describe the effective (or expected) transfer characteristic of the dithered quantizer seen by its input, x, as $e_{\mathbb{E}}(x) = \mathbb{E}\left(e_{N_b\text{-bit}}(x + \Delta t)\right)$.

This can be calculated using

$$e_{\mathbb{E}}(x) = \int_{-\infty}^{+\infty} e_{N_b\text{-bit}}(x + u) \cdot f_{\Delta t}(u) \mathrm{d}u.$$

Substituting for the TDC's transfer characteristic using (3.2) and the Gaussian PDF in (3.20), we obtain that

$$\begin{aligned} e_{\mathbb{E}}(x) &= \left(2^{N_b} - 1\right) - 2 \cdot \sum_{k=-(2^{(N_b-1)}-1)}^{2^{(N_b-1)}-1} F_{\Delta t}(k \cdot t_{\text{res}} - x) \\ &= - \sum_{k=-(2^{(N_b-1)}-1)}^{2^{(N_b-1)}-1} \operatorname{erf}\left(\frac{k \cdot t_{\text{res}} - x}{\sqrt{2} \cdot \sigma_{\Delta t}}\right), \end{aligned} \tag{3.47}$$

also using the Gaussian error function defined in (3.26). When $N_b = 1$, the effective transfer characteristic of a closed-loop BPD can be obtained as

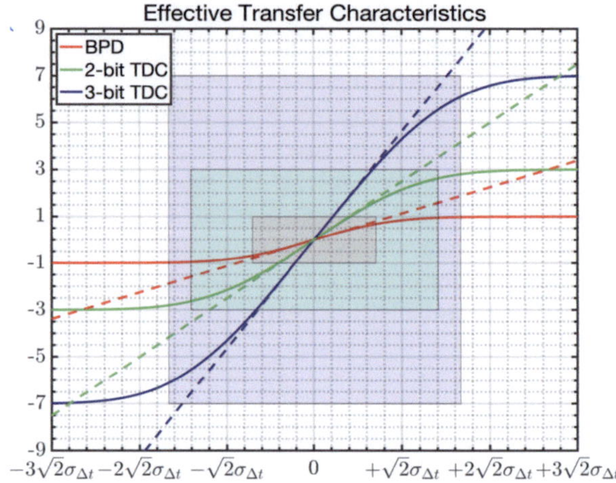

Fig. 3.11 Expected transfer characteristics of a BPD and optimized 2-bit and 3-bit TDCs predicted by (3.48) and (3.47). The shaped areas represent the pseudo-linear operation regions

$$e_{\mathbb{E},\mathrm{BPD}}(x) = \mathrm{erf}\left(\frac{x}{\sqrt{2}\cdot\sigma_{\Delta t}}\right). \tag{3.48}$$

The effective transfer characteristics of a BPD described in (3.4) and the optimized 2-bit and 3-bit TDCs described in (3.2) are plotted in Fig. 3.11.

As shown by the shaded areas, the expected transfer characteristics of the multi-bit TDCs have a symmetric pseudo-linear region that approximately resembles the region accommodating the quantization levels. By recalling Fig. 3.2 and (3.21), the pseudo-linear range for the input, x, can be approximated as

$$\left(-2^{N_b-1}\cdot\hat{\gamma}\cdot\sigma_{\Delta t}, 2^{N_b-1}\cdot\hat{\gamma}\cdot\sigma_{\Delta t}\right),$$

where the optimized normalized resolution $\hat{\gamma}$ will be derived in Chap. 5. The BPD does not have explicit boundaries of quantization levels, and its pseudo-linear region is approximately $(-\sigma_{\Delta t}, \sigma_{\Delta t})$. Referring to Fig. 3.11, note that optimized mid-rise TDCs with more bits typically have a wider pseudo-linear operation region.

For input signals that lie completely within the pseudo-linear region, the TDC functions as a noisy but linear gain element. For example, when using a cubic polynomial to fit the effective transfer characteristic (with the input normalized to $\sigma_{\Delta t}$) in the pseudo-linear region for the three curves in Fig. 3.11, the magnitude of the cubic term is 14.2% of that of the linear term for the BPD case, and 6.3% and 3.8% for the 2-bit and 3-bit TDC cases. Notice that this means that the optimized TDCs with more bits have better linearity within their pseudo-linear region. Figure 3.11 also shows with dashed lines the linearized gains predicted by (3.24) and (3.29); these match the pseudo-linear effective transfer characteristics well. Outside this region, the output of the TDC is compressed and saturates at $\pm 2^{N_b-1}$; hence a compression-type nonlinear distortion is imposed on the input.

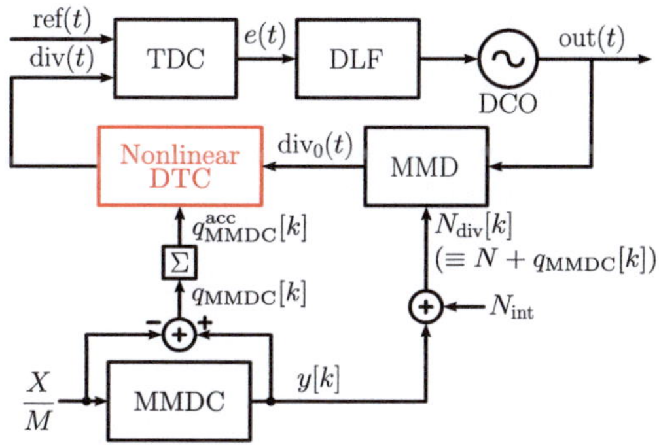

Fig. 3.12 Fractional-N DPLL assisted by a DTC with nonlinearity

3.8.2 Verification

To verify the derived pseudo-linear region and its boundaries for the optimized mid-rise TDCs, we show the results of closed-loop behavioral simulations for DTC-augmented fractional-N operation, while also considering different DTC nonlinearities and MMDC architectures, as shown in Fig. 3.12.

Verification Methodology

We leverage the following facts:

- Due to effects such as variable-loading capacitance and channel-length modulation [23, 24], commonly used "variable-slope" DTCs typically exhibit a "soft" INL that can be approximated by a third-order polynomial that spans about 0.33% of the full scale (FS) [23–36]. After linearity enhancement using techniques such as those described in [23, 24, 27, 29–31], the DTC's INL can be reduced by a factor of about three to four. In this case, the residual INL of the DTC is approximately as illustrated in blue in Fig. 3.13.
- The AQE of the MASH 1-1 MMDC which is commonly used in DPLLs will cause IBS spurs[5] if distorted by a polynomial nonlinearity [17, 37–39]. Therefore, when using a DTC with polynomial nonlinearity to cancel the AQE, the residual error will cause the synthesizer to exhibit IBS spurs. Furthermore, if the TDC can be considered as a linear subsystem, the DTC-induced IBS spurs will be fed through the loop transfer function and appear in the spectrum of the synthesized output signal.
- Stochastic divider controllers, such as [18, 33, 40–42], provide IBS-immunity against polynomial distortion up to certain order. For example, the AQE of an ENOP P1 MMDC [18, 43, 44] is IBS-immune to a cubic nonlinearity. Hence, when

[5] An IBS spur is typically manifest at the frequency $F_{\text{ref}} \cdot X/M$, where F_{ref} is reference frequency, and X and M are the fractional FCW and the modulus of the MMDC.

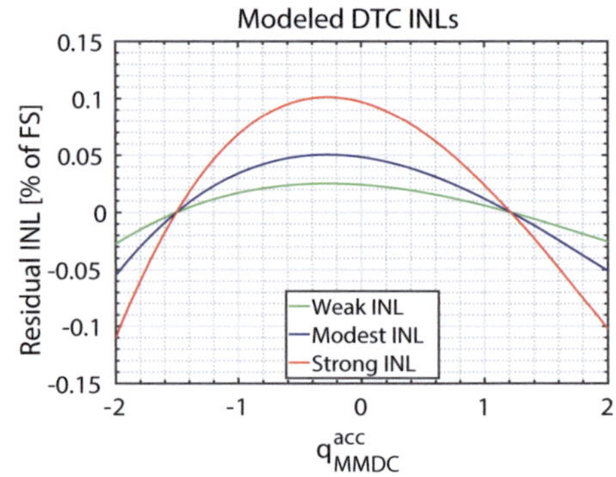

Fig. 3.13 Residual INLs (after linearity enhancement) of the nonlinear DTCs modeled in this subsection to verify the TDC's linearized operation. The modest INL is described as per (3.49) and (3.50)

using the DTC with a polynomial INL to cancel the AQE of such MMDCs, the residual error will be free of IBS spurs. If the DTC's residual error is confined to the pseudo-linear region of the TDC, where the nonlinearity as estimated in Sect. 3.8.1 introduces imperceptible distortion, the IBS-free pattern will be preserved, and the output will be free of IBS spurs. If, on the other hand, the DTC residue exceeds the TDC's pseudo-linear region, the saturation nonlinearity of the TDC will distort the IBS-free pattern and hence regrow spurs at the IBS frequency and its harmonics that will, in turn, be seen at the output of the DPLL.

Therefore, one can vary the type of the MMDC to produce (or not) IBS spurs in the DTC's residual error. Furthermore, one can use range reduction and linearization techniques, such as those described in this section, to tune the magnitude of the modeled DTC's INL, which in turn determines the range of variation of the DTC's residual error.

Results and Discussion

We present behavioral simulations of a type-II first-order DPLL that is parameterized as per Table 3.3. This, as will be discussed in Chap. 5, corresponds to a DPLL that has a jitter-minimized system design. We consider two instances of PD: (i) a BPD and (ii) a 3-bit TDC with an optimized quantization characteristic. Note that, as the linearized gains of the BPD and 3-bit TDC are different, the proportional gain of the DLF (β) is set with different optimized values in order to minimize the system jitter.

As introduced in Sect. 3.6, the original injection of the MMDC's QE in the time domain equals $T_{DCO} \cdot q^{acc}_{MMDC}[k]$. Ideally, a controlled time delay equal to $-T_{DCO} \cdot q^{acc}_{MMDC}[k]$ should be added in the feedback path to cancel this injection. Here, we model the DTC in Fig. 3.12 to have a cubic nonlinear transfer characteristic that is expressed as $-T_{DCO} \cdot f^{\mathcal{N}}_{DTC}\left(q^{acc}_{MMDC}[k]\right)$, where

Table 3.3 Fractional-N DPLL loop and noise parameterization

Parameter	Value
F_{ref}	100 MHz
N_{int}	24
X	227
M	2^{20}
N	$N_{\text{int}} + X/M$
F_{DCO}	$N \cdot F_{\text{ref}}$
K_T	0.145 fs/LSB
β	47.8 (with a BPD), 14.3 (with a 3-bit TDC)
α	$2^{-8} \cdot \beta$
$S_{\phi_{\text{ref}}}$	–155 dBc/Hz
$S_{\phi_{\text{DCO}}}$ (1 MHz)	–108 dBc/Hz

$$\begin{aligned} f_{\text{DTC}}^{\mathcal{N}}(x) &\equiv x + \text{INL}(x) \\ &= \alpha_0 + \alpha_1 \cdot x + \alpha_2 \cdot x^2 + \alpha_3 \cdot x^3. \end{aligned} \tag{3.49}$$

We start by modeling a *"modest"* INL that describes a typical DTC's nonlinearity after linearity enhancement [23], that is:

$$\begin{aligned} \alpha_0 &= +1.938 \cdot 10^{-3}, \alpha_1 = +0.9994, \\ \alpha_2 &= -1.016 \cdot 10^{-3}, \alpha_3 = +1.593 \cdot 10^{-4}. \end{aligned} \tag{3.50}$$

Substituting this into (3.49) yields the INL shown in blue in Fig. 3.13. The INL is such that, in this case, the range of variation of the residual error after cancellation by the DTC matches the width of the pseudo-linear operation of the optimized 3-bit TDC (shown in blue in Fig. 3.11).

The first row of Fig. 3.14 shows the simulated power spectra of the output PN considering a MASH 1-1 (a) and an ENOP P1 (b) [18] as the MMDC. Figure 3.14a shows that, when using a MASH 1-1 MMDC, the DTC's residual error exhibits an IBS spur and its harmonics at integer multiples of $F_{\text{ref}} \cdot X/M = 21.65$ kHz. The IBS pattern is preserved by the pseudo-linear operation of the TDC and transferred directly to the output spectrum. Notice that this phenomenon contradicts the assertion made in [19, 45] that, for the DPLLs with a mid-rise TDCs, as long as the DTC's residual error is limited within the stochastic resonance regime established at the input of the TDC, the intrinsic noise within the loop transferred there will destroy any periodic pattern associated with the residual error and consequently guarantee that the DPLL's output spectrum is free of the spurs related to that. On the contrary, the TDC in this regime is in fact linearized and consequently transfers the spurious pattern of the DTC's residual error virtually undistorted. Figure 3.14b shows that, by using an ENOP P1 in place of the MASH 1-1 that is commonly used in DPLLs, the

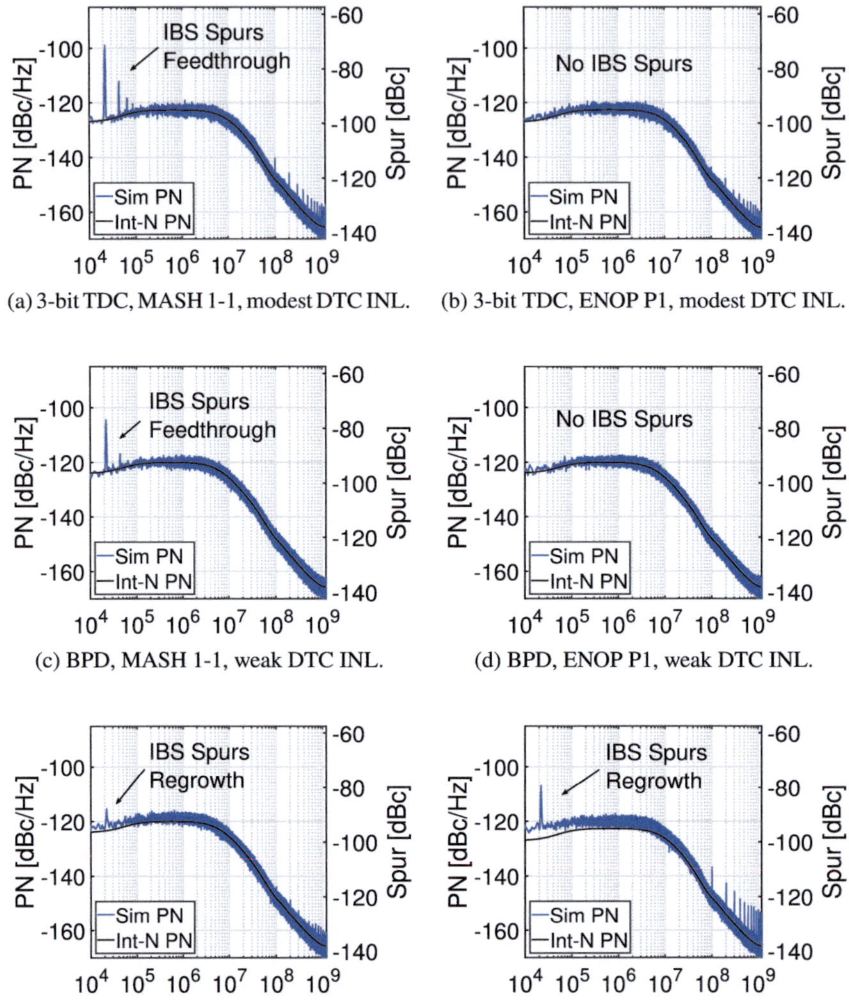

(a) 3-bit TDC, MASH 1-1, modest DTC INL.
(b) 3-bit TDC, ENOP P1, modest DTC INL.
(c) BPD, MASH 1-1, weak DTC INL.
(d) BPD, ENOP P1, weak DTC INL.
(e) BPD, ENOP P1, modest DTC INL.
(f) 3-bit TDC, ENOP P1, strong DTC INL.

Fig. 3.14 Simulated power spectra of the example fractional-N DPLL with different mid-rise TDCs, MMDCs, and DTC nonlinearities. The simulated spectra are compared with that of the integer-N PN (solid black curve) which is predicted using the method that will be described in the next chapter

DTC's residual error is made free of IBS spurs. Hence, after pseudo-linear detection by the TDC, the IBS-immune pattern is preserved and the output is IBS-free as well.

As shown in Fig. 3.11, the pseudo-linear region of a BPD is half that of the 3-bit TDC case. Therefore, we modeled a *"weak"* INL for the DTC by halving the coefficients of the INL in (3.49), as shown in green in Fig. 3.13; this limits the DTC's residual error in the pseudo-linear region of the BPD. The results for the

BPD cases are shown in the second row of Fig. 3.14. As in the 3-bit TDC cases, the linearized operation of the BPD maintains the pattern of the DTC's residual error (either exhibiting IBS spurs or not), which in turn affects the output spectrum. Also in this case, the TDC can be regarded as a linear gain within the loop.

Figure 3.14e shows the case of a BPD while considering a modest DTC INL (described by (3.50) and plotted in blue in Fig. 3.13). In this case, the range of variation of the DTC's residual error exceeds the pseudo-linear region of the BPD. One can observe that, although the DTC's residue *per se* is made IBS-free as a result of driving the cubic DTC's INL with an ENOP P1, IBS spurs in fact regrow in the output spectrum. This is due to the fact that the saturating nonlinearity beyond the pseudo-linear region distorts the otherwise IBS-free pattern of the DTC's residual error. Comparison between Fig. 3.14b and e confirms the superiority of the multi-bit TDC to a BPD due to its larger linearized region of operation.

In Fig. 3.14f, the magnitude of the DTC's INL described by (3.50) is doubled. The *"strong"* INL is plotted in red in Fig. 3.13. In this case, the DTC's residual error exceeds the linearized region of the optimized 3-bit TDC. Hence, similar to Fig. 3.14e, although the DTC's residue is IBS-free, the same IBS regrowth can be observed due to the saturation nonlinearity. The close-in PN level is higher than in the integer-N case; this is due to more noise being contributed following inexact cancellation by the nonlinear DTC, higher TDC's QE, and the reduced effective gain of the TDC which causes a reduction in the loop bandwidth.

In summary, the limited extent of the linearized region of operation associated with mid-rise TDCs entails special considerations for designing and linearizing the DTC and choosing the architecture of the MMDC.

3.9 Conclusion

In this chapter, we have addressed the linearized modeling of mid-rise TDCs—in particular, a BPD and other few-bit TDCs—that are used as phase detectors in DPLLs. Section 3.2 formulates a behavioral model for mid-rise TDCs. In Sect. 3.3, we formulated the MMSE estimation framework that can be used to perform accurate linearized analysis; this predicts the effective gain and the power of the QE of such TDCs in the locked state. This framework takes into account the quantization characteristic of the TDC and the statistical distribution of its input, depending on the operation modes, e.g., integer-N and fractional-N with/without a DTC. Section 3.5 introduces fractional-N operation. In Sects. 3.4 and 3.6, the linearized analysis was further extended to general cases where TDCs with arbitrary equidistant quantization thresholds were considered. All the analytical predictions are confirmed by behavioral simulations in Sect. 3.7. Finally, in Sect. 3.8, we studied the validity limits of the linearized operation of the optimized mid-rise TDCs. A formula for the extent of the pseudo-linear region of operation of such TDCs has been derived explicitly. We have shown that, for input signals that remain within the pseudo-linear region, the TDC in its locked-state functions like a linear gain that preserves and transmits directly to the output spectrum the spectral pattern of its input signal. For input signals that extend beyond the pseudo-linear region, the TDC imposes a saturation-type nonlinear distortion that will regrow fractional spurs. Optimized multi-bit mid-rise TDCs are shown to have a wider linearized region than a commonly used BPD; this is a potential advantage in applications where the pseudo-linear range is important; for example, in DTC-based fractional-N DPLLs.

3.10 Summary

1. In Sect. 3.2, we describe a behavioral model of mid-rise TDCs and their MATLAB models.
2. In Sect. 3.3, we compared and contrasted two state-of-the-art linearized analysis frameworks to calculate the effective gain and QE variance of coarse-resolution TDCs, namely, the MMSE estimator method and small-signal approximation method. A novel simulation method was introduced which (i) confirms that the MMSE estimator always provides estimates that conform to the estimation criterion and (ii) highlights the limitation of the small-signal approach.
3. In Sect. 3.4, the verified and formulated MMSE method is used to estimate the linearized model of a generic N_b-bit TDC operating in integer-N mode, where the steady-state TDC input jitter can be approximated by a Gaussian distribution. The linear gain and QE variance estimated in (3.24) and (3.28) are the same as those derived in [8] using the "small-signal" method. This is a coincidental result due to the special statistical properties of the Gaussian distribution of the input jitter.
4. In Sect. 3.5, we described fractional-N operation enabled by using MASH MMDCs and presented MATLAB models of MASH DDSMs.
5. Section 3.6 derives, using the MMSE estimator, linearized models for TDCs operating in fractional-N with MASH 1-1 and 1-1-1 MMDCs. In the fractional-N cases, the TDC input jitter is overwhelmed by the MMDC's AQE, and hence its distribution is bounded and non-Gaussian. Thus, the MMSE method and "small-signal" method will yield different estimates; the prediction accuracy of the former was confirmed in Sect. 3.7. The estimates obtained using the verified MMSE method are derived explicitly, and these will be used in the system-level analysis and predictions in Chap. 4.
6. In Sect. 3.8, the validity limits of the linearized operation of the mid-rise TDCs were determined. A formula for the "pseudo-linear" region of operation of such TDCs has been derived explicitly; it explains when the "hard quantizer" can be regarded as an almost linear PD and when not.

References

1. N. Da Dalt, Markov chains-based derivation of the phase detector gain in bang-bang PLLs. IEEE Trans. Circuits Syst. II Express Briefs **53**(11), 1195–1199 (2006)
2. M. Zanuso, D. Tasca, S. Levantino, A. Donadel, C. Samori, A.L. Lacaita, Noise analysis and minimization in bang-bang digital PLLs. IEEE Trans. Circuits Syst. II Express Briefs **56**(11), 835–839 (2009)
3. G. Marucci, S. Levantino, P. Maffezzoni, C. Samori, Exploiting stochastic resonance to enhance the performance of digital bang-bang PLLs. IEEE Trans. Circuits Syst. II Express Briefs **60**(10), 632–636 (2013)
4. M.-J. Park, J. Kim, Pseudo-linear analysis of bang-bang controlled timing circuits. IEEE Trans. Circuits Syst. I Regul. Pap. **60**(6), 1381–1394 (2013)
5. S. Levantino, G. Marzin, C. Samori, A.L. Lacaita, A wideband fractional-N PLL with suppressed charge-pump noise and automatic loop filter calibration. IEEE J. Solid-State Circuits **48**(10), 2419–2429 (2013)
6. T.-K. Kuan, S.-I. Liu, A bang bang phase-locked loop using automatic loop gain control and loop latency reduction techniques. IEEE J. Solid-State Circuits **51**(4), 821–831 (2016)
7. H. Xu, A.A. Abidi, Design methodology for phase-locked loops using binary (bang-bang) phase detectors. IEEE Trans. Circuits Syst. I Regul. Pap. **64**(7), 1637–1650 (2017)
8. L. Avallone, M.P. Kennedy, S. Karman, C. Samori, S. Levantino, Jitter minimization in digital PLLs with mid-rise TDCs. IEEE Trans. Circuits Syst. I Regul. Pap. **67**(3), 743–752 (2020)
9. L. Avallone, M. Mercandelli, A. Santiccioli, M.P. Kennedy, S. Levantino, C. Samori, A comprehensive phase noise analysis of bang-bang digital PLLs. IEEE Trans. Circuits Syst. I Regul. Pap. **68**(7), 2775–2786 (2021)
10. H.E. Rowe, Memoryless nonlinearities with gaussian inputs: elementary results. Bell Syst. Tech. J. **61**(7), 1519–1525 (1982)
11. B. Widrow, I. Kollár, Quantization Noise: Roundoff Error in Digital Computation, Signal Processing, Control, and Communications (Cambridge University Press, 2008)
12. M.D. McDonnell, Is electrical noise useful? [point of view]. Proc. IEEE **99**(2), 242–246 (2011)
13. X. Wang, M.P. Kennedy, A random pulse modulation approach to modeling the flicker and white noise of the charge pump of a fractional-N frequency synthesizer. Int. J. Circuit Theory Appl. **50**(4), 1049–1063 (2022)
14. B. Miller, R. Conley, A multiple modulator fractional divider. IEEE Trans. Instrum. Meas. **40**(3), 578–583 (1991)
15. K. Hosseini, M.P. Kennedy, Maximum sequence length MASH digital delta-sigma modulators. IEEE Trans. Circuits Syst. I Regul. Pap. **54**(12), 2628–2638 (2007)
16. K. Hosseini, B. Fitzgibbon, M.P. Kennedy, Observations concerning the generation of spurious tones in digital delta-sigma modulators followed by a memoryless nonlinearity. IEEE Trans. Circuits Syst. II Express Briefs **58**(11), 714–718 (2011)
17. V. Mazzaro, M.P. Kennedy, Spur immunity in MASH-based fractional-N CP-PLLs with polynomial nonlinearities. IEEE Trans. Circuits Syst. I Regul. Pap. **68**(6), 2295–2306 (2021)
18. V. Mazzaro, M.P. Kennedy, A family of $\Delta\Sigma$ modulators with high spur immunity and low folded nonlinearity noise when used in fractional-N frequency synthesizers. IEEE Trans. Circuits Syst. I Regul. Pap. **69**(11), 4499–4509 (2022)
19. D. Tasca, M. Zanuso, G. Marzin, S. Levantino, C. Samori, A.L. Lacaita, A 2.9–4.0-GHz fractional-N digital PLL with bang-bang phase detector and 560-fs_{rms} integrated jitter at 4.5-mW power. IEEE J. Solid-State Circuits **46**(12), 2745–2758 (2011)
20. X. Lu, M.P. Kennedy, Further insights into spur immunity in MASH-based fractional-N CP-PLLs with polynomial nonlinearities, in *2023 30th IEEE International Conference on Electronics, Circuits and Systems (ICECS)* (2023), pp. 1–5
21. S. Levantino, L. Grimaldi, C. Samori, Analysis of adaptive pre-distortion in DTC-based digital fractional-N PLLs, in *2016 12th Conference on Ph.D. Research in Microelectronics and Electronics (PRIME)* (2016), pp. 1–4

22. G. Grimmett, D. Welsh, Probability: An Introduction (Oxford University Press, 2014)
23. J.Z. Ru, C. Palattella, P. Geraedts, E. Klumperink, B. Nauta, A high-linearity digital-to-time converter technique: constant-slope charging. IEEE J. Solid-State Circuits **50**(6), 1412–1423 (2015)
24. Y. Liu, H. Gao, H. Xu, P. Lu, N. Yan, A 10-bit 563-fs step constant-slope digital-to-time converter in 40-nm CMOS with nonlinearity cancellation and range extension techniques. IEEE Trans. Circuits Syst. I Regul. Pap. **71**(2), 526–536 (2024)
25. S. Levantino, G. Marzin, C. Samori, An adaptive pre-distortion technique to mitigate the DTC nonlinearity in digital PLLs. IEEE J. Solid-State Circuits **49**(8), 1762–1772 (2014)
26. H. Liu, D. Tang, Z. Sun, W. Deng, H.C. Ngo, K. Okada, A sub-mW fractional-*N* ADPLL with FOM of -246 dB for IoT applications. IEEE J. Solid-State Circuits **53**(12), 3540–3552 (2018)
27. C. Hwang, H. Park, T. Seong, J. Choi, A 188fs_{rms}-jitter and -243dB-FoM_{jitter} 5.2GHz-ring-DCO-based fractional-N digital PLL with a 1/8 DTC-range-reduction technique using a quadruple-timing-margin phase selector, in *2022 IEEE International Solid- State Circuits Conference (ISSCC)*, vol. 65 (2022), pp. 378–380
28. W. Wu, C.-W. Yao, K. Godbole, R. Ni, P.-Y. Chiang, Y. Han, Y. Zuo, A. Verma, I.S.-C. Lu, S.W. Son, T.B. Cho, A 28-nm 75-fs_{rms} analog fractional-*N* sampling PLL with a highly linear DTC incorporating background DTC gain calibration and reference clock duty cycle correction. IEEE J. Solid-State Circuits **54**(5), 1254–1265 (2019)
29. P. Chen, F. Zhang, Z. Zong, S. Hu, T. Siriburanon, R.B. Staszewski, A 31-μW, 148-fs step, 9-bit capacitor-DAC-based constant-slope digital-to-time converter in 28-nm CMOS. IEEE J. Solid-State Circuits **54**(11), 3075–3085 (2019)
30. A. Santiccioli, M. Mercandelli, L. Bertulessi, A. Parisi, D. Cherniak, A.L. Lacaita, C. Samori, S. Levantino, A 66-fs-rms jitter 12.8-to-15.2-GHz fractional-N bang–bang PLL with digital frequency-error recovery for fast locking. IEEE J. Solid-State Circuits **55**(12), 3349–3361 (2020)
31. W. Wu, C.-W. Yao, C. Guo, P.-Y. Chiang, L. Chen, P.-K. Lau, Z. Bai, S.W. Son, T.B. Cho, A 14-nm ultra-low jitter fractional-N PLL using a DTC range reduction technique and a reconfigurable dual-core VCO. IEEE J. Solid-State Circuits **56**(12), 3756–3767 (2021)
32. H. Park, C. Hwang, T. Seong, Y. Lee, J. Choi, 32.1 a 365fs_{rms}-jitter and -63dBc-fractional spur 5.3 GHz-ring-DCO-based fractional-N DPLL using a DTC second/third- order nonlinearity cancelation and a probability-density-shaping $\Delta\Sigma$M, in *2021 IEEE International Solid- State Circuits Conference (ISSCC)*, vol. 64 (2021), pp. 442–444
33. C. Hwang, H. Park, Y. Lee, T. Seong, J. Choi, A low-jitter and low-fractional-spur ring-DCO-based fractional-N digital PLL using a DTC's second-/third-order nonlinearity cancellation and a probability-density-shaping $\Delta\Sigma$M. IEEE J. Solid-State Circuits **57**(9), 2841–2855 (2022)
34. S.M. Dartizio, F. Tesolin, G. Castoro, F. Buccoleri, L. Lanzoni, M. Rossoni, D. Cherniak, L. Bertulessi, C. Samori, A. L. Lacaita, S. Levantino, 4.3 a 76.7fs-lntegrated-jitter and -71.9dBc in-band fractional-spur bang-bang digital PLL based on an inverse-constant-slope DTC and FCW subtractive dithering, in *2023 IEEE International Solid-State Circuits Conference (ISSCC)* (2023), pp. 3–5
35. M. Rossoni, S.M. Dartizio, F. Tesolin, G. Castoro, R. Dell'Orto, C. Samori, A.L. Lacaita, S. Levantino, 10.1 an 8.75GHz fractional-N digital PLL with a reverse-concavity variable-slope DTC achieving 57.3fs_{rms} integrated jitter and –252.4 dB FoM, in *2024 IEEE International Solid-State Circuits Conference (ISSCC)*, vol. 67 (2024), pp. 188–190
36. D. Xu, Z. Liu, Y. Kuai, H. Huang, Y. Zhang, Z. Sun, B. Liu, W. Wang, Y. Xiong, J. Qiu, W. Madany, Y. Zhang, A.A. Fadila, A. Shirane, K. Okada, "10.3 a 7 GHz digital PLL with cascaded fractional divider and pseudo-differential DTC achieving –62.1 dBc fractional spur and 143.7 fs integrated jitter, in *2024 IEEE International Solid-State Circuits Conference (ISSCC)*, vol. 67 (2024), pp. 192–194
37. D. Banerjee, *PLL Performance, Simulation and Design* (Dog Ear Publishing, Indianapolis, IN, USA, 2006)
38. E. Familier, I. Galton, A fundamental limitation of DC-free quantization noise with respect to nonlinearity-induced spurious tones. IEEE Trans. Signal Process. **61**(16), 4172–4180 (2013)

39. Y. Donnelly, M.P. Kennedy, Prediction of phase noise and spurs in a nonlinear fractional-*N* frequency synthesizer. IEEE Trans. Circuits Syst. I Regul. Pap. **66**(11), 4108–4121 (2019)
40. A. Swaminathan, A. Panigada, E. Masry, I. Galton, A digital requantizer with shaped requantization noise that remains well behaved after nonlinear distortion. IEEE Trans. Signal Process. **55**(11), 5382–5394 (2007)
41. E. Familier, C. Venerus, I. Galton, A class of quantizers with DC-free quantization noise and optimal immunity to nonlinearity-induced spurious tones. IEEE Trans. Signal Process. **61**(17), 4270–4283 (2013)
42. Y. Donnelly, M.P. Kennedy, J. Breslin, S. Tulisi, S. Patil, C. Curtin, S. Brookes, B. Shelly, P. Griffin, M. Keaveney, 4.48-GHz fractional- *N* frequency synthesizer with spurious-tone suppression via probability mass redistribution. IEEE Solid-State Circuits Letters **2**(11), 264–267 (2019)
43. M.P. Kennedy, V. Mazzaro, S. Tulisi, M. Scully, N. McDermott, J. Breslin, 10.4 a 45.5fs-integrated-random-jitter and –75 dBc-integer-boundary-spur BiCMOS fractional-N PLL with suppression of fractional, horn, and wandering spurs, in *2024 IEEE International Solid-State Circuits Conference (ISSCC)*, vol. 67 (2024), pp. 194–196
44. M.P. Kennedy, X. Lu, X. Wang, Insights into architectural spurs in high-performance fractional-*N* frequency synthesizers. IEEE Open J. Solid-State Circuits Soc. **4**, 238–251 (2024)
45. A. Mahmoud, P. Andreani, F. Pepe, A 2.8–3.8-GHz low-spur DTC-based DPLL with a class-D DCO in 65-nm CMOS. IEEE Microwave Wirel. Compon. Lett. **27**(11), 1010–1012 (2017)

Chapter 4
Prediction of System Jitter and Output PN

4.1 Introduction

In the previous chapter, based on MMSE estimation, accurate linearized analysis estimating the effective gain and the power of the QE has been performed for TDCs that feature a hard quantization nonlinearity. In this chapter, the linearized TDC model will be used in DPLL-system-level analysis that yields analytical predictions for the system jitter and output PN. From there, a comprehensive jitter minimization strategy is described in Chap. 5, where its impact on the output PN is discussed.

Recall that DPLLs incorporating a multi-bit TDC as the PD can outperform analog PLLs that use a PFD followed by a CP [1] in design simplicity and power/area consumption, thanks to the full digitization of the PD and loop filter. The DPLL has thus become a popular design option to serve as the frequency synthesizer of choice for many wireless communication and clocking purposes. A major breakthrough for advanced DPLLs has been the adoption of a BPD to minimize the complexity and power/area consumption of the TDC and to achieve a promising trade-off between jitter/noise and power for BBPLLs at the system level [2].

In the integer-N locked state, the sources of random time jitter, i.e., reference clock jitter, TDC's QE, and DCO period jitter, can be referred to the input of the TDC, where they can be approximated by a Gaussian distribution, which can be characterized using the RMS value of the absolute jitter [2–5]. In a BBPLL, this input jitter interacting with the coarse resolution of the BPD inevitably induces a greater QE compared with multi-bit TDCs. The contribution of the BPD's QE can therefore dominate the system's output PN at close-in frequencies (see [3] for example). Avallone et al. [4] have demonstrated that using a TDC with one or two extra bits and an optimized time resolution in place of the BPD can reduce the power of the TDC QE and hence the input-referred jitter, with only a modest increase in the complexity and power consumption.

This chapter first considers integer-N DPLLs with few-bit TDCs and provides an enhanced jitter analysis that outperforms that in [4]. The latter is an extension of

© The Author(s), under exclusive license to Springer Nature Switzerland AG 2026

X. Wang and M. P. Kennedy, *Jitter and Spur Minimization in Fractional-N Digital Frequency Synthesizers*, Analog Circuits and Signal Processing,
https://doi.org/10.1007/978-3-032-27186-0_4

[3] that focused only on BBPLLs. The prediction of the input jitter contributed by the TDC's QE (or the jitter arising from the "limit-cycle regime") in those works is derived using the method introduced in [6]. It will be shown that they lack prediction accuracy, which in turn leads to incorrect jitter minimization. This motivates us to find an alternative method to perform a more accurate input jitter analysis. To date, [7] provides the most accurate jitter analysis framework for a BBPLL, which originates from the discrete-time method used in [8]. Instead of integrating the power spectrum using a continuous-time system model [3–5], the analysis framework in [7] derives the RMS input jitter by solving its difference equation using a MRDT BBPLL model. This chapter extends the analysis to the multi-bit TDC case and uses behavioral simulations to verify the accuracy of the prediction.

In this chapter, the MRDT model for the DPLL incorporating a multi-bit TDC introduced in Chap. 2 will first be linearized. It is then used to derive accurate predictions for the system's input and output jitters and output PN in the following sections. Section 4.2 presents integer-N mode predictions, where the RMS value measured from the simulated jitter confirms that the accuracy of the prediction outperforms the counterpart predicted by [4]. The NTFs that predict the PSD of the output PN in relation to the contributions from the inherent jitter/noise sources will be provided in Sect. 4.2.4. Section 4.3 covers the predictions for a fractional-N DPLL (without DTC-based QE cancellation). This case sets the worst-case jitter/PN performance limit for DPLLs operating in the fractional-N mode.

4.2 Integer-*N* Prediction

Recall from the analysis in previous chapters that a generic N_b-bit TDC has the transfer characteristic expressed as (3.2), which is depicted as Fig. 3.2. The TDC quantizer can be modeled as a linear gain element with gain K_{TDC} to whose output is added the QE, $q_{\text{TDC}}[k]$, as in (3.5). Adopting the MMSE estimator formulated in Chap. 3, the linearized gain and the variance of the QE of the TDC, considering the probability distribution of the input jitter being zero-mean Gaussian in the integer-N case, are derived in (3.24) and (3.28).

The linearized TDC model can be substituted in place of the TDC's behavioral model into the MRDT system model, as shown in Fig. 2.1. The z-domain integer-N DPLL system model is shown in Fig. 4.1.

In the remainder of this section, jitter analysis based on this system model will be performed. In Sect. 4.2.1, the RMS input jitter, $\sigma_{\Delta t}$, of the DPLL with a multi-bit TDC will be derived as an explicit function of the parameter γ in (3.21). Notice that, at first glance, the relationship between γ and $\sigma_{\Delta t}$ might seem recursive in the sense that the normalized TDC resolution (γ) is set with respect to a known RMS jitter ($\sigma_{\Delta t}$), yet setting a particular value for γ varies the linearized gain and power of the QE of the TDC which, in turn, can change the value of $\sigma_{\Delta t}$. However, as will be shown in Sect. 4.2.2, starting with a specified γ value, together with some known loop parameters, the corresponding $\sigma_{\Delta t}$ value can be predicted (calculated). Then,

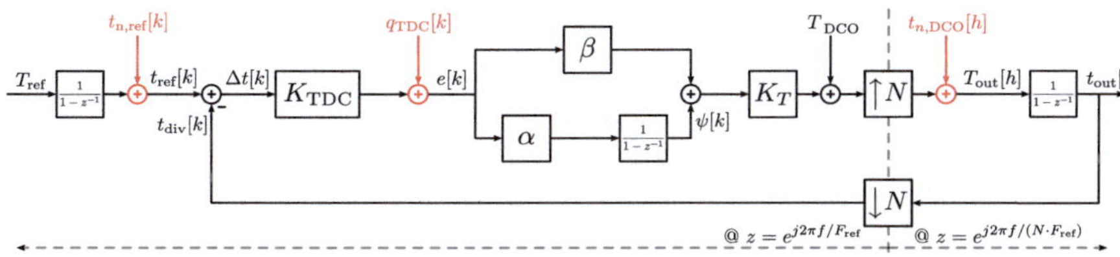

Fig. 4.1 z-domain model for a MRDT DPLL system, where injections of the jitter/noise sources are marked in red

substituting γ and $\sigma_{\Delta t}$ into (3.21), the absolute TDC resolution (t_{res}) can be set; using this in the behavioral simulation, the DPLL will achieve a simulated system jitter that matches the prediction precisely.

Furthermore, it will be proven in Chap. 5 that, to minimize the overall system jitter, there exists an optimal value of γ for each value of $N_b = 1, 2, 3, \ldots$. The optimal γ values will be derived and tabulated there, together with a jitter-minimization-oriented design procedure which also takes into account optimization of the PN performance.

4.2.1 *Input Jitter*

In this subsection, analysis of the time jitter referred to the system's input will be provided. The output jitter analysis will be performed in later subsections.

Following the model in Fig. 4.1, rearranging (2.13), and substituting the relationships in (2.9) and (2.10) yields a recurrent equation for the timestamp of the divided signal which can be expressed as

$$t_{\text{div}}[k+1] = t_{\text{div}}[k] + \sum_{h=N\cdot k+1}^{N\cdot(k+1)} \left\{ \underbrace{K_T\left(\beta \cdot e[k] + \psi[k]\right)}_{\text{control via DLF}} + \underbrace{T_{\text{DCO}} + t_{n,\text{DCO}}[h]}_{\text{DCO period with jitter in (2.5)}} \right\}. \tag{4.1}$$

Recall that $\psi[k] = \psi[k-1] + \alpha \cdot e[k]$ is the output of the integral path of the DLF. As the DPLL's stability requires that $\alpha \ll \beta$ [9, 10], $\psi[k]$ in (4.1) can be neglected without sacrificing the accuracy of the jitter transfer analysis.

Substituting the TDC-related expressions in (3.1) and (3.5) into (4.1), one can derive the difference equation of the input time difference (jitter) shown as

$$\Delta t[k+1] = \Delta t[k] + (t_{\text{ref}}[k+1] - t_{\text{ref}}[k]) - \sum_{h=N\cdot k+1}^{N\cdot(k+1)} \left\{ K_{\text{TDC}} \cdot \beta \cdot K_T \cdot \Delta t[k] + \beta \cdot K_T \cdot q_{\text{TDC}}[k] + T_{\text{DCO}} + t_{n,\text{DCO}}[h] \right\}. \tag{4.2}$$

In integer-N mode, the divide ratio N is a constant integer. Therefore, the sum operator in (4.2) can be replaced by multiplication by N for the terms in the sum that are independent of the index h. Substituting the definition of synchronous reference jitter from (2.2) and recalling that $T_{\text{ref}} = N \cdot T_{\text{DCO}}$ in the locked state, (4.2) can be simplified and rearranged as

$$\Delta t[k+1] = (1-X)\cdot \Delta t[k] + \left(t_{n,\text{ref}}[k+1] - t_{n,\text{ref}}[k]\right) - \left\{ \frac{X}{K_{\text{TDC}}} \cdot q_{\text{TDC}}[k] + \sum_{h=N\cdot k+1}^{N\cdot(k+1)} t_{n,\text{DCO}}[h] \right\}, \tag{4.3}$$

which uses the shorthand notation,

$$X = K_{\text{TDC}} \cdot \beta \cdot K_T \cdot N. \tag{4.4}$$

Solving the difference equation (4.3) yields

$$\Delta t[k+1] = (1-X)^{k+1} \cdot \Delta t[0] + t_{n,\text{ref}}[k+1] - \sum_{i=0}^{k} (1-X)^{(k-i)} \left\{ X \cdot t_{n,\text{ref}}[i] + \frac{X}{K_{\text{TDC}}} \cdot q_{\text{TDC}}[i] + \sum_{h=Ni+1}^{N(i+1)} t_{n,\text{DCO}}[h] \right\}. \tag{4.5}$$

For a stable loop, $0 < X < 2$ is always satisfied [7]. Therefore, in steady state, where $k \to \infty$, the first term on the right-hand side vanishes, which reflects the ability of a DPLL to close the initial phase/time gap at the input.

The noise sources are assumed to be zero-mean, ergodic, and uncorrelated with each other [7, 11]. In particular, the absence of correlation between the predicted TDC's QE and the other sources is guaranteed by the MMSE estimation. Ergodicity allows one to take (time-domain) variances on both sides of (4.5), whilst noticing that the last term represents a sum of N independent DCO period jitters, to calculate the (ensemble) input jitter power as a sum of the powers of the noise/jitter sources, which is

$$\sigma_{\Delta t}^2 = \sigma_{t_{n,\text{ref}}}^2 + \sum_{i=0}^{k} (1-X)^{2(k-i)} \left\{ X^2 \sigma_{t_{n,\text{ref}}}^2 + \frac{X^2}{K_{\text{TDC}}^2} \cdot \sigma_{Q_{\text{TDC}}}^2 + N \cdot \sigma_{t_{n,\text{DCO}}}^2 \right\}. \tag{4.6}$$

Approaching the steady state, $0 < X < 2$ guarantees the convergence of the geometric series in (4.6), hence $\lim_{k\to\infty} \sum_{i=0}^{k} (1-X)^{2(k-i)} = \frac{1}{X(2-X)}$ which, when substituted into (4.6), yields

$$\sigma_{\Delta t}^2 = \frac{2}{2-X} \cdot \sigma_{t_{n,\text{ref}}}^2 + \frac{X}{2-X} \cdot \frac{\sigma_{Q_{\text{TDC}}}^2}{K_{\text{TDC}}^2} + \frac{N}{X(2-X)} \cdot \sigma_{t_{n,\text{DCO}}}^2. \tag{4.7}$$

To simplify the analysis, we use the shorthand notations:

$$Y = \sqrt{\frac{2}{\pi}} \cdot \left(1 + 2 \cdot \sum_{k=1}^{2^{(N_b-1)}-1} e^{-\frac{k^2\gamma^2}{2}}\right), \tag{4.8}$$

and

$$Z = (2^{N_b} - 1)^2 - 8 \cdot \sum_{k=1}^{2^{(N_b-1)}-1} k \cdot \mathrm{erf}\left(\frac{k \cdot \gamma}{\sqrt{2}}\right). \tag{4.9}$$

Equations (3.24) and (3.28) then become

$$K_{\mathrm{TDC}} = \frac{Y}{\sigma_{\Delta t}}, \tag{4.10}$$

and

$$\sigma^2_{Q_{\mathrm{TDC}}} = Z - Y^2, \tag{4.11}$$

which simplify the variance of the input-referred TDC's QE in (4.7) as

$$\sigma^2_{Q_{\mathrm{in,TDC}}} \equiv \frac{\sigma^2_{Q_{\mathrm{TDC}}}}{K^2_{\mathrm{TDC}}} = \left(\frac{Z}{Y^2} - 1\right) \cdot \sigma^2_{\Delta t}. \tag{4.12}$$

Note that Y has the dimension of LSB, and that the Z has the dimension of LSB^2. Comparing the expressions of the power of the TDC's QE in (3.11) with (4.11), note that the quantities Z and Y^2 have physical meanings: Z represents the power (variance) of the TDC's output (E in (3.11)), and Y^2 represents the power of the linear portion of the output ($\hat{K}_{\mathrm{TDC}} \cdot \sigma_{\Delta t}$ in (3.11)).

Substituting (4.10) into (4.4), together with (4.12) into (4.7), yields a quadratic equation for the RMS input jitter as

$$\sigma^2_{\Delta t} - \left(\frac{N\beta K_T \cdot Z}{2 \cdot Y} + \frac{\sigma^2_{t_n,\mathrm{DCO}}}{2 \cdot \beta K_T \cdot Y}\right) \cdot \sigma_{\Delta t} - \sigma^2_{t_n,\mathrm{ref}} = 0,$$

the positive root of which can be represented as

$$\sigma_{\Delta t} = \frac{\eta}{2} + \sqrt{\left(\frac{\eta}{2}\right)^2 + \sigma^2_{t_n,\mathrm{ref}}}, \tag{4.13}$$

where

$$\eta = \underbrace{\frac{N\beta K_T \cdot Z}{2 \cdot Y}}_{\text{From TDC QE}} + \underbrace{\frac{\sigma^2_{t_n,\mathrm{DCO}}}{2 \cdot \beta K_T \cdot Y}}_{\text{From DCO jitter}}. \tag{4.14}$$

Note that, in the special case of a BPD, $N_b = 1$ and $\gamma = \infty$, giving $Y = \sqrt{\frac{2}{\pi}}$ and $Z = 1$, which, when substituted into (4.14), yield

$$\eta_{\text{BPD}} = \sqrt{\frac{\pi}{2}} \cdot \left(\frac{N\beta K_T}{2} + \frac{\sigma_{t_n,\text{DCO}}^2}{2 \cdot \beta K_T} \right). \tag{4.15}$$

The prediction for η in (4.14) is a general case for DPLLs using generic N_b-bit TDCs. The special case, η_{BPD} in (4.15), corresponds to the most accurate prediction to date in [7], which focuses exclusively on the analysis of BBPLLs.

4.2.2 Comparison of η Predicted in (4.14) Versus That in [4]

The input jitter analysis for DPLLs with multi-bit TDCs conducted in [4] omits the contribution from the reference jitter. Hence, its derived input jitter is limited to the prediction of the value of η, which contains only the contributions from the TDC's QE and DCO jitter, or, using the language of [4], the jitters derived in the "limit-cycle" and "random-noise" regimes. Nevertheless, the accuracy of its prediction of η can be compared against that of (4.14). Using the terminology of this section, [4] predicts that

$$\eta_{[4]} = \frac{N\beta K_T}{\sqrt{3}} \cdot \left[(2^{N_b} - 1) - 2 \cdot \sum_{k=1}^{2^{(N_b-1)}-1} \text{erf}\left(\frac{k \cdot \gamma}{\sqrt{2}} \right) \right] + \frac{\sigma_{t_n,\text{DCO}}^2}{2 \cdot \beta K_T \cdot Y}, \tag{4.16}$$

whose prediction of the contribution from the DCO jitter (the second term) is the same as its counterpart in (4.14), yet its prediction of the TDC's QE's contribution is fundamentally different from (4.14).

To compare and contrast the predictions in this work and that in [4], closed-loop behavioral simulations are performed for DPLLs parameterized with the loop and noise settings used in [7, Sect. III], which are tabulated in Table 4.1.[1] For a fair comparison with [4], what is different from Table 4.1 is that the reference source is assumed to be free of jitter, and hence the PSD of the reference jitter $S_{\phi_{\text{ref}}} = -\infty$ and $\sigma_{t_{n,\text{ref}}} = 0$. The cases where a BPD, 2-bit, and 3-bit TDCs are used as the PD for the DPLLs are simulated. For the latter two cases, $\gamma = 1.08$ and 0.66 are chosen, as suggested by [4], to mitigate the input jitter. Substituting the γ value, (4.14) and (4.16) predict the values of η for the two methods, which is equivalent to $\sigma_{\Delta t}$ without reference jitter. The values of γ and $\sigma_{\Delta t}$ are substituted into (3.21) to select the absolute TDC resolution used in the simulations.

The standard deviation of the simulated sequence of the TDC input samples, $\Delta t[k]$, is measured. The comparisons tabulated in Table 4.2 show that, for all three

[1] The relationships between $S_{\phi_{\text{ref}}}$ & $\sigma_{t_{n,\text{ref}}}$ and $S_{\phi_{\text{DCO}}}(\Delta f)$ & $\sigma_{t_{n,\text{DCO}}}$ are derived in (4.25) and (4.27).

Table 4.1 Integer-N DPLL loop and noise parameterization

Parameter	Value
F_{ref}	100 MHz
N	24
F_{DCO}	$N \cdot F_{\text{ref}}$
β	70
α	$2^{-8} \cdot \beta$
K_T	0.145 fs/bit
$S_{\phi_{\text{ref}}}$	–155 dBc/Hz ($\sigma_{t_{n,\text{ref}}} = 283$ fs)
$S_{\phi_{\text{DCO}}}$ (1 MHz)	–108 dBc/Hz ($\sigma_{t_{n,\text{DCO}}} = 33.9$ fs)

Table 4.2 Comparison of η predicted by [4] versus work and measured from simulations, assuming a jitter-free reference

TDC	[4] by (4.16)	This work by (4.14)	Simulation
BPD	211 fs	223 fs	222 fs
2-bit	252 fs	267 fs	267 fs
3-bit	368 fs	396 fs	399 fs

Table 4.3 Comparison of predicted and simulated $\sigma_{\Delta t}$

TDC	No DCO jitter		Reference and DCO jitters	
	Prediction	Simulation	Prediction	Simulation
BPD	369 fs	369 fs	416 fs	416 fs
2-bit	423 fs	423 fs	446 fs	447 fs
3-bit	528 fs	529 fs	543 fs	544 fs

TDC cases, the input jitters predicted here using (4.14) outperform their counterparts by [4] in the sense that the predictions using the former match the simulated results much better. The reason that the prediction in the latter lacks accuracy is that the jitter component contributed by the TDC's QE is underestimated in (4.16) of [4].

To verify further the accuracy of the prediction using (4.13), the analysis is extended to the cases where (i) the DCO is jitter-free, hence $S_{\phi_{\text{DCO}}} = -\infty$ and $\sigma_{t_{n,\text{DCO}}} = 0$, and (ii) the reference and DCO both exhibit jitter as parameterized in Table 4.1. The results tabulated in Table 4.3 verify the prediction accuracy of this work for these cases.[2]

Accurate prediction of RMS input jitter is crucial in terms of ascribing the correct linearized gain and power of the QE to the TDC. In particular, note that (3.24) and

[2] Readers should not be confused by the increase of RMS input jitter in Tables 4.2 and 4.3, when extra numbers of bits are used in the TDC. This is because, in these comparisons, the loop parameterization is not yet optimized corresponding to the TDC change. The loop optimization will be performed in Chap. 5.

(3.28) include $\sigma_{\Delta t}$. These will be used directly for linearized analysis of DPLLs at the system level. As simulations confirm the accuracy of the proposed prediction of the input jitter in all the cases of reference and DCO jitter combinations and with all the tested types of TDCs, the output jitter analysis and jitter minimization in the next chapter will be performed based on the expressions (4.13) and (4.14).

4.2.3 Output Jitter

The RMS *output* jitter of the DPLL that functions as a frequency synthesizer is typically of more interest than the DPLL's *input* jitter. In general, the output jitter is the one that ought to be minimized. References [3, 4] ignore the reference jitter in their derivation. In particular, the output jitter is claimed to be quantitatively equivalent to the TDC-input jitter. However, as reference jitter is indeed inevitable, the relationship between the system's input and output jitters, both of which are statistically dependent on the reference jitter, may not be that straightforward. In this subsection, the analysis of the output jitter is performed to reveal its relationship with the input one.

The analysis begins by noticing that the variance of the jitter of the divided signal ($t_{\text{div}}[k]$) is the same as that of the output signal ($t_{\text{out}}[h]$). By inspecting Fig. 4.1, assuming the divider *per se* does not add jitter to the transitions, the former sequence is just a decimated and compressed version of the latter that is uniformly sampled for every N elements, where N is the integer divide ratio of the feedback divider. Hence, the RMS jitter of $t_{\text{div}}[k]$ is equivalent to the RMS output jitter. The cycle-wise jitter associated with the $t_{\text{div}}[k]$ sequence is defined as the deviation of the kth sample from the expected value as

$$t_{\text{n,div}}[k] = t_{\text{div}}[k] - \mathbb{E}\left(t_{\text{div}}[k]\right) = t_{\text{div}}[k] - k \cdot T_{\text{ref}}, \tag{4.17}$$

where the DPLL locks the divided frequency to the reference frequency.

Substituting (3.1), (2.1), and (4.17) into (4.3), after some manipulations, yields a difference equation of $t_{\text{n,div}}[k]$ as

$$t_{\text{n,div}}[k+1] = (1-X) \cdot t_{\text{n,div}}[k] + X \cdot t_{n,\text{ref}}[k] + \frac{X}{K_{\text{TDC}}} \cdot q_{\text{TDC}}[k] + \sum_{h=N\cdot k+1}^{N\cdot(k+1)} t_{n,\text{DCO}}[h],$$

solving which and taking variances of both sides following the procedure detailed in Sect. 4.2.1, the variance of the jitter of the divided signal is given by

$$\sigma^2_{t_{\text{n,div}}} = \frac{X}{2-X} \cdot \sigma^2_{t_{n,\text{ref}}} + \frac{X}{2-X} \cdot \frac{\sigma^2_{Q_{\text{TDC}}}}{K^2_{\text{TDC}}} + \frac{N}{X(2-X)} \cdot \sigma^2_{t_{n,\text{DCO}}}. \tag{4.18}$$

Substituting (4.18) into (4.7) yields

$$\sigma_{\Delta t}^2 = \sigma_{t_{\text{n,div}}}^2 + \sigma_{t_{n,\text{ref}}}^2. \tag{4.19}$$

Although the randomly varying divided signal (represented by the first term on the right-hand side) is dependent on the reference jitter (represented by the second term) because it contains the latter transferred to its position, these two variables are statistically uncorrelated when referred to the input of the DPLL because the variance of the sum of these variables equals the sum of the variances of the variables. The RMS output jitter can then be expressed as

$$\sigma_{t_{\text{n,out}}} = \sqrt{\sigma_{\Delta t}^2 - \sigma_{t_{n,\text{ref}}}^2} = \sqrt{\frac{\eta^2}{2} + \eta\sqrt{\left(\frac{\eta}{2}\right)^2 + \sigma_{t_{n,\text{ref}}}^2}}. \tag{4.20}$$

The jitter minimization will be performed directly on $\sigma_{t_{\text{n,out}}}$ in Chap. 5.

4.2.4 Noise Transfer Functions for Output PN

In this subsection, prediction of the output PN of integer-N DPLLs is broken down into the contributions from the noise sources introduced above. Based on this prediction, the impact of jitter minimization on the system's PN signature as a guideline to derive the optimal design methodology will be discussed in Chap. 5.

The multi-rate system model shown in Fig. 4.1 can be further linearized using a linear time-invariant (LTI) model, as shown in Fig. 4.2, where $z = e^{j2\pi f/(N\cdot F_{\text{ref}})}$.[3] The phase perturbations injected by the reference jitter, TDC's QE, and DCO jitter are marked in red, and these are transferred to the overall output PN level via their respective NTFs. Unlike Fig. 4.1, which refers the white Gaussian DCO period jitter to the DCO input, Fig. 4.2 models the DCO jitter referred to its output for simplicity of analysis; in this way the DCO-output-referred PN has a −20 dB/dec slope in its power spectrum.

For the conciseness of presentation, the feed-forward (or open-loop) gain of the system is represented as

$$G_{\text{FF}}(z) = K_{\text{TDC}} \cdot \left(\beta + \frac{\alpha}{1 - z^{-N}}\right) \cdot K_T \cdot \frac{1 - z^{-N}}{1 - z^{-1}} \cdot \frac{1}{1 - z^{-1}}, \tag{4.21}$$

---

[3] A *time-variant* model that takes into account the downsampling operation of the feedback divider, which, in principle, imposes a folding effect on the out-of-band PN, has been reported in [7]. That modeling approach is more accurate, but it complicates the DPLL model and analysis. As pointed out in [12], the slight increase in the far-off PN level caused by the time-invariant approximation is negligible for most DPLL systems. Therefore, the feedback divider is modeled as an LTI divide-by-N module in this section to simplify the analysis. The feedback division operation inherently contributes a unit delay as well, as shown in Fig. 4.2, according to [7, Fig. 5]; this models the delay transferring from $t_{\text{div}}[k+1]$ in (2.13) to $t_{\text{div}}[k]$ in Fig. 2.1.

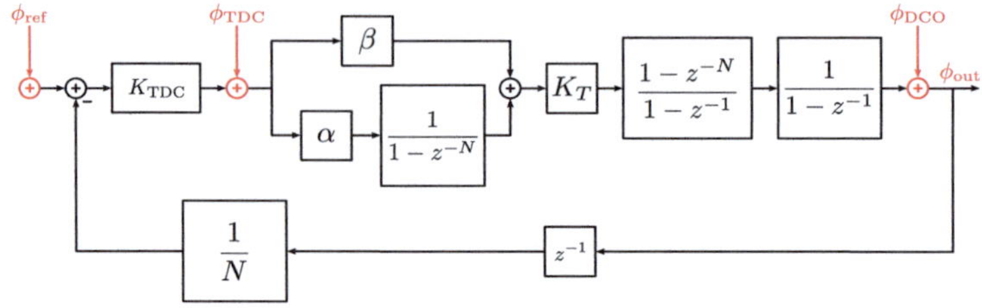

Fig. 4.2 Phase-domain LTI model of an integer-N DPLL system for PN analysis

and the feedback gain as

$$G_{\text{FB}}(z) = \frac{1}{N} \cdot z^{-1}. \tag{4.22}$$

Hence, the loop gain of the system is

$$G_{\text{loop}}(z) = G_{\text{FF}}(z) \cdot G_{\text{FB}}(z). \tag{4.23}$$

$S_{\phi_{\text{out}}}$ Due to Reference Jitter

The system output PN PSD due to the (assumed) white reference PN is

$$S_{\phi_{\text{out,ref}}}(f) = \left| \frac{G_{\text{FF}}(z)}{1 + G_{\text{loop}}(z)} \right|^2 \cdot S_{\phi_{\text{ref}}}, \tag{4.24}$$

where f in this section represents the offset frequency in relation to the center output frequency. The PSD of the PN of the reference jitter is related to the RMS reference jitter via

$$S_{\phi_{\text{ref}}} = (2\pi)^2 \, F_{\text{ref}} \cdot \sigma^2_{t_{n,\text{ref}}}. \tag{4.25}$$

$S_{\phi_{\text{out}}}$ Due to TDC's QE

The system output PN PSD contributed by the TDC QE can be expressed as

$$S_{\phi_{\text{out,TDC}}}(f) = \left| \frac{1}{K_{\text{TDC}}} \cdot \frac{G_{\text{FF}}(z)}{1 + G_{\text{loop}}(z)} \right|^2 \cdot S_{\phi_{\text{TDC}}},$$

where $S_{\phi_{\text{TDC}}} = (2\pi F_{\text{ref}})^2 \cdot S_{Q_{\text{TDC}}}$ is the phase-domain PSD of the TDC's QE transferred from the PSD of TDC's QE, which is $S_{Q_{\text{TDC}}} = \sigma^2_{Q_{\text{TDC}}}/F_{\text{ref}}$ as the TDC's QE is randomized in the stochastic resonance (SR) regime so that it has a white power spectrum and its noise power is uniformly distributed in the bandwidth, F_{ref} [7]. $\sigma^2_{Q_{\text{TDC}}}$ is presented in (4.11). Together, it can be rewritten that

$$S_{\phi_{\text{out,TDC}}}(f) = (2\pi)^2\, F_{\text{ref}} \cdot \left| \frac{G_{\text{FF}}(z)}{1 + G_{\text{loop}}(z)} \right|^2 \cdot \sigma^2_{Q_{\text{in,TDC}}}, \tag{4.26}$$

where the variance of TDC-input-referred QE, $\sigma^2_{Q_{\text{in,TDC}}}$, is expressed in (4.12).

$S_{\phi_{\text{out}}}$ Due to DCO Jitter

The −20 dB/dec PSD of the DCO-output-referred PN due to the DCO jitter is expressed as

$$S_{\phi_{\text{DCO}}}(f) = S_{\phi_{\text{DCO}}}(\Delta f) \cdot \left(\frac{2\pi(\Delta f)}{N \cdot F_{\text{ref}}} \right)^2 \cdot \frac{1}{\left|1 - z^{-1}\right|^2},$$

where the parameterization, $S_{\phi_{\text{DCO}}}(\Delta f)$, represents the value of the PSD referred to a specific offset frequency Δf [7]. For example, $S_{\phi_{\text{DCO}}}(1\ \text{MHz}) = -108$ dBc/Hz specified in Table 4.1. $S_{\phi_{\text{DCO}}}(\Delta f)$ can be calculated by

$$S_{\phi_{\text{DCO}}}(\Delta f) = \frac{(N \cdot F_{\text{ref}})^3}{(\Delta f)^2} \cdot \sigma^2_{t_n,\text{DCO}}, \tag{4.27}$$

The system output PN PSD due to the DCO period jitter can be expressed as

$$S_{\phi_{\text{out,DCO}}}(f) = \left| \frac{1}{1 + G_{\text{loop}}(z)} \right|^2 \cdot S_{\phi_{\text{DCO}}}(f). \tag{4.28}$$

Finally, the PSD of the DPLL's output PN is a linear combination of the PSDs contributed by the uncorrelated noise sources in (4.24), (4.26), and (4.28), namely

$$S_{\phi_{\text{out}}}(f) = S_{\phi_{\text{out,ref}}}(f) + S_{\phi_{\text{out,TDC}}}(f) + S_{\phi_{\text{out,DCO}}}(f).$$

The above NTFs are used to predict the DPLL's output PN in Chap. 5, which is then compared with the behaviorally simulated counterpart.

4.2.5 Loop Bandwidth

Notice that the NTFs for the reference jitter and TDC's QE contain the lowpass term

$$\left| \frac{G_{\text{FF}}(z)}{1 + G_{\text{loop}}(z)} \right|^2, \tag{4.29}$$

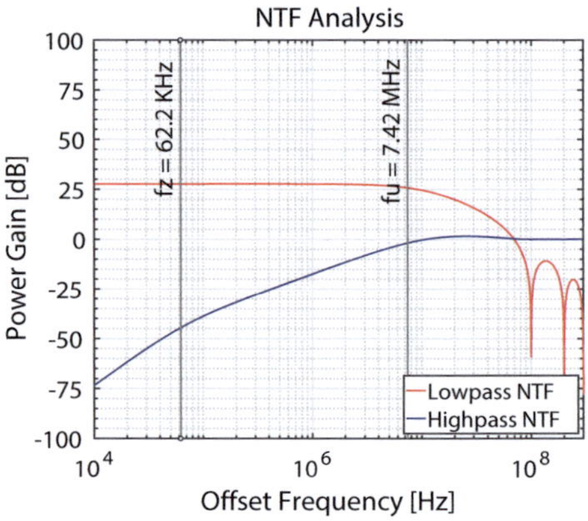

Fig. 4.3 PN NTFs in (4.29) and (4.30) for the system modeled in Sect. 4.2.2

whilst that for the DCO jitter is defined by the highpass filter

$$\left|\frac{1}{1+G_{\text{loop}}(z)}\right|^2. \tag{4.30}$$

Plots of (4.29) and (4.30) are shown in Fig. 4.3.

To facilitate the frequency-domain analysis for the NTFs, the narrowband approximation is considered, where $f \ll F_{\text{ref}}$, which yields $z^{-1} \approx 1 - j2\pi f/\left(N \cdot F_{\text{ref}}\right)$ and $z^{-N} \approx 1 - j2\pi f/F_{\text{ref}}$. In this way, the zero frequency contributed by the proportional-integral-based DLF with a transfer function, $\beta + \alpha/\left(1 - z^{-N}\right)$, can be simplified as

$$f_z \approx \frac{F_{\text{ref}}}{2\pi} \cdot \frac{\alpha}{\beta}. \tag{4.31}$$

Moreover, the term associated with α in (4.21) can be neglected as $\alpha \ll \beta$. As $2\pi f/(N \cdot F_{\text{ref}}) \ll 1$, it can further approximated that $z^{-1} \approx 1$ in (4.22) for the feedback path [13]. With these approximations, the NTFs in (4.21)–(4.23) can be simplified in the frequency domain as

$$G_{\text{FF}}(f) \approx \frac{K_{\text{TDC}} \cdot \beta \cdot K_T \cdot N^2 \cdot F_{\text{ref}}}{j2\pi f}, \tag{4.32}$$

$$G_{\text{FB}}(f) \approx \frac{1}{N},$$

$$G_{\text{loop}}(f) \approx \frac{K_{\text{TDC}} \cdot \beta \cdot K_T \cdot N \cdot F_{\text{ref}}}{j2\pi f}. \tag{4.33}$$

The corner frequency can be found by calculating the unity-gain frequency, where $\left|G_{\text{loop}}(f_u)\right| = 1$, as it indeed manifests as the pole in the denominators of (4.29) and (4.30). Here,

$$f_u \approx \frac{K_{\mathrm{TDC}} \cdot \beta \cdot K_T \cdot N \cdot F_{\mathrm{ref}}}{2\pi}, \tag{4.34}$$

which is commonly referred to as the "loop bandwidth" of the DPLL.

Substituting (4.32) and (4.33) into (4.29), the lowpass NTFs for the PNs of the reference and TDC become a lowpass filter with a DC gain of N^2, and a -20 dB/dec roll-off beyond f_u. On the other hand, substituting (4.31) and (4.33) into (4.30) yields the NTF for the DCO PN as a highpass filter with a $+40$ dB/dec slope below f_z, $+20$ dB/dec slope between f_z and f_u, and a unity highpass gain. The plotted z-domain NTFs using (4.29) and (4.30) for the system modeled in Table 4.1 are shown in Fig. 4.3, where the DC gain of (4.29) is $10\log_{10}(24^2) = 27.6$ dB, $f_z = 62.2$ kHz and $f_u = 7.42$ MHz as per (4.31) and (4.34). Note that, due to the discrete-time nature of the lowpass NTF in (4.29), high-frequency lobes are present in the out-of-band spectrum instead of a constant-slope roll-off.

MATLAB Script for Post Simulation Analysis in Locked State

In this chapter, we show the MATLAB script in Listing 4.1 that performs linearized analysis of the MRDT-modeled PLL system in the locked state. For simplicity of presentation, in the script, we show the analysis model that reports and plots the key figures predicted in this chapter, namely when the loop operates in the integer-N mode.

Listing 4.1 MATLAB script for the locked state analysis of simulated MRDT model of integer-N DPLLs with a mid-rise TDC.

```
%%%%%%%%%%%%%%%%%%%%%%%%%%%%%%%%%%%%%%%%%%%%%%%%%%%%%%%%%%%%%%%
%                   DATA TRIM TO LOCKED STATE                 %
%%%%%%%%%%%%%%%%%%%%%%%%%%%%%%%%%%%%%%%%%%%%%%%%%%%%%%%%%%%%%%%

t_div=t_div(ceil(N)+2:end-ceil(N));
Deltat=Deltat(ceil(N)+2:end-ceil(N));
t_out=t_out(ceil(N)+2:floor(end*0.95));

tstart=ncycles/F_ref*1/4;

index_start_in=find(t_div>tstart);
index_start_in=index_start_in(1);
index_start_out=find(t_out>tstart);
index_start_out=index_start_out(1);

Deltat_locked=Deltat(index_start_in:end-1);
t_out_locked=t_out(index_start_out:end-1);

%%%%%%%%%%%%%%%%%%%%%%%%%%%%%%%%%%%%%%%%%%%%%%%%%%%%%%%%%%%%%%%
%                 SIMULATED RESULTS ACQUISITION               %
%%%%%%%%%%%%%%%%%%%%%%%%%%%%%%%%%%%%%%%%%%%%%%%%%%%%%%%%%%%%%%%

sigma_Deltat_sim = std(Deltat_locked)

[F_mean,t_n_rms,phi_n_rms,f_PSD_out,PSD_phi_out] = acquisition(
    t_out_locked);

%%%%%%%%%%%%%%%%%%%%%%%%%%%%%%%%%%%%%%%%%%%%%%%%%%%%%%%%%%%%%%%
%                      PREDICTION MODELS                      %
%%%%%%%%%%%%%%%%%%%%%%%%%%%%%%%%%%%%%%%%%%%%%%%%%%%%%%%%%%%%%%%
```

```

%% Integer-N Jitter prediction
% Optimum gamma for TDC, can be changed
if Nb_TDC == 2
    gamma_TDC = 0.996;
elseif Nb_TDC == 3
    gamma_TDC = 0.588;
end
if Nb_TDC == 1
    eta = sqrt(pi/2)*N*beta*KT/2 + sqrt(pi/2)*sigma_t_n_DCO^2/(2*
        beta*KT);
    sigma_Deltat_int_pred = eta/2+sqrt((eta/2)^2+sigma_t_n_ref^2)
    K_TDC = sqrt(2/pi)/sigma_Deltat_int_pred;
    sigma_Q_TDC = sqrt(1 - 2/pi);
    sigma_Q_in_TDC = sqrt(pi/2-1)*sigma_Deltat_int_pred;
else
    syms k
    Y = eval(sqrt(2/pi)*(1+2*symsum(exp(-k^2*gamma_TDC^2/2),k,1,2^(
        Nb_TDC-1)-1)));
    Z = eval((2^Nb_TDC-1)^2-8*symsum(k*erf(k*gamma_TDC/sqrt(2)),k
        ,1,2^(Nb_TDC-1)-1));
    eta = N*beta*KT*Z/(2*Y) + sigma_t_n_DCO^2/(2*beta*KT*Y);
    sigma_Deltat_int_pred = eta/2+sqrt((eta/2)^2+sigma_t_n_ref^2)
    K_TDC = Y / sigma_Deltat_int_pred;
    sigma_Q_TDC = sqrt(Z - Y^2);
    sigma_Q_in_TDC = sqrt((Z/Y^2-1)*sigma_Deltat_int_pred^2);
end

%% Zero frequency
fz = alpha/beta*F_ref/(2*pi);
%% Unit-gain frequency
fu = K_TDC*F_ref*N*beta*KT/2/pi;

%% PN prediction
f             = f_PSD_out;
z_1           = exp(-1i*2*pi*f/(N*F_ref));
% Noise PSDs at source
S_phi_TDC     = (2*pi)^2*F_ref*sigma_Q_in_TDC^2;
S_phi_DCO     = 10^(S_phi_DCO_Deltaf/10)*((2*pi *1e6)/(N*F_ref))^2./
    abs((1-z_1)).^2;
S_phi_ref     = 10^(S_phi_ref/10);
% Loop transfer functions
G_FF          = K_TDC*KT*(beta+alpha./(1-z_1.^N)).*(1-z_1.^N)./(1-
    z_1).^2;
G_FB          = z_1/N;
G_loop        = G_FF.*G_FB;
G             = G_loop./(1+G_loop);
% Output characteristic NTFs
G_out_LP      = abs(G./G_FB).^2;
G_out_HP      = abs(G./G_loop).^2;
% Output PN NTFs
G_out_ref     = G_out_LP;
G_out_DCO     = G_out_HP;
G_out_TDC     = G_out_LP;
% Output PN contributions
S_phi_out_ref = S_phi_ref .* G_out_ref;
S_phi_out_DCO = S_phi_DCO .* G_out_DCO;
S_phi_out_TDC = S_phi_TDC .* G_out_TDC;
S_phi_out     = S_phi_out_ref + S_phi_out_DCO + S_phi_out_TDC;

%%%%%%%%%%%%%%%%%%%%%%%%%%%%%%%%%%%%%%%%%%%%%%%%%%%%%%%%%%%%%%
%                        RESULT PLOTS                        %
%%%%%%%%%%%%%%%%%%%%%%%%%%%%%%%%%%%%%%%%%%%%%%%%%%%%%%%%%%%%%%
```

```
88
89 % Plot characteristic NTFs per se
90 figure('Position',[0 0 900 700]+100)
91 semilogx(f,10*log10(G_out_LP),'r','linewidth',3);
92 hold on
93 semilogx(f,10*log10(G_out_HP),'b','linewidth',3);
94 xlim([1e4, 3e8]);
95 ylim([-100, 100]);
96 legend('Lowpass NTF','Highpass NTF')
97 grid on
98 x1 = xline(fz,'-',{'fz'},'HandleVisibility','off');
99 x1.LineWidth=3;
100 x1.FontSize=20;
101 x1.LabelHorizontalAlignment = 'left';
102 x2 = xline(fu,'-',{'fu'},'HandleVisibility','off');
103 x2.LineWidth=3;
104 x2.FontSize=20;
105 x2.LabelHorizontalAlignment = 'left';
106 title('NTF Analysis')
107 ax=gca;
108 xlabel(ax,'Offset Frequency [Hz]');
109 ylabel(ax,'Power Gain [dB]');
110 ax.LineWidth=3;
111 ax.FontSize=20;
112
113 % Plot simulated vs. predicted PNs
114 figure('Position',[0 0 900 700]+100)
115 semilogx(f,10*log10(PSD_phi_out),'Color',[0.5 0.5 0.5],'linewidth'
        ,3);
116 hold on
117 semilogx(f,10*log10(S_phi_out_ref),'r','linewidth',3);
118 semilogx(f,10*log10(S_phi_out_DCO),'b','linewidth',3);
119 semilogx(f,10*log10(S_phi_out_TDC),'g','linewidth',3);
120 semilogx(f,10*log10(S_phi_out),'k','linewidth',3);
121 ax=gca;
122 ax.LineWidth=3;
123 ax.FontSize=20;
124 xlim([2e4 inf])
125 ylim([-170 -90]);
126 xlabel(ax,'Offset Frequency [Hz]');
127 ylabel(ax,'Output Phase Noise [dBc/Hz]');
128 legend('Total Sim','Ref Pred','DCO Pred','TDC Pred','Total Int-N
        Pred','Location','southwest')
129 grid on
130 title('Output Phase Noise Power Spectrum')
```

The script consists of four sections. In the first section, from Lines 5 to 17, we regard the last three quarters of the simulated time as when the loop operates in the locked state, and trim the signals of interest to the specified range. In particular, we focus on the locked state input time difference (or input jitter) $\Delta t[k]$ and the locked state timestamps of the synthesized output signal, $t_{\text{out}}[h]$.

In the next section, i.e., Lines 23 and 25, we calculate the standard deviation of the simulated $\Delta t[k]$ (or RMS input jitter), $\sigma_{\Delta t_{\text{sim}}}$, and the PSD of the simulated output PN, $S_{\phi_{\text{out,sim}}}(f)$, by processing $t_{\text{out}}[h]$ with the `acquisition` function in Listing 2.7.

In the prediction section, Lines 31–83, the analytical predictions for the system jitter and PN that are made in this section are calculated. Specifically, we calculate the RMS input jitter, $\sigma_{\Delta t_{\text{pred}}}$, predicted in (4.13), its intermediate value, η, as per (4.14) and (4.15), the linearized TDC gain, K_{TDC}, as per (4.10), the RMS TDC's output and input-referred QEs, $\sigma_{Q_{\text{TDC}}}$ and $\sigma_{Q_{\text{in,TDC}}}$, as per (4.11) and (4.12), and the loop's zero frequency and loop bandwidth, f_z and f_u, analyzed in (4.31) and (4.34). The prediction of the output PN as a linear combination of the PSDs of the uncorrelated PN contributors transferred via their NTFs, i.e., (4.24), (4.26), and (4.28), are also calculated there.

Finally, in the last section of the script from Line 89, we show the code that illustrates the characteristic lowpass and highpass NTFs, as shown in Fig. 4.3, and also that plots the matched simulated and predicted PNs, such as those that will be shown in Sect. 4.4, which confirm the prediction accuracy of our work.

4.3 Fractional-*N* (Without DTC) Prediction

A fractional-*N* DPLL, such as that shown in Fig. 1.2, uses a MASH DDSM, commonly of order greater than one, e.g., MASH 1–1 or 1–1–1, to control the MMD in the feedback path in order to produce an output frequency that is a non-integer multiple of the reference frequency. Switching from the integer-*N* mode to the fractional-*N* mode, the QE arising from the MMDC presents at the input of the TDC, and is more significant than the input jitter contributed by the intrinsic noise/jitter sources [14]. This reduces the linearized gain of the TDC drastically due to the dependence of the TDC gain on the statistical properties of the input jitter. Intuitively, this can be understood as the result of the TDC's input range being increased, whilst the output range of the TDC saturates to its top and bottom quantization levels. Consequently, apart from the contribution from the DDSM noise, the suppressed effective gain severely degrades the output PN performance in the locked state [15]. This scenario is regarded in this book as the *worst-case* performance limit of a fractional-*N* DPLL, because it produces the most output PN.

Advanced fractional-*N* DPLLs typically use techniques such as a DTC [15] to reduce the MMDC's noise prior to the TDC in order to mitigate its negative impact on the PN performance. Ideally, a DTC with a linear gain can cancel the DDSM noise entirely and recover the integer-*N* PN, which can be regarded as the theoretical *best-case* performance limit of the fractional-*N* DPLL. However, in reality, the DTC exhibits nonidealities such as quantization, polynomial, and element mismatch nonlinearities, and hence can only mitigate the output PN to a level between the worst and best performance limits. It is useful to derive explicit predictions for these performance limits in order to analyze the effectiveness of techniques to mitigate fractional-*N* nonlinearity-induced PN and spurs.

Analysis of an integer-*N* DPLL, corresponding to the "best-case limit" in this section, has been widely studied in the literature, e.g., [4, 5, 7, 8]. Prior to this work, Avallone et al. [7] had provided the most accurate prediction for the input jitter and output PN based on a discrete-time system model of BBPLLs. In Sect. 4.2, their analysis was extended to DPLLs with a generic multi-bit TDC. However, the scope so far has been limited to integer-N operation. The fractional-N PN was discussed briefly by Xu and Abidi [5]. However, due to the continuous-time nature of the model used in [5], it lacks accuracy in terms of PN prediction [7]. More importantly, [5] determines the fractional-N's TDC gain as a function of the RMS input jitter, assuming that the MMDC's QE-dominated input jitter is Gaussian distributed; this assumption is incorrect, as shown in [16]. An improved TDC linearized analysis that takes into account the interaction between the TDC characteristic and the distribution of the MMDC's AQE was provided in Sect. 3.6. Furthermore, both [5, 7] are limited to the analysis of a BBPLL with a BPD. PN prediction in the case of fractional-N DPLLs using a generic multi-bit TDC, i.e., the "worst-case limit" will be provided here.

This section compares the PN performance limits when the same DPLL operates in its integer-N and fractional-N modes. The MRDT modeling framework is adopted and extended to analyze the DPLL with a multi-bit TDC operating in fractional-N mode. Together with the linearized analysis of the TDC, a linearized system model and output PN prediction are presented. The predicted PN, taking into account the contributions from reference clock jitter, TDC's QE, DCO period jitter, and MMDC's QE, is verified by the behaviorally simulated counterpart.

4.3.1 Fractional-N Input Jitter Transfer Mechanism

The input jitter analysis starts with the multi-rate z-domain model shown in Fig. 4.4 for the fractional-N DPLL, which is updated from Fig. 2.3 by linearizing the TDC as in (3.5). As per the model, a recurrent equation for the timestamp of the divided signal can be expressed as

$$t_{\mathrm{div}}[k+1] = t_{\mathrm{div}}[k] + \sum_{h=N_{\mathrm{div}}[k]\cdot k+1}^{N_{\mathrm{div}}[k]\cdot(k+1)} \left\{ \underbrace{K_T\left(\beta \cdot e[k] + \psi[k]\right)}_{\text{control via DLF}} + \underbrace{T_{\mathrm{DCO}} + t_{n,\mathrm{DCO}}[h]}_{\text{DCO period with jitter}} \right\}. \tag{4.35}$$

Note that, in contrast to (4.1) for the integer-N mode, the lower and upper limits of the sum operator both contain the cycle-wise varying divide ratio, $N_{\mathrm{div}}[k]$. Similar to the analysis after (4.1), $\psi[k]$ in (4.35) can be neglected without sacrificing the accuracy of the jitter transfer analysis.

Substituting (3.5) into (4.35) and some manipulations yield the difference equation as

$$\begin{aligned}\Delta t[k+1] = {} & \Delta t[k] + \left(t_{\mathrm{ref}}[k+1] - t_{\mathrm{ref}}[k]\right) - \\ & \sum_{h=N_{\mathrm{div}}[k]\cdot k+1}^{N_{\mathrm{div}}[k]\cdot(k+1)} \left\{K_{\mathrm{TDC}}\beta K_T \cdot \Delta t[k] + \beta K_T \cdot q_{\mathrm{TDC}}[k] + T_{\mathrm{DCO}} + t_{n,\mathrm{DCO}}[h]\right\},\end{aligned} \tag{4.36}$$

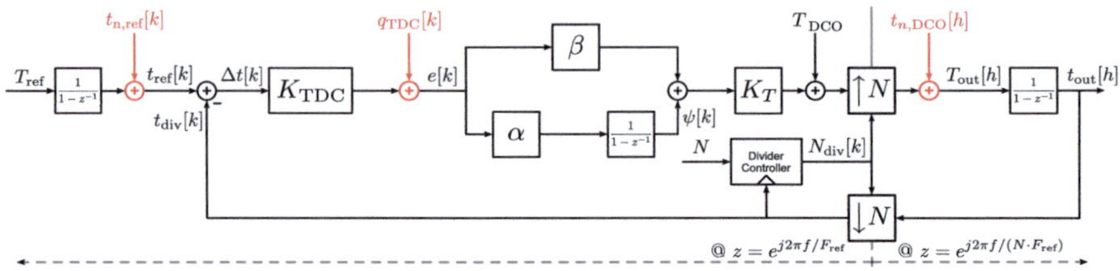

Fig. 4.4 z-domain model of a MRDT fractional-N DPLL system. The injected noise/jitter sources are marked in red

the sum in which contains $N_{\mathrm{div}}[k]$ elements. Note that $N_{\mathrm{div}}[k]$ produced by the MMDC can be expressed as

$$N_{\mathrm{div}}[k] \triangleq N + q_{\mathrm{MMDC}}[k].$$

Substituting the expression of the reference period with synchronous jitter in (2.2) into (4.36), extracting the T_{DCO} term from the sum, and recalling that $T_{\mathrm{ref}} = N \cdot T_{\mathrm{DCO}}$ in the locked state, Eq. (4.36) can be simplified to

$$\Delta t[k+1] = (1 - K_{\mathrm{TDC}}\beta K_T N) \cdot \Delta t[k] - q_{\mathrm{MMDC}}[k] \cdot T_{\mathrm{DCO}} + \text{"jitter terms contributed by intrinsic sources"}. \tag{4.37}$$

Equation (4.37) contains the jitter terms contributed by the TDC's QE and reference and DCO jitters; these are quantitatively negligible compared to the $q_{\mathrm{MMDC}}[k] \cdot T_{\mathrm{DCO}}$ term, and hence can be omitted. Furthermore, as derived and verified in Chap. 3, the fractional-N operation suppresses the TDC gain by a factor of about 1000 compared to the integer-N case, which pushes to zero the coefficient $K_{\mathrm{TDC}}\beta K_T N$, which was optimized within the interval (0,1) in the integer-N mode in order to guarantee loop stability. Therefore, in the steady state ($k \to \infty$), with the effect of the initial time difference $\Delta t[0]$ decaying to zero, (4.37) becomes

$$\Delta t[k+1] \approx -T_{\mathrm{DCO}} \cdot \sum_{i=0}^{k} q_{\mathrm{MMDC}}[i], \tag{4.38}$$

which indicates that the fractional-N input time difference is dominated by the waveform of the MMDC's AQE scaled by T_{DCO}.

Finally, taking the RMS values of both sides of (4.38) and substituting the standard deviation for the MASH DDSM's AQE [14] yields the fractional-N RMS input jitter as

$$\sigma_{\Delta t} \approx \begin{cases} \sqrt{\frac{1}{12}} \cdot T_{\text{DCO}}, & \text{for EFM1}, \\ \sqrt{\frac{1}{6}} \cdot T_{\text{DCO}}, & \text{for MASH 1–1}, \\ \sqrt{\frac{1}{2}} \cdot T_{\text{DCO}}, & \text{for MASH 1–1–1}, \\ & \vdots \end{cases}$$

The linearized gain and the variance of the QE of a generic N_b-bit TDC, taking as its input the AQE of a MASH 1–1 or 1–1–1 DDSM, have been derived explicitly in Chap. 3, and these will be used directly in the linear analysis and PN predictions in the remainder of this section.

4.3.2 Fractional-N PN Transfer Functions

The multi-rate system model shown in Fig. 4.4 can be linearized using an LTI model similar to Fig. 4.2, where $z = e^{j2\pi f/(N \cdot F_{\text{ref}})}$. The QE arising from the MMDC is injected into the feedback path as per the jitter transfer analyzed in Sect. 4.3.1, as shown in Fig. 4.5. Note that the MMDC's output QE, $q_{\text{MMDC}}[k]$, equals the QE injected into the final stage of the MASH, $q_l[k]$, that is lth-order highpass filtered by the DDSM's NTF, where l is the order of the MASH DDSM.

For conciseness of presentation, the feed-forward (or open-loop), feedback, and loop gains of the system are represented by (4.21), (4.22), and (4.23). Note that the NTFs of the output PNs contributed by the uncorrelated noises of the reference, TDC, and DCO in fractional-N mode are the same as Eqs. (4.24), (4.26), and (4.28).

The PSD of the MMDC's noise injected into the loop can be expressed as

$$T_{\text{DCO}}^2 \cdot \left| \frac{1}{1 - z^{-N}} \right|^2 \cdot S_{\phi_{\text{MMDC}}}(f)$$

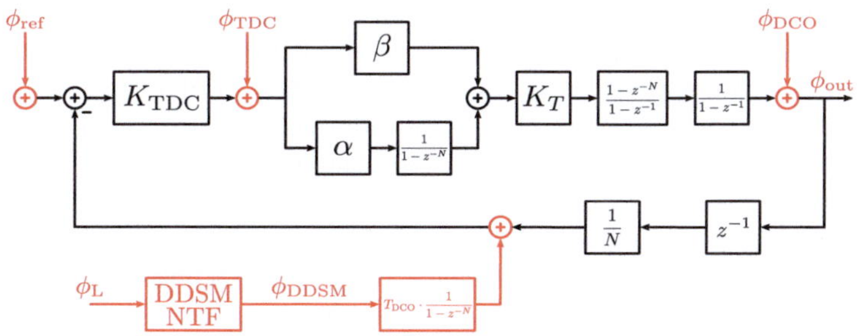

Fig. 4.5 LTI system model of a fractional-N DPLL for PN analysis. The injected noise/jitter sources are marked in red

where $S_{\phi_{\text{MMDC}}}(f) = \left|\left(1 - z^{-N}\right)^l\right|^2 \cdot S_{\phi_l}$ is the PSD at the output of the MMDC. $S_{\phi_l} = (2\pi F_{\text{ref}})^2 \cdot S_{Q_l}$ represents the PSD of the QE injected into the lth stage transferred to the phase domain, where $S_{Q_l} = \sigma_{Q_l}^2 / F_{\text{ref}}$ because the QE is white and uniformly distributed in its bandwidth and $\sigma_{Q_l}^2 = 1/12$ is the variance of the QE produced by a 1-bit quantizer. Finally, the system's output PN due to the MMDC's QE can be expressed as

$$S_{\phi_{\text{out,MMDC}}}(f) = \left|\frac{-G_{\text{FF}}(z)}{1 + G_{\text{loop}}(z)}\right|^2 \cdot \left|\frac{T_{\text{DCO}}}{1 - z^{-N}}\right|^2 \cdot S_{\phi_{\text{MMDC}}}(f). \tag{4.39}$$

The PSD of the DPLL's output PN is a linear combination of the PSDs contributed by each of the uncorrelated noise sources in (4.24), (4.26), (4.28), and (4.39), namely

$$S_{\phi_{\text{out}}}(f) = S_{\phi_{\text{out,ref}}}(f) + S_{\phi_{\text{out,TDC}}}(f) + S_{\phi_{\text{out,DCO}}}(f) + S_{\phi_{\text{out,MMDC}}}(f).$$

The above NTFs are used to predict the fractional-N DPLL's output PN in the next section, where it is then compared with its behaviorally simulated counterpart.

4.4 Verification for PN Predictions

To verify the accuracy of the analysis for both the integer-N and fractional-N output PNs, behavioral simulations using the MRDT model are performed, the PSDs of whose output PNs are extracted in Fig. 4.6 to compare with the analytical predictions in this chapter. The simulations use BPD, 2-bit and 3-bit TDCs as the PD. The model parameters are those used in Table 4.1. For the 2-bit and 3-bit TDCs, time resolutions of 1.08 and 0.66 times the integer-N RMS input jitter are chosen, which minimizes the jitter for these cases [4]. A MASH 1–1–1 MMDC is used to implement fractional-N operation; this randomizes the spectrum of the TDC's QE sufficiently. The nominal divide ratio is $N = 24 + 3277/2^{16}$.

In each case, as shown in Fig. 4.6, the prediction matches the simulation. Switching from the integer-N to the fractional-N mode, with the parameters of the loop and the intrinsic noise sources unchanged, the injected MMDC's QE significantly degrades the K_{TDC}, which in turn reduces the unity-gain frequency f_u of (4.23). Note that f_u is the pole frequency of all the denominators of (4.24), (4.26), (4.28), and (4.39); reducing the value of K_{TDC} consequently has the effect of reducing the loop bandwidth.

The reduced bandwidth does not change the close-in PN level contributed by the reference jitter, but effectively reduces the corner frequency of the lowpass filter. The DCO's PN is highpass filtered; hence it contributes more noise due to the reduced bandwidth. The TDC's PN contribution is lowpass filtered. Its close-in PN level is determined by the power of the input-referred TDC's QE, $S_{\phi_{\text{TDC}}}/K_{\text{TDC}}^2$. Switching to the fractional-N mode, the input-referred TDC's QE is increased dramatically as

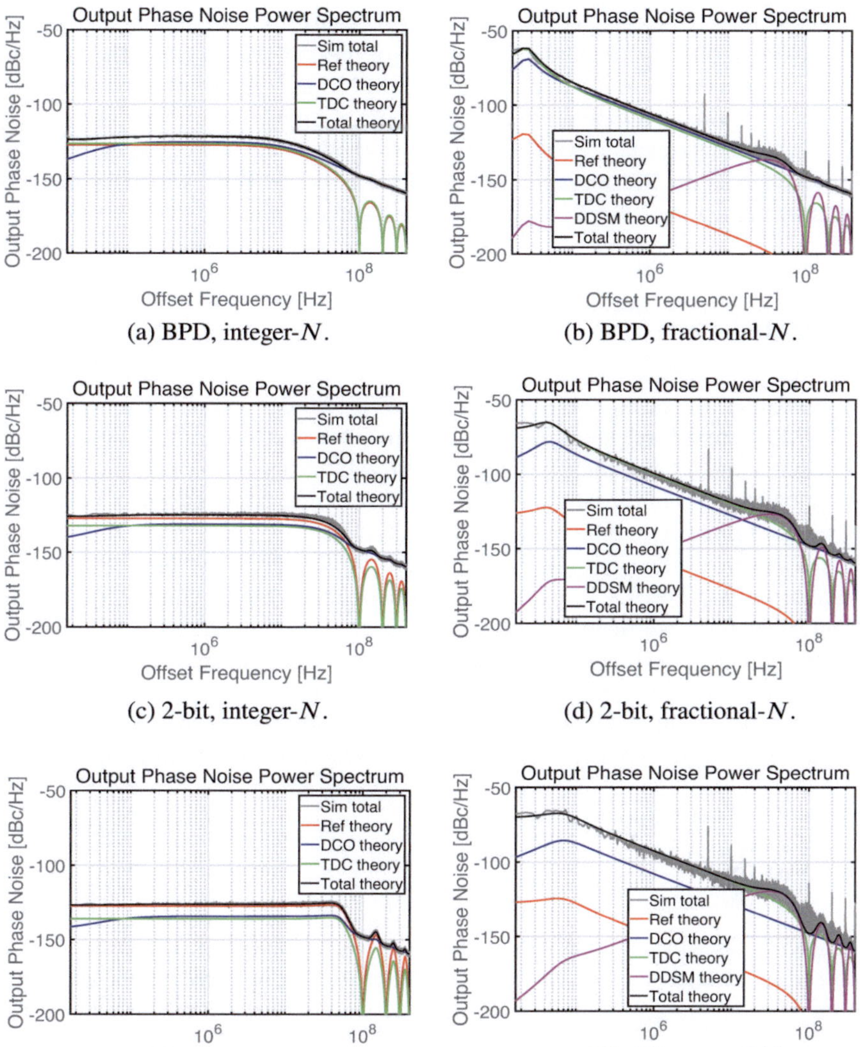

(a) BPD, integer-N.
(b) BPD, fractional-N.
(c) 2-bit, integer-N.
(d) 2-bit, fractional-N.
(e) 3-bit, integer-N.
(f) 3-bit, fractional-N.

Fig. 4.6 Simulated versus predicted PN PSDs for DPLLs with a mid-rise TDC. A MASH 1–1–1 MMDC is selected to facilitate fractional-N operation. Sim total (gray) is the simulated output PN; Ref theory (red), DCO theory (blue), TDC theory (green), and DDSM theory (magenta) are the analytically predicted PNs contributed by reference and DCO jitters, TDC's and MMDC's QEs; Total theory (black) is the sum of all the predicted components its value is mainly determined by the saturation of the top and bottom quantization levels overwhelmed by the much larger span of the TDC input jitter. The MMDC's QE *per se* is manifest as a bump in the output PN spectrum. Furthermore, DDSM-facilitated fractional-N operation exhibits spurs at integer and fractional multiples of the reference frequency, 100 MHz, in this example. This is because the MMDC's QE overwhelms the SR operation region of the mid-rise TDC, which saturates the TDC's dynamic range and drives it into a limit cycle again.

4.5 Conclusion

In this chapter, the accurately linearized TDC model derived in Chap. 3 is used in the DPLL-system-level analysis. The system jitter and PN in the integer-N and fractional-N modes are analytically predicted, and the accuracy of the predictions is confirmed by behaviorally simulated results. The closed-form integer-N predictions will be used directly in the next chapter to derive an optimum jitter-minimization-oriented system design strategy. Fractional-N analysis describes the worst-case performance limit of DPLLs. It revealed the phenomenon of the significantly degraded PN/jitter performance of the DPLL due to the injection of the MMDC's QE into the loop. This demonstrates that *a priori* QE cancellation techniques, such as using a DTC, that cancel the MMDC's QE before its injection are necessary in order to recover the potentially best-case integer-N performance of the loop. Modeling, analysis, and design enhancement for DTCs will be discussed from Chap. 6.

4.6 Summary

1. Based on the linear models of mid-rise coarse-resolution TDCs derived in the previous chapter, the closed-form predictions of the system's input and output jitters in the integer-N mode have been derived in Sect. 4.2 of this chapter. The accuracy of predictions in this chapter is demonstrated to be superior to that of their counterparts made in [4]. The NTFs relating the noise sources within the loop to the output PN were also reported.
2. Section 4.3 provides jitter/PN prediction for fractional-N operation. The much deteriorated jitter and PN performances, when DTC-based QE cancellation is not in use, confirm the need to use DTCs in such DPLLs.
3. Finally, Sect. 4.4 confirms the accuracy of PN predictions made in this book by matching the prediction and simulation results.

References

1. X. Wang, M.P. Kennedy, A random pulse modulation approach to modeling the flicker and white noise of the charge pump of a fractional-N frequency synthesizer. Int. J. Circuit Theory Appl. **50**(4), 1049–1063 (2022)
2. S. Levantino, G. Marzin, C. Samori, A.L. Lacaita, A wideband fractional-N PLL with suppressed charge-pump noise and automatic loop filter calibration. IEEE J. Solid-State Circuits **48**(10), 2419–2429 (2013)
3. M. Zanuso, D. Tasca, S. Levantino, A. Donadel, C. Samori, A.L. Lacaita, Noise analysis and minimization in bang-bang digital PLLs. IEEE Trans. Circuits Syst. II Express Briefs **56**(11), 835–839 (2009)
4. L. Avallone, M.P. Kennedy, S. Karman, C. Samori, S. Levantino, Jitter minimization in digital PLLs with mid-rise TDCs. IEEE Trans. Circuits Syst. I Regul. Pap. **67**(3), 743–752 (2020)
5. H. Xu, A.A. Abidi, Design methodology for phase-locked loops using binary (bang-bang) phase detectors. IEEE Trans. Circuits Syst. I Regul. Pap. **64**(7), 1637–1650 (2017)
6. N. Da Dalt, A design-oriented study of the nonlinear dynamics of digital bang-bang PLLs. IEEE Trans. Circuits Syst. I Regul. Pap. **52**(1), 21–31 (2005)
7. L. Avallone, M. Mercandelli, A. Santiccioli, M.P. Kennedy, S. Levantino, C. Samori, A comprehensive phase noise analysis of bang-bang digital PLLs. IEEE Trans. Circuits Syst. I Regul. Pap. **68**(7), 2775–2786 (2021)
8. T.-K. Kuan, S.-I. Liu, A bang bang phase-locked loop using automatic loop gain control and loop latency reduction techniques. IEEE J. Solid-State Circuits **51**(4), 821–831 (2016)
9. V. Kratyuk, P.K. Hanumolu, U.-K. Moon, K. Mayaram, A design procedure for all-digital phase-locked loops based on a charge-pump phase-locked-loop analogy. IEEE Trans. Circuits Syst. II Express Briefs **54**(3), 247–251 (2007)
10. N. Da Dalt, Linearized analysis of a digital bang-bang PLL and its validity limits applied to jitter transfer and jitter generation. IEEE Trans. Circuits Syst. I Regul. Pap. **55**(11), 3663–3675 (2008)
11. G. Marucci, S. Levantino, P. Maffezzoni, C. Samori, Exploiting stochastic resonance to enhance the performance of digital bang-bang PLLs. IEEE Trans. Circuits Syst. II Express Briefs **60**(10), 632–636 (2013)
12. Y. Hu, T. Siriburanon, R.B. Staszewski, Multirate timestamp modeling for ultralow-jitter frequency synthesis: a tutorial. IEEE Trans. Circuits Syst. II Express Briefs **69**(7), 3030–3036 (2022)
13. F.M. Gardner, *Phaselock Techniques* (John Wiley & Sons, 2005)
14. S. Levantino, L. Grimaldi, C. Samori, Analysis of adaptive pre-distortion in DTC-based digital fractional-N PLLs, in *2016 12th Conference on Ph.D. Research in Microelectronics and Electronics (PRIME)*, pp. 1–4 (2016)
15. D. Tasca, M. Zanuso, G. Marzin, S. Levantino, C. Samori, A.L. Lacaita, A 2.9–4.0-GHz fractional-N digital PLL with bang-bang phase detector and 560-fs_{rms} integrated jitter at 4.5-mW power. IEEE J. Solid-State Circuits **46**(12), 2745–2758 (2011)
16. V. Mazzaro, M.P. Kennedy, Spur immunity in MASH-based fractional-N CP-PLLs with polynomial nonlinearities. IEEE Trans. Circuits Syst. I Regul. Pap. **68**(6), 2295–2306 (2021)

Chapter 5
Enhanced Jitter Minimization for Integer-N DPLLs and Its Impact on Output PN

5.1 Introduction

This chapter considers DPLLs with mid-rise TDCs and provides a jitter minimization strategy that outperforms that in [1]. Previous papers on jitter minimization, such as [1, 2], aim to minimize the DPLL's *input jitter* that is assumed to be equivalent to the output jitter as the reference jitter is omitted in their derivations. A comprehensive jitter minimization procedure for the *output jitter* of DPLLs is introduced, which is based on the DPLL's output jitter which was derived in Chap. 4.

Furthermore, despite the jitter minimization described in [1], its influence on the system's output PN, which is another key performance metric for DPLLs used as radiofrequency synthesizers, has not been studied prior to this; this chapter addresses this impact analytically. Accurate predictions for the output PN components have been provided in Chap. 4 that are verified against behavioral simulation results. It will be concluded that, after jitter minimization, TDCs with more bits can reduce significantly the PN components induced by the TDC and DCO. A methodology designed to *mask* the output PN induced by the TDC by that induced by the reference jitter using a minimum number of TDC bits is provided, the result of which is that the system's output PN is dominated by the contributions from the reference and DCO jitters only.

This chapter is arranged as follows. Section 5.2 proposes a minimization strategy for the output jitter by optimizing the loop parameters and the time resolution of the TDC. Inspecting these, Sect. 5.3 discusses the impact of the proposed jitter minimization strategy on the system's output PN. Section 5.4 discusses the best choice of TDC bits under type-I and type-II loop configurations. Section 5.5 presents a jitter-minimization-oriented design procedure. The methodology is verified by prediction versus simulation results in Sect. 5.6. Finally, conclusions are presented in Sect. 5.7.

© The Author(s), under exclusive license to Springer Nature Switzerland AG 2026

X. Wang and M. P. Kennedy, *Jitter and Spur Minimization in Fractional-N Digital Frequency Synthesizers*, Analog Circuits and Signal Processing,
https://doi.org/10.1007/978-3-032-27186-0_5

5.2 Jitter Minimization

Mid-rise TDCs with a few more bits can be used in place of a BPD in order to mitigate the QE resulting from the latter's coarse quantization and, in turn, to minimize the PN and jitter that it contributes. For example, 2-bit and 3-bit mid-rise TDCs were used in [3, 4], where they were referred to as "optimal-threshold TDCs". Unlike a BPD that has only one quantization threshold at zero, the quantization thresholds of multi-bit TDCs must be carefully designed to minimize the QE optimally. Accurately derived optimal values of the quantization thresholds play an important role; for example, when auto-calibration schemes are used to calibrate the thresholds adaptively [3], the theoretically optimal quantization thresholds need to be predefined as target values, to which the auto-calibrated thresholds will converge. Minimization of the multi-bit TDC's QE by optimizing the choice of quantization thresholds has been described by Avallone et al. [1].

Two input jitter minimization schemes were proposed in [1] that scale the normalized TDC resolution, γ, to find the best resolution that minimizes the target quantities. Namely, the first of these is a "local optimization" that aims to minimize the power of the TDC's QE, whilst the other is a "global optimization" that aims to minimize the RMS system jitter directly. This section provides an enhanced jitter minimization strategy for the DPLL based on the jitter expressions derived in the previous chapter. It will be demonstrated that the difference that existed between the two minimization schemes in [1] is due to its flawed prediction of the input jitter that was discussed in Chap. 4. In fact, the two schemes in [1] are substantially identical, in the sense that, if the correct expressions are used, the quantities to be minimized by them are the same.

Another limitation of [1] is that it only considers TDCs with equidistant quantization thresholds. In this section, we will extend the analysis to the arbitrary-quantization-threshold case and show that multi-bit TDCs with optimized equidistant thresholds can yield both the optimum QE and jitter minimization.

5.2.1 Loop Dynamic Optimization

To minimize the RMS output jitter, $\sigma_{t_{\mathrm{n,out}}}$, derived in Chap. 4, the minimization first determines the optimal loop parameterization, namely, the optimal setting of the DLF proportional gain β. A necessary condition for the optimal value, $\hat{\beta}$, is when it makes the first-order partial derivative of the $\sigma_{t_{\mathrm{n,out}}}$ with respect to β equal to zero, that is

$$\frac{\partial \sigma_{t_{\mathrm{n,out}}}}{\partial \beta} = 0. \tag{5.1}$$

Substituting (4.14) and (4.20) into (5.1) yields

$$\left[\eta+\sqrt{\left(\frac{\eta}{2}\right)^2+\sigma_{t_{n,\mathrm{ref}}}^2}+\frac{\eta^2}{4}\left(\left(\frac{\eta}{2}\right)^2+\sigma_{t_{n,\mathrm{ref}}}^2\right)^{-\frac{1}{2}}\right]$$
$$\cdot\left(\frac{\eta^2}{2}+\eta\sqrt{\left(\frac{\eta}{2}\right)^2+\sigma_{t_{n,\mathrm{ref}}}^2}\right)^{-\frac{1}{2}}\cdot\frac{\partial\eta}{\partial\beta}=0.$$

As all the preceding terms are positive, the only real-valued solution can be found when

$$\frac{\partial\eta}{\partial\beta}=0, \tag{5.2}$$

with which the second-order derivative $\partial^2\sigma_{t_{n,\mathrm{out}}}/\partial\beta^2$ can be proven to be positive. Together, these yield the necessary and sufficient conditions for $\hat{\beta}$ that makes $\partial\eta/\partial\beta=0$ and minimizes the output jitter. Similarly, it is trivial to show that the value $\hat{\beta}$ also minimizes the system's input jitter expressed in (4.13).

Substituting (4.14) to solve (5.2), the optimal loop setting can be found as

$$\hat{\beta}=\frac{\sigma_{t_{n,\mathrm{DCO}}}}{K_T\sqrt{N\cdot Z}}. \tag{5.3}$$

This value, substituted into (4.14), in fact equalizes the amounts of jitter contributed by the TDC's QE and the DCO period jitter. This is the sufficient condition for SR operation that randomizes and disrupts limit-cycle behavior of the TDC in the locked state [5], and also yields a spur-free output PN spectrum, as will be shown in Sect. 5.6. The optimal value $\hat{\eta}$ is thus expressed as

$$\hat{\eta}=\sqrt{\frac{Z}{Y^2}}\cdot\sqrt{N}\cdot\sigma_{t_{n,\mathrm{DCO}}}, \tag{5.4}$$

recalling the definitions of Z and Y from Chap. 4.

5.2.2 *TDC Time Resolution Optimization*

To reduce the input and output jitters further, one can vary the time resolution, γ, of the N_b-bit TDC so that $\hat{\eta}$ in (5.4), that is readily optimized with respect to the loop dynamics, will be ultimately minimized. Therefore, with $\sqrt{N}$ and $\sigma_{t_{n,\mathrm{DCO}}}$ fixed, the expression $\sqrt{Z/Y^2}$ needs to be minimized.

It is interesting to note that, when $\sqrt{Z/Y^2}$ is minimized by choosing the best γ value, the quantity $\left(Z/Y^2-1\right)$ is also minimized exactly, which is the coefficient in $\sigma_{Q_{\mathrm{in,TDC}}}^2=\left(Z/Y^2-1\right)\cdot\sigma_{\Delta t}^2$ representing the power of the TDC's input-referred QE. This implies that the global optimization result, $\hat{\gamma}$, that minimizes the system jitters is effectively the same as the local optimization result that minimizes the TDC's QE.

In other words, the optimal TDC resolution $\hat{\gamma}$ should be the one that simultaneously minimizes the DPLL's input jitter in (4.13), output jitter in (4.20), and the TDC-input-referred QE in (4.12). On the other hand, although [1] defines the TDC's QE term, "K_q", to be the same as $\left(Z/Y^2 - 1\right)$ using this work's terminology, the system jitter represented by "K_J" in that work is derived from the flawed prediction in (4.16). This results in the global optimization—"K_J versus γ" and the local optimization—"K_q versus γ" performed in [1] being fundamentally different, which we have now shown to be incorrect.

In this subsection, we will provide a strategy to choose the optimum quantization thresholds of the TDC that minimize the TDC's QE and the system jitter of the synthesizer. The strategy is derived based on the integer-N assumption, where the steady-state input jitter approximates a zero-mean Gaussian distribution. As mentioned in Chap. 3, in the fractional-N operation of such synthesizers, effective DTC-based cancellation is typically used, which results in the input jitter remaining Gaussian and the derived strategy remaining optimum.

To improve the versatility of this analysis, we generalize the study to N_b-bit TDCs with arbitrary (nonuniformly distributed) quantization thresholds. We will conclude with a strategy for QE and jitter minimization in the case of mid-rise TDCs with an arbitrary number of bits.

Generalized Analysis Considering Arbitrary Quantization Thresholds

Instead of the equidistant quantization thresholds assumed in Chap. 3 that are $0, \pm t_{\text{res}}, \pm 2 \cdot t_{\text{res}}, \ldots, \pm\left(2^{N_b-1} - 1\right) \cdot t_{\text{res}}$, here we assume that the quantization thresholds are nonuniformly distributed and have arbitrary values that equal $t_{\text{res},k}$, $\forall k \in \left\{k \in \mathbb{Z} \,\middle|\, -2^{N_b-1} + 1 \leq k \leq 2^{N_b-1} - 1\right\}$. In this case, the normalized kth threshold is updated from (3.21) to become

$$\gamma_k = \frac{t_{\text{res},k}}{\sigma_{\Delta t}}, \quad \forall k \in \left\{k \in \mathbb{Z} \,\middle|\, -2^{N_b-1} + 1 \leq k \leq 2^{N_b-1} - 1\right\}.$$

Following the MMSE estimation procedure from (3.22) to (3.28), one can still obtain the expressions for the TDC's linearized gain ($\hat{K}_{\text{TDC}} = Y/\sigma_{\Delta t}$) and the powers of its QE ($\sigma^2_{Q_{\text{TDC}}} = Z - Y^2$) and input-referred QE ($\sigma^2_{Q_{\text{in,TDC}}} = \left(Z/Y^2 - 1\right) \cdot \sigma^2_{\Delta t}$) in the forms of Eqs. (4.10), (4.11), and (4.12). In the case of arbitrary quantization thresholds, the shorthand notations are updated as

$$Y = \sqrt{\frac{2}{\pi}} \cdot \sum_{k=-2^{(N_b-1)}+1}^{2^{(N_b-1)}-1} e^{-\frac{\gamma_k^2}{2}}, \tag{5.5}$$

and

$$Z = \left(2^{N_b} - 1\right)^2 - 4 \cdot \sum_{k=-2^{(N_b-1)}+1}^{2^{(N_b-1)}-1} k \cdot \operatorname{erf}\left(\frac{\gamma_k}{\sqrt{2}}\right). \tag{5.6}$$

When minimizing the representative quantity, Z/Y^2, by best choosing the TDC's resolution, the TDC's QE power in (4.12) and the system output jitter in (5.4) will be minimized simultaneously. Recall from Chap. 4 that Z represents the power of the TDC's output, and Y^2 represents the power of the linear portion of the output. Hence, minimizing the ratio Z/Y^2 in fact maximizes the linear portion of the TDC's output, which is equivalent to minimizing the quantization noise component. Furthermore, as will be shown in this subsection, with more bits used in the TDC, the optimized value of Z/Y^2 approaches unity. Substituting $Z/Y^2 \to 1$ into (4.12) reflects the effect of nullifying the TDC's QE with improved TDC resolution.

As per (5.5) and (5.6), Z/Y^2 is a complicated function that depends on a total of $(2^{N_b} - 1)$ different variables—$\gamma_k, \forall k \in \{k \in \mathbb{Z} \,|\, -2^{N_b-1} + 1 \le k \le 2^{N_b-1} - 1\}$. In the next subsection, we will perform jitter minimization for this generalized case.

Optimized Minimization

To find the optimal setting for each γ_k threshold, we notice that the optimized TDC design that minimizes the QE/jitter exists when the partial derivatives of the ratio (Z/Y^2) with respect to all the γ_k settings are simultaneously zero, namely

$$\frac{\partial\left(\frac{Z}{Y^2}\right)}{\partial\gamma_k} = 0, \forall k \in \left\{k \in \mathbb{Z} \,\middle|\, -2^{N_b-1} + 1 \le k \le 2^{N_b-1} - 1\right\};$$

this equation can be rearranged as

$$\frac{Y^2 \cdot \frac{\partial Z}{\partial\gamma_k} - 2 \cdot Y \cdot Z \cdot \frac{\partial Y}{\partial\gamma_k}}{Y^4} = 0. \tag{5.7}$$

Taking the derivative of (5.6) gives

$$\frac{\partial Z}{\partial\gamma_k} = -4 \cdot \sqrt{\frac{2}{\pi}} \cdot k \cdot e^{-\frac{\gamma_k^2}{2}},$$

and the derivative of (5.5) is

$$\frac{\partial Y}{\partial\gamma_k} = -\sqrt{\frac{2}{\pi}} \cdot \gamma_k \cdot e^{-\frac{\gamma_k^2}{2}}.$$

Substituting these into (5.7) returns the optimum setting for the kth quantization threshold, that is

$$\gamma_k = 2 \cdot \frac{Y}{Z} \cdot k, \forall k \in \left\{k \in \mathbb{Z} \,\middle|\, -2^{N_b-1} + 1 \le k \le 2^{N_b-1} - 1\right\}. \tag{5.8}$$

Notice that this means that the optimum quantization thresholds of the TDC must have values equal to $0, \pm 2\frac{Y}{Z}, \pm 4\frac{Y}{Z}, \pm 6\frac{Y}{Z}, \ldots, \pm \left(2^{N_b} - 2\right) \frac{Y}{Z}$. Given that the common factor, Y/Z, is the same for all the terms, this indeed indicates that the *equidistant-quantization-threshold* TDC with a normalized input resolution equal to $2 \cdot Y/Z$ will achieve QE minimization and result in the minimized jitter.

Verification

To verify the optimality of the derived TDC quantization thresholds for jitter minimization, we perform a numerical search for the optimal TDC quantization resolution that minimizes the target ratio Z/Y^2. For simplicity of analysis, the numerical search is performed considering the case of equidistant quantization thresholds. The result will show that the searched optimal thresholds match their theoretically derived counterparts (5.8) exactly.

We define a dimensionless variable $K = \sqrt{Z/Y^2}$ as a function of the normalized TDC's equidistant quantization threshold γ in (3.21) using the equidistant-quantization notations of Y and Z in (4.8) and (4.9), which becomes

$$K(\gamma) = \sqrt{\frac{\pi}{2}} \cdot \sqrt{\frac{(2^{N_b} - 1)^2 - 8 \sum_{k=1}^{2^{(N_b-1)}-1} k \cdot \mathrm{erf}\left(\frac{k\gamma}{\sqrt{2}}\right)}{\left(1 + 2 \cdot \sum_{k=1}^{2^{(N_b-1)}-1} e^{-\frac{k^2\gamma^2}{2}}\right)^2}}. \tag{5.9}$$

We will use numerical methods to find the optimal $\hat{\gamma}$ values for TDCs with different numbers of bits, N_b.

It is worth noticing that, for TDCs with different numbers of bits $N_b = 1, 2, 3, \ldots$, calculating $K(\gamma)$ with extreme γ settings yields the maximum value as

$$K_{\max} = \lim_{\gamma \to 0} K(\gamma) = \lim_{\gamma \to \infty} K(\gamma) = \sqrt{\frac{\pi}{2}}.$$

Meanwhile, notice that using a large number of bits for the TDC can reduce the TDC's QE referred to its input, which theoretically reduces the latter to zero as the number of bits approaches infinity. This implies that, expressing K as a function of N_b and the corresponding optimal $\hat{\gamma}$, the non-negative $\sigma^2_{Q_{\text{in,TDC}}}$ in (4.12) can be minimized with

$$K_{\min} = \lim_{N_b \to \infty} K\left(N_b, \hat{\gamma}\right) = 1.$$

Moderate values of $0 < \gamma < \infty$ and N_b keep the value of K within the range $(1, \sqrt{\pi/2}]$, which in turn minimizes the values of $\hat{\eta}$, $\sigma_{\Delta t}$, and $\sigma_{t_{\text{n,out}}}$.

A numerical search for the optimal $\hat{\gamma}$ and $\hat{K}$ pairs by sweeping γ in (5.9) was performed for $N_b = 1, 2, \ldots, 6$, and the results are shown graphically in Fig. 5.1 and summarized in Table 5.1. A BPD, where $N_b = 1$, does not have an explicit γ,

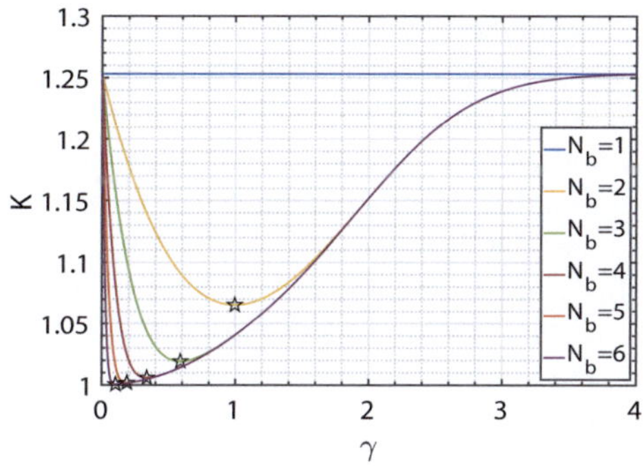

Fig. 5.1 Calculated K with swept γ by (5.9) for $N_b = 1, 2, \ldots, 6$. Stars represent the optimal value of γ, denoted $\hat{\gamma}$, and its corresponding $\hat{K}$ for each case that are tabulated in Table 5.1

Table 5.1 Comparison of optimal γ searched versus derived for $N_b = 1, 2, \ldots, 6$

N_b	Numerically searched				Calculated by (5.8)
	$\hat{\gamma}$	$\hat{K}$	$\hat{Y}$	$\hat{Z}$	$2 \cdot Y/Z$
1	N/A	1.253	0.798	1	N/A
2	0.996	1.065	1.770	3.554	0.996
3	0.588	1.019	3.277	11.151	0.588
4	0.336	1.006	5.884	35.036	0.336
5	0.188	1.002	10.598	112.772	0.188
6	0.104	1.001	19.201	369.432	0.104

but $K = \sqrt{\pi/2}$ in this case, which is represented by the horizontal line at the top of the figure. The plots are consistent with the analytical predictions for $K_{\max}$ and $K_{\min}$. The optimal $\hat{\gamma}$ and $\hat{K}$ pairs are tabulated in Table 5.1, from which one can see that, with more bits used by the TDC, the optimal value $\hat{K}$ is reduced from $\sqrt{\pi/2}$ toward 1, where $N_b = 2$ or 3 already reduces $\hat{K}$ to a level that is close to 1, which is the theoretical ideal case for a "linear" TDC that generates no QE.

Furthermore, as introduced in Sect. 5.2.1, the optimal loop parameter $\hat{\beta}$ balances the values of the TDC's QE and DCO terms within Eq. (4.14), which can ultimately be minimized by the optimal $\hat{\gamma}$, as discussed in this subsection. However, these jitter minimization practices cannot alter the jitter contributed by the reference jitter, i.e., the $\sigma^2_{t_n,\mathrm{ref}}$ term in the expressions for the system input and output jitters, (4.13) and (4.20). This fact limits the jitter minimization effect when the reference clock is the dominant contributor to the jitter.

Finally, substituting the optimal $\hat{Z}$ values from Table 5.1 into (5.3) yields the optimal setting for the loop dynamics, $\hat{\beta}$. In this case, the minimized values for the DPLL's input and output jitters can be predicted using (4.13) and (4.20), where the $\hat{\eta}$ value is calculated by substituting $\hat{\gamma}$ and $\hat{K}$ in Table 5.1 into (5.4).

In Table 5.1, we also compare the numerically searched optimal γ settings with the predictions calculated using (5.8). The table shows that the numerically searched optimal resolutions for equidistant TDCs match exactly the analytically calculated result in this section, i.e., the brute force searched results are in fact $2 \cdot Y/Z$ in (5.8). We will confirm the optimality of this resolution in Sect. 5.6, which also shows a relative insensitivity to the TDC's mismatch nonlinearity.

To summarize, in this section we extended the MMSE analysis from equidistant-threshold TDCs in Chap. 3 to TDCs with arbitrarily distributed thresholds. We confirmed analytically that the optimal TDC architecture indeed has equidistant quantization thresholds. The QE-minimization-oriented TDC design strategy indicates that the optimal normalized input resolution should equal $2 \cdot Y/Z$ in (5.8). Examples for $N_b = 1, 2, \ldots, 6$ are provided in Table 5.1.

5.3 Impacts of Jitter Minimization on PN

The jitter minimization strategy described in this chapter varies (i) the loop dynamics, i.e., β in (5.3), (ii) the estimated TDC linearized gain, K_{TDC} in (4.10), and (iii) the TDC's QE, e.g., $\sigma^2_{Q_{\text{in,TDC}}}$ in (4.12). Together, these affect the DPLL's output PN performance. This section analyzes separately the impacts of the jitter minimization on the contributions to the output PN from the reference jitter, the TDC's QE, and the DCO jitter, considering which some design rules of thumb for efficient PN mitigation can be derived. Finally, the section concludes with a design procedure to achieve low-jitter and low-PN DPLL systems.

The jitter minimization strategy proposed in this chapter optimizes the values of the TDC's linearized gain and the DLF's proportional gain, which alter the loop dynamics simultaneously. In particular, the change of the loop bandwidth is critical. Substituting the expressions for K_{TDC} and $\hat{\beta}$ in (4.10) and (5.3) into (4.34) yields the loop bandwidth as in

$$f_u = \sqrt{\frac{Y^2}{Z} \cdot \frac{\sigma_{t_{n,\text{DCO}}}}{\sigma_{\Delta t}} \cdot \frac{\sqrt{N}}{2\pi}} \cdot F_{\text{ref}}. \tag{5.10}$$

Since all the fractions on the right-hand side are dimensionless, the dimension of (5.10) is defined by F_{ref}, and hence is Hz, as expected. After jitter minimization, one can observe that, within (5.10),

- the RMS input-referred jitter, $\sigma_{\Delta t}$, in the denominator was minimized;
- as the factor $K = \sqrt{Z/Y^2}$ in (5.9) was minimized from the maximum value of $K = \sqrt{\pi/2}$ in the BPD case toward 1 as $N_b \to \infty$, interestingly, the term $\sqrt{Y^2/Z}$ must be maximized from $\sqrt{2/\pi} \approx 0.8$ toward 1;
- all the other terms remain unchanged.

Altogether, one can conclude that, compared to a BBPLL system with β optimized for the BBPLL case, the jitter minimization via (i) increasing N_b, (ii) optimizing β for the new N_b choice, and (iii) optimizing γ increases the loop bandwidth, but only slightly. In general, when the overall output jitter is guaranteed to be minimized, a broadened loop bandwidth is a desirable feature in terms of tracking speed, loop stability, etc.

Compared to before jitter minimization, the impact of the strategy on the contributions to the output PN from each of the noise/jitter sources individually can be observed as below.

5.3.1 Reference-Jitter-Induced PN

This strategy does not change the power of the reference jitter represented by the white PSD, $S_{\phi_{\text{ref}}}$, or the DC gain of the NTF in (4.24). Therefore, the magnitude of the close-in PN level of $S_{\phi_{\text{out,ref}}}(f)$ will remain unchanged. The bandwidth of the close-in PN level is larger. However, as introduced above, the change in the bandwidth is insignificant and can be neglected.

5.3.2 TDC-QE-Induced PN

As with the reference analysis above, the lowpass NTF's DC gain is unchanged, whilst its bandwidth is changed slightly. Referring to (4.26), as the white power of TDC-input-referred QE as expressed in (4.12) is minimized, the entire level of the PN contributed by the TDC's QE will be reduced. This is because, the more TDC bits are used, the closer the Z/Y^2 is pushed to 1, and hence the less is the power of the TDC's QE. Substituting (4.12) into (4.26), one can extrapolate that, in the limit, a TDC with an infinite number of bits contributes zero PN, which corresponds to a conceptually linear TDC that does not generate quantization noise.

5.3.3 DCO-Jitter-Induced PN

Jitter minimization does not change the intrinsic power of the DCO jitter that is represented by the $S_{\phi_{\text{DCO}}}(f)$ with a -20 dB/dec slope in (4.28). In the meantime, as the corner frequency of the highpass filtering NTF is pushed higher, the PN contributed by the DCO will be better suppressed.

Finally, it can be concluded that jitter minimization performed in this chapter can increase the loop bandwidth modestly, which is desirable when the overall output jitter is guaranteed to be minimized. The optimized bandwidth rebalances slightly the PN contributions from the reference and DCO jitters, in the sense that the reference-induced PN is changed insignificantly, whilst the DCO-induced PN is better suppressed. The overall PN level contributed by the TDC's QE is reduced by using extra TDC bits if the loop dynamics and TDC resolution are optimized accordingly.

5.4 Optimal Choice of Number of Bits Used for TDC

As increasing the number of bits used by the TDC reduces its contributed output PN level, it is useful to derive a criterion that determines how many bits should be used to best mitigate the TDC-QE-induced PN. Conventionally, in the PN power spectrum, once the magnitude of the TDC-induced PN contribution is overwhelmed by the contributions of the reference's and DCO's jitters, the TDC-induced level will be *spectrally masked* by the latter, in which case one can argue that the TDC-induced PN has been *"canceled completely"*. This section derives some design rules of thumb considering two separate scenarios, i.e., when $f_z \ll f_u$ and when $f_z < f_u$.

The intrinsic PN contributions from the reference jitter and TDC's QE are both assumed white, and the NTFs for them are both defined by the same lowpass filter in (4.29). Therefore, their contributions to the output PN have the same lowpass filtered shape. On the other hand, the intrinsic DCO-output-referred PN has a -20 dB/dec slope which, when passed through the highpass NTF of the type-II DPLL as discussed in Chap. 4, yields an output PN spectrum that has $+20$ dB/dec, flat, and -20 dB/dec slopes in the three regions separated by f_z and f_u. Loop stability requires that $f_z < f_u$ [6]. One can observe that, due to different settings of the α/β ratio, the distance between f_z and f_u varies. In some of the reported systems, e.g., [7] and [8, Fig. 6], $f_z \ll f_u$. In this case, the DCO-induced output PN has a broad flat level at close-in frequencies, which contributes to the overall in-band PN level together with the reference and TDC PNs. Note that the overall output PN in this case is similar to that of a type-I DPLL that does not have an integral path in the DLF, and hence $f_z = 0$ [9]. Meanwhile, some other systems, e.g., [2, 10], push f_z close to f_u. In this way, the DCO-induced PN is suppressed much more and exhibits peaking near the corner frequency. Hence, the close-in phase noise level is only determined by the reference and TDC PNs in this case. In this chapter, we call the former configuration the "$f_z \ll f_u$" case and the latter the "$f_z < f_u$" case.

The output PN power spectra, together with the breakdown of the PN contributions in these two cases, following the above analysis, are illustrated in Fig. 5.2. To simplify the analysis, the levels of the flat regions in the power spectrum induced by the reference jitter, DCO jitter, and TDC QE are represented as S_{ref}, S_{DCO}, and S_{TDC}. Their

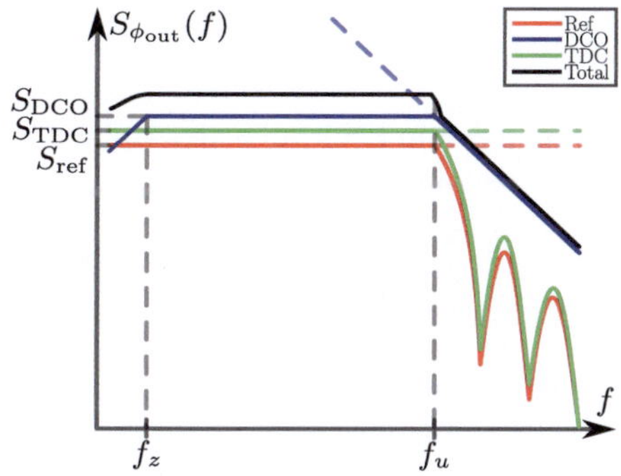

(a) $f_z \ll f_u$.

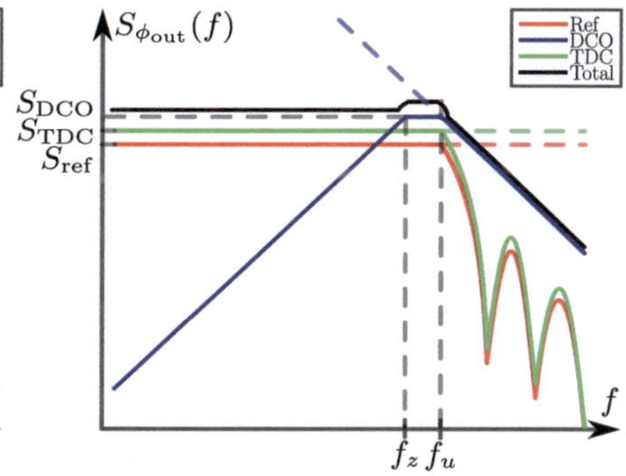

(b) $f_z < f_u$.

Fig. 5.2 Illustration of the output PN power spectra in the two scenarios. Ref (red), DCO (blue), and TDC (green) curves represent the output PNs contributed by the reference jitter, DCO jitter, and TDC QE. Total (black) represents the overall PN

magnitudes are quantitatively derived using the RMS jitters elaborated in previous discussion as follows:

- substitute (4.25) into (4.24) and recall that the DC gain of the NTF expressed by (4.29) is N^2; the PSD value of the flat region of the output PN induced by the reference jitter can be derived as

$$S_{\text{ref}} = (2\pi)^2 \cdot N^2 \cdot F_{\text{ref}} \cdot \sigma^2_{t_{n,\text{ref}}}; \tag{5.11}$$

- similarly, recall (4.26), the level of the flat region induced by the TDC's QE is

$$S_{\text{TDC}} = (2\pi)^2 \cdot N^2 \cdot F_{\text{ref}} \cdot \sigma^2_{Q_{\text{in,TDC}}}, \tag{5.12}$$

  which is equivalent to, as per (4.12),

$$S_{\text{TDC}} = (2\pi)^2 \cdot N^2 \cdot F_{\text{ref}} \cdot \left(K^2 - 1\right) \cdot \sigma^2_{\Delta t}, \tag{5.13}$$

  where K is defined in (5.9);
- notice that, between f_z and f_u, $\left|\frac{1}{1+G_{\text{loop}}(z)}\right|^2 \approx (f/f_u)^2$, substituting which, together with (5.10) and (4.27), into (4.28) yields the level of the flat region induced by the DCO jitter as

$$S_{\text{DCO}} = (2\pi)^2 \cdot N^2 \cdot F_{\text{ref}} \cdot K^2 \cdot \sigma^2_{\Delta t}. \tag{5.14}$$

Note that one can discard the common coefficient $(2\pi)^2 \cdot N^2 \cdot F_{\text{ref}}$ that appears in all the expressions, (5.11)–(5.14), when comparing their magnitudes.

5.4.1 Case 1: $f_z \ll f_u$

As illustrated in Fig. 5.2a, in this case, the output close-in PN level is determined by the contributions from all three PN sources. One can observe the following:

- for S_{TDC} versus S_{DCO}: interestingly, comparing (5.13) and (5.14), as the inequality of $\left(K^2 - 1\right) \cdot \sigma_{\Delta t}^2 < K^2 \cdot \sigma_{\Delta t}^2$ always holds, this implies that the DCO-induced output PN always exceeds that induced by the TDC, as long as the jitter minimization is performed correctly. Note that this relationship is true for TDCs with any number of bits, and is even true for the BPD that generates the worst-case QE;
- for S_{TDC} versus $S_{\mathrm{DCO}} + S_{\mathrm{ref}}$: taking into account the reference jitter in (5.11), one can see that the PN level contributed by the TDC's QE must be masked by that contributed by both the reference and DCO jitters as $\left(K^2 - 1\right) \cdot \sigma_{\Delta t}^2 < K^2 \cdot \sigma_{\Delta t}^2 + \sigma_{t_{n,\mathrm{ref}}}^2$ is always satisfied;
- for S_{TDC}, S_{DCO}, and S_{ref}: by the proposed jitter minimization, K^2 and $\sigma_{\Delta t}^2$ are minimized, which lowers the output PN level contributed by the DCO and TDC simultaneously, yet has no effect on that from the reference jitter. Interestingly, one can deduce that the jitter minimization has an impact on the PN level that is *very similar* to that on the input/output jitters. As discussed in Sect. 5.2.2, the proposed jitter minimization strategy minimizes the jitters contributed by the DCO and TDC, but has no effect on the reference-induced jitter. Explicitly, the close-in output PN level is defined by (5.11), (5.13), and (5.14) as $(2\pi)^2 \cdot N^2 \cdot F_{\mathrm{ref}} \left[\left(2 \cdot K^2 - 1\right) \cdot \sigma_{\Delta t}^2 + \sigma_{t_{n,\mathrm{ref}}}^2\right]$;
- for S_{ref} versus S_{DCO}: furthermore, recall (4.19) that $\sigma_{\Delta t}^2 = \sigma_{t_{\mathrm{n,out}}}^2 + \sigma_{t_{n,\mathrm{ref}}}^2$, hence $\sigma_{\Delta t}^2 > \sigma_{t_{n,\mathrm{ref}}}^2$ always holds. As K^2 is minimized within the range of $(1, \pi/2]$, comparing (5.11) and (5.14), one can see that $\sigma_{t_{n,\mathrm{ref}}}^2 < K^2 \cdot \sigma_{\Delta t}^2$ is always satisfied. Therefore, no matter how much the DCO-induced PN is minimized, its magnitude is reduced toward the reference-induced level and can never be lower than the latter. In an extremely jitter-minimized case using a large number of bits, the magnitude of the overall close-in PN level is about twice S_{ref}.

5.4.2 Case 2: $f_z < f_u$

In this case, the close-in output PN level is only contributed by the PNs induced by the reference and the TDC, which, as discussed already, have the same shape. As the DCO-induced output PN is greatly suppressed just by relocating the f_z position, this configuration is considered a better solution for enhanced PN performance. Note that the relationship of S_{ref} versus S_{DCO} discussed in the last item in Case 1 still exists, which implies that the peaking in the DCO PN contribution will always exceed the reference-induced PN level. In order to have the TDC-induced PN *masked* by the reference-induced one, a useful inequality can be concluded as a design criterion. By comparing (5.11) and (5.12), one needs to ensure that

$$\sigma^2_{Q_{\text{in,TDC}}} < \sigma^2_{t_{n,\text{ref}}}, \tag{5.15}$$

which can be accomplished by adopting more bits for the TDC followed by a correct jitter minimization. The inequality indicates that, as long as the power of the TDC-input-referred jitter is less than that of the reference jitter, the close-in PN will be dominated solely by the reference PN, as if the TDC is *"ideal"* and contributes no noise. Note that this PN minimization can be achieved by using a relatively small number of bits for the TDC, where one can already almost *"completely cancel"* the TDC-QE-induced PN from the output spectrum. This gives an advantage when using a DPLL in place of an analog CP-PLL, whose CP exhibits inevitable physical noises that cannot be easily mitigated [11]. Using additional bits for the TDC beyond this point is meaningless, as it only consumes more area and power.

5.5 Complete Jitter-Minimization-Oriented Design Procedure

In order to achieve this optimal PN performance, insight into the inequality (5.15) can be gained by rewriting it as $\left(K^2 - 1\right) \cdot \sigma^2_{\Delta t} < \sigma^2_{t_{n,\text{ref}}}$. The optimal value $\hat{K}$ can be obtained from Table 5.1, where $\sigma^2_{\Delta t}$ is derived in the step for (4.13), and $\sigma^2_{t_{n,\text{ref}}}$ is a known characteristic of the reference source. The design procedure proposed in this work is summarized in the flowchart shown in Fig. 5.3; one can follow this procedure to achieve a type-II DPLL which has a minimized system jitter and an optimized PN performance that is dominated by the contributions from the reference and DCO jitters only, whilst using a minimum number of bits in the TDC.

Listing 5.1 shows the MATLAB script that programs the proposed jitter-minimization-oriented design procedure to determine the optimum setting for the loop and TDC in a DPLL with mid-rise TDCs. Note that, for presentation simplicity, we show here the basic loop and noise parameterization as per Table 4.1. The values in Lines 4–8 can be customized to be used for the design of other loops.

Listing 5.1 MATLAB script encoding the jitter-minimization-oriented design procedure proposed in Fig. 5.3.

```
1  clear all
2  close all
3  %% Basic parameters
4  F_ref            = 100e6;
5  N                = 24;
6  KT               = 0.145e-15;
7  sigma_t_n_ref    = 283e-15;
8  sigma_t_n_DCO    = 33.9e-15;
9
10 %% DLF ratio?
11 R = 2^-8; % ratio of alpha to beta
```

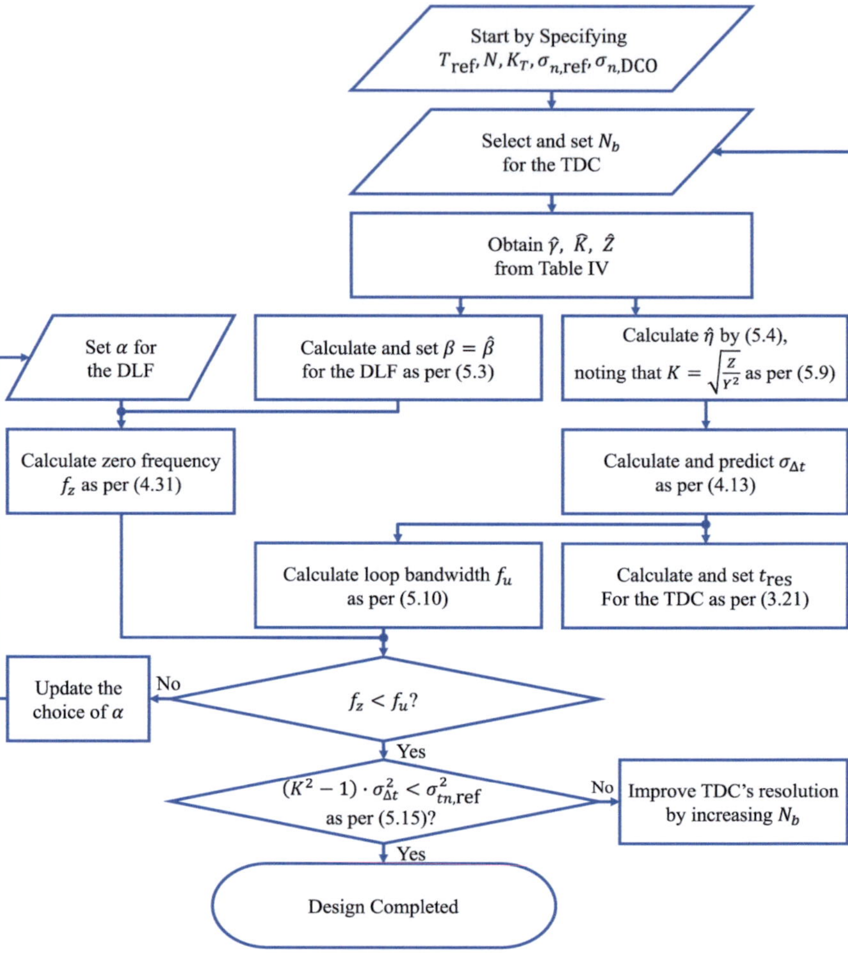

Fig. 5.3 Proposed jitter-minimization-oriented design procedure for a type-II DPLL whose output PN is dominated by the reference and DCO noises only

```
%% Number of TDC bits?
Nb_TDC = 1;

%% Table 5.3
if Nb_TDC == 1
    gamma = 0;
    K = sqrt(pi/2);
    Z = 1;
elseif Nb_TDC == 2
    gamma = 0.996;
```

```
    K = 1.065;
    Z = 3.554;
elseif Nb_TDC == 3
    gamma = 0.588;
    K = 1.019;
    Z = 11.151;
end

%% Calculation
beta                = sigma_t_n_DCO/KT/sqrt(N*Z)
fz                  = F_ref/2/pi*R
eta                 = K* sqrt(N)*sigma_t_n_DCO;
sigma_Deltat_pred = eta/2 + sqrt((eta/2)^2 + sigma_t_n_ref^2)
t_res               = gamma*sigma_Deltat_pred
fu                  = 1/K * sigma_t_n_DCO/sigma_Deltat_pred*sqrt(N)/(2*pi)*F_ref
if fz<fu
    fprintf(1,'\nZero frequency is less than loop bandwidth, good.\n');
    if (K^2-1)*sigma_Deltat_pred^2<sigma_t_n_ref^2
        fprintf(1,'\nTDC masked by ref, completed!\n');
    else
        fprintf(1,'\nTDC not masked by ref, try increasing TDC bits!\n');
    end
else
    fprintf(1,'\nZero frequency is greater than loop bandwidth, choose another R.\n');
end
```

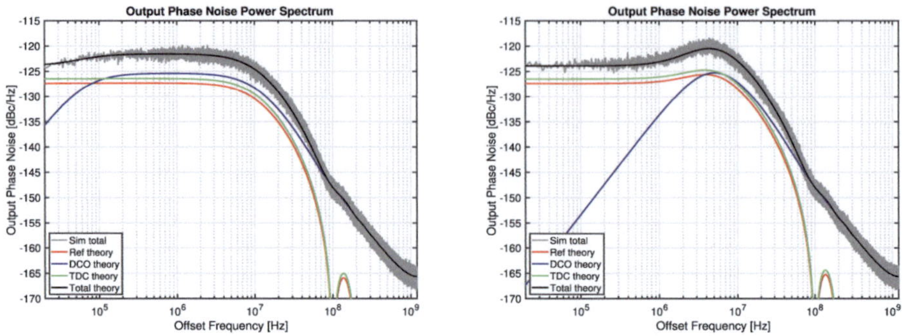

(a) $f_z \ll f_u$. (b) $f_z < f_u$.

Fig. 5.4 Simulated versus predicted PN PSDs for a BBPLL *before* jitter minimization. Sim total (gray) is the simulated output PN; Ref theory (red), DCO theory (blue), TDC theory (green) are the analytically predicted PNs contributed by reference and DCO jitters, and TDC QE; Total theory (black) is the sum of all the predicted

5.6 Verification

To verify the analysis and predictions in this chapter, the MRDT DPLL system using the parameters presented in Table 4.1 is simulated. The BBPLL described in [8] with $\beta = 70$ and $\alpha = 2^{-8} \cdot \beta$ is chosen. The PSD of the output PN is plotted in Fig. 5.4a, which is the same as [8, Fig. 6c, right]. In this case, the resultant bandwidths are $f_z = 62.2$ kHz and $f_u = 7.42$ MHz. Note that this β value, and hence f_u, is not optimized for jitter minimization purposes. The relationship between f_z and f_u corresponds to the $f_z \ll f_u$ case discussed. To configure the $f_z < f_u$ case, the other parameters are kept unchanged and $\alpha = 40 \cdot 2^{-8} \cdot \beta$ is set; this yields $f_z = 2.49$ MHz which is much closer to f_u. The reconfigured output PN is shown in Fig. 5.4b. Figure 5.4 demonstrates the representative cases classified in Sect. 5.4. Case (i) of Table 5.2[1] shows the key system values that are predicted and then compared with their simulated counterparts. Note that, in Table 5.2, the predicted and simulated results match very well.

5.6.1 Analysis of Effect of Jitter and PN Mitigation

For the two cases shown in Fig. 5.4, jitter minimization is performed by setting the optimal loop bandwidth using $\hat{\beta}$ calculated with (5.3) and increasing TDC $N_b = 2, 3$ with the optimal values of $\hat{\gamma}$ in Table 5.1. The columns of Fig. 5.5 show the output

[1] The simulated RMS input jitter, $\sigma_{\Delta t_{\text{sim}}}$ in Table 5.2, is acquired following the method introduced in Chap. 2, i.e., by taking the standard deviation of the simulated samples of the locked-state input time difference, Δt.

Table 5.2 Predicted values and simulated input jitter for the DPLLs of Figs. 5.4 and 5.5

Case	TDC	$\sigma_{Q_{\text{TDC}}}$ [bit]	$\sigma_{Q_{\text{in,TDC}}}$ [fs]	K_{TDC} [bit/s]	β	f_u [MHz]	$\sigma_{\Delta t_{\text{pred}}}$ [fs]	$\sigma_{\Delta t_{\text{sim}}}$ [fs]
(i)	BPD (with $\beta = 70$) [8]	0.603	314	$1.92 \cdot 10^{12}$	70	7.42	416	416
(ii)	BPD (with $\hat{\beta}$)	0.603	306	$1.97 \cdot 10^{12}$	47.8	5.20	405	407
(iii)	2-bit (with $\hat{\beta}$)	0.650	141	$4.60 \cdot 10^{12}$	25.3	6.44	385	385
(iv)	3-bit (with $\hat{\beta}$)	0.646	74.9	$8.62 \cdot 10^{12}$	14.3	6.82	380	380
(v)	3-bit (w/ $\beta = 90\% \cdot \hat{\beta}$)	0.646	75.0	$8.61 \cdot 10^{12}$	12.9	6.13	381	381
(vi)	3-bit (w/ $\beta = 110\% \cdot \hat{\beta}$)	0.646	75.0	$8.61 \cdot 10^{12}$	15.7	7.49	381	381

PNs after the jitter minimization process for the representative $f_z \ll f_u$ and $f_z < f_u$ cases, respectively. The key values are tabulated in Cases (ii)–(iv) in Table 5.2.

The following observations are made:

- comparing the BPD cases before and after jitter minimization: as recorded in Cases (i) and (ii) of Table 5.2, jitter minimization by optimizing the loop bandwidth helps to reduce the system jitter from 416 to 405 fs;
- considering increased N_b with jitter minimization: in terms of jitter minimization, one can see that the proposed strategy reduces the RMS TDC-input-referred QE, $\sigma_{Q_{\text{in,TDC}}}$, and the RMS input jitter, $\sigma_{\Delta t}$, simultaneously. In terms of PN performance, the strategy increases the loop bandwidth modestly, as expected. The TDC-QE-induced PN level is reduced from above to below the reference-induced level. Note that the reference-induced noise level is determined by $\sigma_{t_{n,\text{ref}}} = 283$ fs. Referring to the fourth column of Table 5.2, $N_b \geq 2$ makes $\sigma_{Q_{\text{in,TDC}}} < \sigma_{t_{n,\text{ref}}}$. As per (5.15), this in turn brings the TDC-induced level below the reference-induced level. The larger the number of bits, the lower is the TDC-induced PN level. With jitter minimization, the flat region of the DCO-induced PN level decreases as well. However, as predicted in Sect. 5.4, it cannot be reduced to be lower than the reference-induced level. Hence, for the $f_z \ll f_u$ case with jitter minimization, the overall close-in PN level is determined by the PN contributed by the DCO and reference;
- considering the $f_z < f_u$ case: in terms of the key values, the only difference from the $f_z \ll f_u$ case is that $f_z = 2.49$ MHz now. In this case, the strategy successfully minimizes the TDC's QE and system jitters as well. The close-in output PN level is only determined by the reference and TDC PNs, and is lower compared with the $f_z \ll f_u$ case. Increasing N_b, together with correct jitter minimization, the TDC-QE-induced level can be reduced significantly. In Fig. 5.5b, the TDC-induced level contributes more to the overall PN level than the reference one, whereas in Fig. 5.5f,

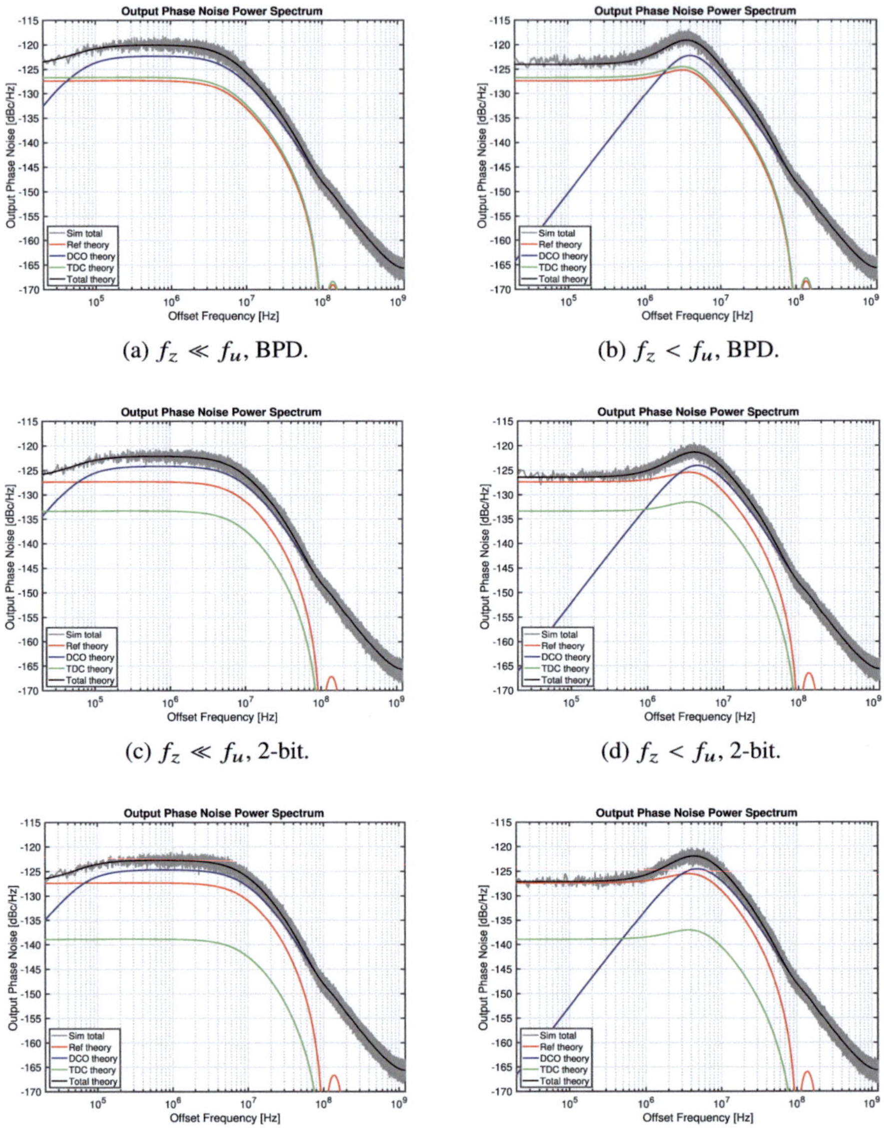

(a) $f_z \ll f_u$, BPD.

(b) $f_z < f_u$, BPD.

(c) $f_z \ll f_u$, 2-bit.

(d) $f_z < f_u$, 2-bit.

(e) $f_z \ll f_u$, 3-bit.

(f) $f_z < f_u$, 3-bit.

Fig. 5.5 Simulated versus predicted PN PSDs for DPLLs *after* jitter minimization. Sim total (gray) is the simulated output PN; Ref theory (red), DCO theory (blue), TDC theory (green) are the analytically predicted PNs contributed by reference and DCO jitters, and TDC QE; Total theory (black) is the sum of all the predicted the TDC-PN is overwhelmingly masked by the reference-PN so that the close-in level is dominated by the reference PN *per se*. Similar to the previous observation, $N_b \geq 2$ satisfies the inequality in (5.15) which ensures that the reference-induced PN level is dominant. Notice that, as predicted in Sect. 5.4, the DCO-induced PN always has a narrow flat region whose magnitude can be reduced, yet can never be lower than the reference-induced level;

- finally, as discussed in Sect. 5.2.1, the output spectra are free of spurs due to the fact that the TDC is operating in the SR regime.

Quantitatively, the predicted RMS input jitters match the simulated counterparts for all the cases presented, which confirms the accuracy and effectiveness of the albeit idealized analysis in this work.

5.6.2 Sensitivity of the Optimized Designs to Variations, Mismatch and Nonlinearity

To assess the optimality of the loop bandwidth set by applying $\hat{\beta}$, recalling (4.34), DPLLs with a 3-bit TDC are predicted and simulated where the loop bandwidth is slightly increased (or reduced) by setting a value for β that is 10% greater (or less) than $\hat{\beta}$. The predicted and simulated key values of these cases are tabulated in Cases (v) and (vi) of Table 5.2. Compared with Case (iv), which describes the 3-bit-TDC DPLL that is optimized as per the proposed jitter minimization strategy, one can observe that making the loop bandwidth slightly narrower or wider in fact degrades the TDC-input-referred QE and the system jitter, but only slightly. This indicates that the loop bandwidth set by $\hat{\beta}$ derived in this work is indeed an optimized one. It also confirms that the value of β is insensitive to small variations in the sense that having a value of β that deviates slightly from $\hat{\beta}$ only degrades the system jitter marginally (by 1 fs in simulations).

The sensitivity of the derived optimal setting to the TDC's resolution is also assessed, for which two typical cases using a 2-bit TDC are simulated as examples. Table 5.3 tabulates the simulated system jitters. The TDCs in the simulations have zero as the center threshold and positive and negative quantization thresholds that symmetrically or asymmetrically mismatch the optimal resolution, as shown in Fig. 5.6. Symmetrically mismatched thresholds are 10% wider ($\pm t_{\text{res}}^{(\text{b})}$ in Fig. 5.6a) or narrower ($\pm t_{\text{res}}^{(\text{c})}$ in Fig. 5.6a) than the optimal. Asymmetrically mismatched positive and negative thresholds in Case (d) of Table 5.3 are 10% right offset from the optimal ($\pm t_{\text{res}}^{(\text{d})}$ in Fig. 5.6b); those in Case (e) are 5% right and 10% left offset from the optimal ($\pm t_{\text{res}}^{(\text{e})}$ in Fig. 5.6b). The asymmetric mismatch can also be regarded as the TDC exhibiting a static nonlinearity. The mismatched cases once again confirm the optimality of the optimized TDC's resolution, whilst the slightly degraded jitter performances show that the proposed TDC setting is relatively insensitive to mismatch and the TDC's nonlinearity.

Table 5.3 Simulated input jitter of 2-bit-TDC DPLLs with $\hat{\beta}$ and ideal (optimized) or nonideal TDC thresholds

Case	Feature of γ	t_{res}	$\sigma_{\Delta t_{\text{simulation}}}$ [fs]
(a)	Optimum	$\hat{t_{\text{res}}} = \hat{\gamma}/\sigma_{\Delta t}$	384.9
(b)	Symmetric mismatch	$110\% \cdot \hat{t_{\text{res}}}$	385.4
(c)		$90\% \cdot \hat{t_{\text{res}}}$	385.9
(d)	Asymmetric mismatch (nonlinearity)	$\hat{t_{\text{res}}} + 10\% \left\lvert \hat{t_{\text{res}}} \right\rvert$	385.3
(e)		$-\widehat{t_{\text{res}}} - 10\% \lvert\widehat{t_{\text{res}}}\rvert$ & $\widehat{t_{\text{res}}} + 5\% \lvert\widehat{t_{\text{res}}}\rvert$	385.2

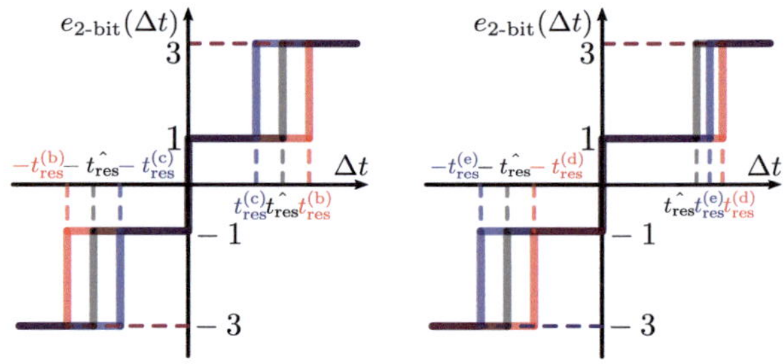

(a) Symmetric mismatch. (b) Asymmetric mismatch.

Fig. 5.6 Illustration of 2-bit TDCs with nonideal thresholds described in Cases (b)–(e) in Table 5.3

5.7 Conclusion

This chapter is based on the comprehensive analysis in Chap. 4 of the input and output jitters of DPLLs with mid-rise TDCs with an accuracy which outperforms that of a recent analysis of this architecture in [1]. A more complete jitter minimization strategy has been provided. This shows that TDCs with equidistant quantization thresholds that are optimized according to our strategy are able to achieve optimum jitter minimization at both the TDC and DPLL levels. The impact of jitter minimization on the DPLL's output PN performance is studied, considering the characteristics of the loop's NTFs and bandwidth that are changed by jitter minimization. As per the case studies for different loop dynamics, this chapter provides insights, design rules of thumb, and a detailed procedure that assist efficient jitter and PN mitigation for DPLLs used as frequency synthesizers. The analysis and design in this work are confirmed with closed-loop behavioral simulations. Chap. 7 will show that this design optimization strategy also yields the optimized loop design for fractional-N DPLLs using DTCs, when the DTCs' nonlinearities are memoryless, such as in the case for the most significant sources of nonlinearity—hard quantization, soft integral, and element mismatch nonlinearities—studied in Chap. 6.

5.8 Summary

1. In Sect. 5.2, the jitter minimization method for an integer-N DPLLs using a mid-rise TDC with a specific number of bits is proposed. This is achieved by (i) finding the optimum value of the coefficient of the proportional path of the DLF, β, using (5.3), which sets the optimum loop bandwidth, and (ii) finding the optimum setting for the time resolution of the TDC recorded in Table 5.1.
2. Fig. 5.1 shows that (i) using more bits in the TDC within the range of the central-limited input noise mitigates the overall jitter, and (ii) the marginal mitigation effect by using more TDC bits is diminishing. Motivated by this observation, Sect. 5.3 investigates the impact of jitter minimization on the output PN spectrum.
3. Sect. 5.4 introduces the design rule of thumb for the optimal choice of the number of bits used in the TDC.
4. Following the above discussion, Sect. 5.5 proposes a novel jitter-minimization-oriented DPLL design procedure in Fig. 5.3, which helps to achieve simultaneously (i) a minimized system jitter, (ii) using the minimum hardware (number of bits) in the TDC, (iii) masking the TDC's PN contribution below the reference jitter's contribution so that the output PN is dominated by the inevitable reference and DCO's jitters only.
5. Sect. 5.6 confirms the effectiveness of our method, which also demonstrates the robustness of the design strategy against variations, mismatch, and TDC nonlinearity.

References

1. L. Avallone, M.P. Kennedy, S. Karman, C. Samori, S. Levantino, Jitter minimization in digital PLLs with mid-rise TDCs. IEEE Trans. Circuits Syst. I Regul. Pap. **67**(3), 743–752 (2020). (March)
2. M. Zanuso, D. Tasca, S. Levantino, A. Donadel, C. Samori, A.L. Lacaita, Noise analysis and minimization in bang-bang digital PLLs. IEEE Trans. Circuits Syst. II Express Briefs **56**(11), 835–839 (2009). (Nov)
3. T. Seong, Y. Lee, S. Yoo, J. Choi, A 320-fs RMS jitter and – 75-dBc reference-spur ring-DCO-based digital PLL using an optimal-threshold TDC. IEEE J. Solid-State Circuits **54**(9), 2501–2512 (2019). (Sep.)
4. H. Park, C. Hwang, T. Seong, J. Choi, A low-jitter ring-DCO-based fractional-N digital PLL with a 1/8 DTC-range-reduction technique using a quadruple-timing-margin phase selector. IEEE J. Solid-State Circuits **57**(12), 3527–3537 (2022). (Dec)
5. G. Marucci, S. Levantino, P. Maffezzoni, C. Samori, Exploiting stochastic resonance to enhance the performance of digital bang-bang PLLs. IEEE Trans. Circuits Syst. II Express Briefs **60**(10), 632–636 (2013). (Oct)
6. H. Xu, A.A. Abidi, Design methodology for phase-locked loops using binary (bang-bang) phase detectors. IEEE Trans. Circuits Syst. I Regul. Pap. **64**(7), 1637–1650 (2017). (July)
7. S. Jang, S. Kim, S.-H. Chu, G.-S. Jeong, Y. Kim, D.-K. Jeong, An optimum loop gain tracking all-digital PLL using autocorrelation of bang–bang phase-frequency detection. IEEE Trans. Circuits Syst. II Express Briefs **62**(9), 836–840 (2015). (Sep.)

8. L. Avallone, M. Mercandelli, A. Santiccioli, M.P. Kennedy, S. Levantino, C. Samori, A comprehensive phase noise analysis of bang-bang digital PLLs. IEEE Trans. Circuits Syst. I Regul. Pap. **68**(7), 2775–2786 (2021). (July)
9. R.B. Staszewski, P.T. Balsara, Phase-domain all-digital phase-locked loop. IEEE Trans. Circuits Syst. II Express Briefs **52**(3), 159–163 (2005). (March)
10. S. Levantino, G. Marzin, C. Samori, A.L. Lacaita, A wideband fractional-N PLL with suppressed charge-pump noise and automatic loop filter calibration. IEEE J. Solid-State Circuits **48**(10), 2419–2429 (2013). (Oct)
11. X. Wang, M.P. Kennedy, A random pulse modulation approach to modeling the flicker and white noise of the charge pump of a fractional-N frequency synthesizer. Int. J. Circuit Theory Appl. **50**(4), 1049–1063 (2022)

Chapter 6
Analysis of DTC-Assisted Fractional-*N* DPLL

6.1 Introduction

As studied in Chaps. 3 and 4, when using an MMDC to modulate the instantaneous divide ratio of the MMD in order to facilitate fractional-*N* operation [1], the AQE arising from the MMDC will be injected directly into the loop (see Eq. (4.37)); this degrades the overall PN/jitter performance of the frequency synthesizer.

This injection has an additional negative impact on DPLLs that use a few-bit TDC; as derived in Eq. (4.38), the injected MMDC's AQE overwhelms the TDC's input jitter in the locked state, which drastically reduces the effective gain of the TDC and causes much more TDC-induced QE (see Table 3.2). Consequently, the DPLL's output PN and jitter are significantly deteriorated (see Fig. 4.6).

In CP-PLLs [2, 3] and some DPLLs [4, 5], the aforementioned QE can be mitigated by adopting higher-order lowpass loop filters and/or "digiphase" cancellation techniques that match and subtract the QE *after* the PD. In CP-PLLs, it is uncommon to zero the QE *before* the PD operation while this is essential in many DPLLs. For example, in DPLLs that use a mid-rise TDC as the PD [6–8], including BBPLLs that use a 1-bit BPD [9–11], such *a posteriori* approaches are inadequate. This is partly due to the fact that the output resolution is limited by the coarse detection of the TDC. Furthermore, as mentioned above, when the unmitigated MMDC's QE is presented at the input of the TDC, the effective gain of the TDC is severely degraded; this in turn drastically shrinks the loop bandwidth and deteriorates the PN and jitter of the DPLL in the locked state.

Zanuso et al. [5] first proposed an *a priori* QE cancellation method for fractional-*N* DPLLs that introduces a phase interpolator, which is now commonly referred to as a "DTC", before the TDC in order to mitigate the MMDC's QE *before* its injection into the loop [12, 13]. Tasca et al. [10] extended this idea and proposed the first DTC-enhanced fractional-*N* BBPLL. It has now become common practice to include a DTC to accompany the MMD. A representative DTC-based fractional-*N* DPLL is illustrated in Fig. 1.5, the MRDT model of which is shown in Fig. 2.4.

© The Author(s), under exclusive license to Springer Nature Switzerland AG 2026

X. Wang and M. P. Kennedy, *Jitter and Spur Minimization in Fractional-N Digital Frequency Synthesizers*, Analog Circuits and Signal Processing,
https://doi.org/10.1007/978-3-032-27186-0_6

The core component of the DTC is a digitally controlled delay (DCD) that, similar to the digiphase scheme, is controlled by the MMDC's AQE. The DCD introduces a corrective time (equivalently phase) delay in the feedback (and/or reference) path which can ideally match and cancel the concerned QE entirely. In the ideal case, the DTC-based feedback path can be regarded as a "true fractional-N divider". In this way, the primary concerns about nonlinearity in the loop are transferred from the TDC to the DTC.

In the ideal case, i.e., if the DTC has infinite resolution and perfect linearity, the deterministic time-domain jitter contributed by the MMDC's QE can be reproduced and zeroed exactly. Reducing the time differences between the reference and divided edges with the DTC, before presenting them to the TDC, allows the DTC-enhanced fractional-N synthesizer to achieve, in principle, the steady-state PN and jitter performance limits corresponding to integer-N operation. Note that the integer-N performances were predicted in Chap. 4, which can be optimized as per the strategy proposed in Chap. 5; these have been verified against behavioral simulations.

However, the effectiveness of practical DTCs suffers from various sources of nonlinearity associated with the DTC *per se*. The nonlinearities result in inexact cancellation of the concerned QE; a residual error after nonlinear-DTC-based cancellation will still be injected into the loop; this limits the ability of the DTC-based cancellation strategy to recover the integer-N performance.

In this chapter, the function, design, and modeling approach of the DTCs will be introduced. In Sect. 6.2, we will elaborate simplified DTC macromodels that take different types of nonlinearities into account and will discuss the separability and linear combination of their distinct noise and spurious contributions to the overall output spectra. Section 6.3 provides insights into the generation and transfer mechanisms of DQI spurs at source and in the output PN. The dependence of DQI spurs on the MMDC and DTC-related parameters will be revealed. In Sect. 6.4, we will reveal the phenomenon of fractional spurs and PN induced by the random element mismatch (REM) of the DCD using typical REM profiles associated with both the non-segmented and segmented DTCs reported in the literature. Section 6.5 compares two state-of-the-art DTC segmentation and control methods and studies the behavior of DTC-based cancellation in relation to the calibration method used inside the DTC.

6.2 Analysis of Nonideal DTCs

In a practical DTC, in addition to the inevitable thermal noise contributed by the physical devices in the DCD [14–16], the DTC exhibits three major types of nonlinearity, which cause inexact QE cancellation and thereby induce excess output PN and spurs, namely

1. the *hard quantization nonlinearity* due to the DTC's limited resolution; this results in a special type of DTC-quantization-induced (DQI) spurs, which we will show in this section and analyze in detail in the next section;

2. the *soft integral nonlinearity* associated with the transfer characteristic of the DCD due to its architecture that causes excess PN and IBS spurs [17–19]; and
3. the *random element mismatch* amongst the delay cells within the DCD due to inevitable process-voltage-temperature (PVT) variations.

Interestingly, DTC-block and PLL-system-level analyses will show that the spectral signatures of the PN and spurs induced by these nonlinearities are distinguishable in the spectrum; the overall PN and spurious response resembles a linear combination of three distinct families of PN and spurs. In this section, we will analyze and distinguish these three nonidealities and present DTC macromodels that take all of them into account; we will explain intuitively the reasons for the separability and combination of the underlying nonlinear phenomena.

6.2.1 Ideal DTC

We start by reviewing the rationale of the DTC-based cancellation in the *ideal*, i.e., perfect cancellation, case. The behavioral model of the feedback path of the DTC-enhanced fractional-N frequency synthesizer in Fig. 1.5 is shown in Fig. 6.1a.[1]

As introduced in Chap. 1, recalling (1.2), fractional-N synthesis is realized by modulating with an MMDC the instantaneous divide ratio of the MMD as $N_{\text{div}}[k] = N_{\text{int}} + y[k]$ during the kth reference clock cycle, where N_{int} is the integer part of the nominal divide ratio (N), and $y[k]$ is the MMDC's output. $y[k]$ quantizes the fractional value X/M, where the integer values X and M are the fractional FCW and the MMDC's modulus, respectively. The modulation operation generates the QE of the MMDC,

$$q_{\text{MMDC}}[k] \triangleq y[k] - \frac{X}{M},$$

which, when substituted into (1.2), yields

$$N_{\text{div}}[k] \triangleq N + q_{\text{MMDC}}[k].$$

This indicates that, throughout modulation by the MMDC, apart from the desired ratio (N), the MMDC's QE ($q_{\text{MMDC}}[k]$) is injected into the loop.

As derived in Chap. 4, this QE is accumulated cycle-by-cycle and is scaled by the oscillation period of the DCO, T_{out}, as in

$$t^{\text{acc}}[k+1] = T_{\text{out}} \cdot q^{\text{acc}}_{\text{MMDC}}[k], \tag{6.1}$$

[1] The DCD in the DTC can be segmented, as shown in Fig. 6.1b. Analysis of segmented DTCs will be provided in the following sections.

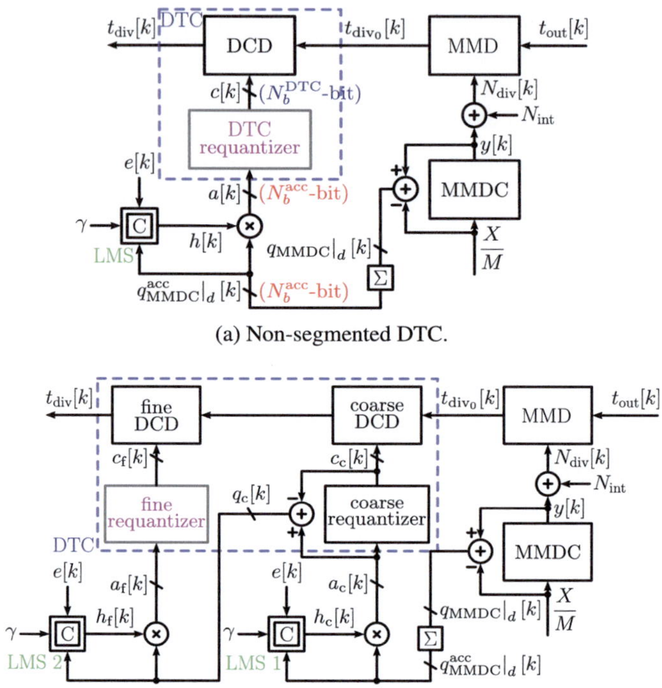

(a) Non-segmented DTC.

(b) Segmented DTC.

Fig. 6.1 Behavioral models of the DTC-based MMDC QE cancellation

which, if not canceled, is injected at the TDC's input to adversely affect the system jitter, where, recalling (3.37), $q_{\text{MMDC}}^{\text{acc}}[k] = \sum_{n=0}^{k} q_{\text{MMDC}}[n]$ represents the accumulated MMDC's QE.

In DTC-based fractional-N PLLs, the DTC is used to produce a time delay that ideally equals

$$t_{\text{DTC}}[k+1] = -T_{\text{out}} \cdot q_{\text{MMDC}}^{\text{acc}}[k], \tag{6.2}$$

which, when added after the MMD, can cancel $t^{\text{acc}}[k+1]$ in (6.1) entirely, before reaching the TDC. In this case, Fig. 6.1 can be regarded as an "ideal fractional divider".[2]

[2] Note that such a fractional divider could also be implemented in an analog PLL in place of the conventional uncompensated MMD [15, 20–23].

6.2.2 Practical DTC

In a practical implementation, this is realized by first extracting the digital code that represents $q^{acc}_{MMDC}[k]$; this can be expressed as

$$q^{acc}_{MMDC}|_d[k] = M \cdot q^{acc}_{MMDC}[k]. \tag{6.3}$$

Note that $q^{acc}_{MMDC}[k]$ is the classical definition of the MMDC's AQE, which is a fractional value with resolution equal to $1/M$, and $q^{acc}_{MMDC}|_d[k]$ is its digital representation with unity resolution.[3] $q^{acc}_{MMDC}|_d[k]$ then controls the DTC, whose core component is a DCD, as shown in Fig. 6.1.

Least-Mean-Square (LMS) Auto-Calibration

LMS adaptive filtering is typically used to calibrate the gain of the DTC in order to match the original injection of the MMDC's QE to cover the DCO's tuning range in the presence of PVT variations [24, 25]; this is represented by the double-framed "C" block in Fig. 6.1.

The LMS algorithm is designed to converge to the optimum value of the gain, $\hat{h} = \lim_{k\to\infty} h[k]$, by correlating

$$h[k] = \sum_{i=0}^{k} \lambda \cdot e[i] \cdot q^{acc}_{MMDC}[i],$$

where $e[i]$ is the error signal output by the TDC as shown in Fig. 1.5 and λ is the LMS step size. In principle, the LMS algorithm delivers an optimum DTC gain which ensures that the ensembles of the error signal, e, and q^{acc}_{MMDC} are uncorrelated, hence giving the best QE cancellation across PVT variations. In the segmented configuration (Fig. 6.1b), the gains of the two sections are auto-calibrated with two separate LMS correlators.

Note that the LMS does not alter the word length or equivalent resolution when mapping $q^{acc}_{MMDC}|_d[k]$ to the signal $a[k]$ that feeds the DTC requantizer.

MATLAB Modeling Approach for LMS-Based DTC Gain Autocalibration

Because the innovative usage of LMS algorithms for run-time calibration is not the focus of this book, we briefly recall a method that can be used to adaptively train the DTC gain within our MRDT system model [26].

---

[3] This notation method will be explained in detail in Chap. 7, where the method to determine quantization-related parameters of the DTC will be discussed.

Listing 6.1 shows the MATLAB model of the LMS adaptive filter discussed above. This correlates two input signals `a` and `b` to train a gain factor `gain`, which converges to an optimum value that minimizes the cross correlation between them.

Listing 6.1 MATLAB behavioral model of the LMS adaptive filter for autocalibration.

```
function gain = LMS(a,b,g_1,lambda)
    %% LMS adaptive filtering
    product = a * b;
    gain = g_1 + lambda * product;
end
```

When used in the MRDT simulation in Listing 2.1, the `DTC` function should not be invoked before the iteration in Line 57; instead, a cycle-wise DTC corrective time delay should be generated within each iteration, where the DTC gain is calibrated cycle by cycle. Namely, Listing 6.2 should be added before Line 86 modeling the DTC-based cancellation in the feedback path.

Listing 6.2 MATLAB code used to calibrate the gain of the DTC in every simulated iteraton.

```
%% LMS correlator
gain(k) = LMS(q_acc(k),e(k),gain(k-1)
   ,2^-10/M);
%% DTC correction calibrated by adaptive
   DTC gain
t_DTC(k) = DTC_ON * DTC(q_acc(k)*gain(k),1/
   F_DCO,M,res_DTC,DTC_quant_ON,
   DTC_nonlin_ON,alpha0,alpha1,alpha2,
   alpha3);
```

In segmented DTCs, each of the DTC branches should be calibrated by its own LMS correlator; this is shown in Fig. 6.1b.

Inherent Quantization in a Finite-Resolution DTC

In practice, the DTC's DCD is typically designed to have relatively low (about 10-bit) equivalent resolution in order to limit hardware consumption and control complexity [10, 21, 27]. On the other hand, commercial frequency synthesizers, e.g., [28, 29], use very large modulus (M) in the MMDC in order to achieve fine-resolution synthesis; this results in its AQE that controls the DTC having very high, e.g., 35-bit, equivalent resolution that is much finer than what can be provided with the DTC.

This resolution reduction from the DTC's N_b^{acc}-bit *digital control word*, $a[k]$, to an N_b^{DTC}-bit control code, $c[k]$, that controls the N_b^{DTC}-bit DCD to yield an *analog time delay output*, cannot be realized without a digital-to-digital requantization operation, that we consider to be performed at the control input of the DTC, *in addition to* the digital-to-analog conversion in the DCD. This is indicated by the "DTC requantizer" block in Fig. 6.1a. As will be discussed in Sect. 6.3, the requantization operation can be a simple N_b^{acc}-to-N_b^{DTC}-bit truncation or a digital Δ-Σ requantization. The requantization causes inexact cancellation of the MMDC's QE and consequently injects a residual error into the loop that degrades the system's noise performance. In this chapter, we call the residual error after cancellation caused by the *hard quantization nonlinearity* in the DTC, the "DTC's QE".

As we will reveal in Sect. 6.3, regardless of the DTC's configuration and control methods, i.e., a non-segmented or segmented DCD controlled by normal and/or DDSM-based requantizers [10, 12, 13, 15, 21, 30–33], as long as there is a resolution gap between the DTC control input and time output, DQI spurs prevail and will be present in the DPLL's output spectrum. This contradicts the claim made in [34] that the residual error due to quantization by the DTC has a white power spectrum and that, in the case where the MMDC has an order higher or equal to two, the residual error is free of spurs, regardless of any gain mismatch associated with the DTC. For notational simplicity, we will refer to these DTC-quantization-induced spurs as "DQI spurs".

It will be shown that the DQI spurs transferred to the DPLL's output can degrade the spectral purity of the output more than the more widely studied fractional spurs [17, 18]. A family of DTC-enhancement techniques called input-dithered quantization (IDQ) which, if used in the DTC, can provide immunity to the DQI spurs, will be described in detail in Chap. 7.

Nonlinear DCD

Apart from the aforementioned hard quantization nonlinearity, another source of DTC's nonlinearity that causes inexact DTC-based cancellation and produces noise and spurs is the inevitable *soft* INL associated with the DCD's transfer characteristic. For example, in commonly used "variable-slope" DTCs, the DCD consists of a number of inverter stages with an RC network inserted in the middle comprising a resistor and a digitally controlled capacitor bank to provide a tunable time delay, as shown in Fig. 6.2 [35].

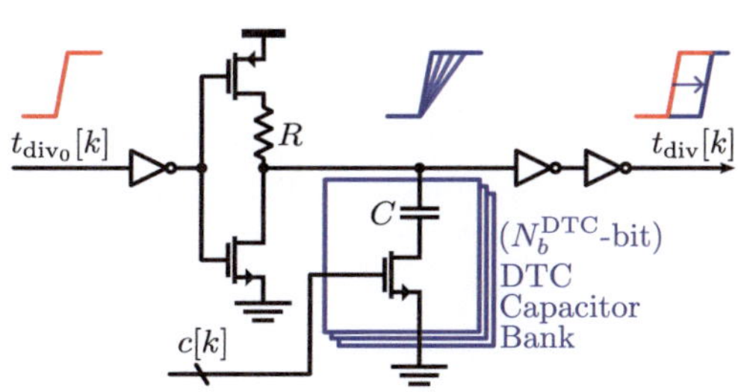

Fig. 6.2 Commonly used variable-slope DCD

As analyzed in [35, Sect. II] and [36, Sect. II], due to effects such as variable-loading capacitance and channel-length modulation, the DCD exhibits a prominent concave-shaped[4] INL. It is known that the DDSM's AQE interacting with a nonlinear transfer function yields spurious tones such as a prominent IBS and excess folded noise [17–19, 37, 38].

REM in DCD

The third source of nonlinearity is the inevitable REM between the delay cells within the DCD due to PVT variations. Once manufactured, the element mismatch profile of the DCD is fixed, which can be regarded as a static nonlinear function that is exercised by the MMDC's AQE cycle after cycle; this will in turn induce yet another cluster of fractional spurs and PN, as will be shown in this section.

Note that, the aforementioned DTC linearity enhancement techniques typically aim to mitigate the smooth polynomial INL curve of the DCD whilst recognizing that, compared to this, the DCD's REM has a more limited range of variation, e.g., about $1/10^3$ of the DR of the DCD [35], hence reducing its relative adverse contribution. We will show in this section and in Sect. 6.4 that, when the DCD's INL is not compensated, the REM contributes equally detrimental fractional spurs and PN/jitter power. In the *best case scenario*, where the polynomial INL has been minimized, the residual INL, namely the intact REM of the DCD, can actually dominate the spur and PN profile induced by the DCD's nonlinearity.

In following subsections, we will provide DTC macromodels that take these three nonlinearities into account. The resulting noise and spurious responses will be analyzed and distinguished. The macromodels also help to provide insight into the mechanisms that lead to the separability of their spectral signatures and cause the overall PN spectrum to be a linear combination of them.

6.2.3 DTC Macromodels

We start by simplifying the behavioral model of the PLL's feedback path in Fig. 6.1a to focus only on the original injection and DTC-based cancellation of the accumulated QE, i.e., $t^{\mathrm{acc}}[k+1]$ in (6.1) and $t_{\mathrm{DTC}}[k+1]$ in (6.2). The simplified behavioral model is presented in Fig. 6.3a. As shown here, the DCD with a nominal physical unit delay τ_{DCD} is calibrated by the gain $h[k]$ trained by the LMS algorithm to match the original injection. $f_{\mathrm{DCD}}^{\mathcal{N}}(\cdot)$ includes the inevitable nonlinearities, i.e., polynomial and REM INLs, of the delay line.

---

[4] In this book, a "concave" function refers to a function whose derivative, in the interval of interest, is monotonically decreasing. An example *concave* INL is shown by solid black curve in Fig. 6.4, which has a "∩" shape in the dynamic range.

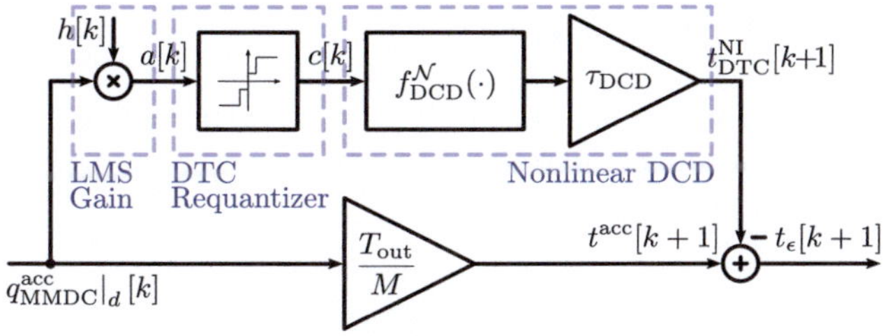

(a) Simplified behavioral model.

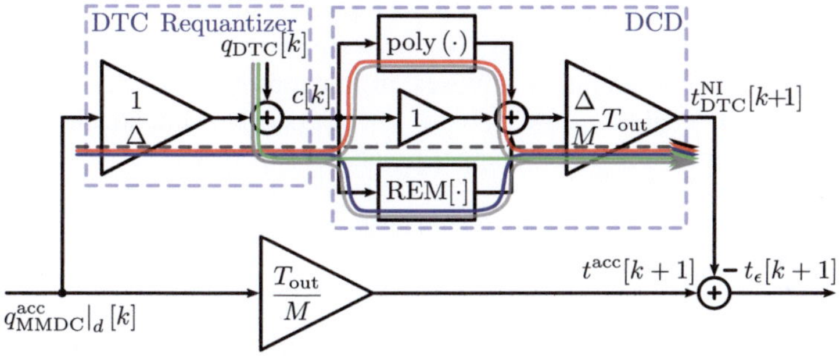

(b) Nonideal model annotated with the signal flows.

Fig. 6.3 DTC models simplified from the behavioral model in Fig. 6.1a focusing on the QE cancellation mechanism with a quantized nonlinear DTC

LMS Calibration

In steady state, when the loop is locked and the LMS algorithm has settled, $h[k]$ will converge toward a fixed value to match the signal gains of the original injection and DTC compensation so that $t^{\mathrm{acc}}[k] + t_{\mathrm{DTC}}[k] \to 0$. Therefore, by noting that the Δ-to-1 DTC requantizer has an inherent gain of $1/\Delta$, we simplify our analysis by assuming that the nominal delay of the DCD, $\tau_{\mathrm{DCD}} = \frac{\Delta}{M} \cdot T_{\mathrm{out}}$. In this case, the LMS gain will converge toward unity, in order to have the DTC cancellation match the original injection with a gain equaling T_{out}/M (as will be discussed below). In other words,

$$a[k] = q^{\mathrm{acc}}_{\mathrm{MMDC}}|_d[k], \tag{6.4}$$

and we can omit the LMS correlator from the analysis and regard $q^{\mathrm{acc}}_{\mathrm{MMDC}}|_d[k]$ as the input to the simplified model describing DTC-based cancellation. The model can be further refined as in Fig. 6.3b to show the nonidealities in detail.

Original Injection of MMDC's AQE

Recalling (6.4) in steady state and substituting (6.3) into (6.1), the original injection is simplified as a linear gain factor of T_{out}/M that transfers $q_{\text{MMDC}}^{\text{acc}}|_d[k]$ to $t^{\text{acc}}[k+1]$, as shown in the lower branches of Fig. 6.3.

Inherent Quantization by DTC Requantizer

The Δ-to-1 DTC requantizer, where Δ is its input resolution, as shown in Fig. 6.3a, takes as input $q_{\text{MMDC}}^{\text{acc}}|_d[k]$ and yields $c[k]$; both signals have unity resolution. Effectively, the quantization operation imposes an inherent gain of $1/\Delta$ to $q_{\text{MMDC}}^{\text{acc}}|_d[k]$ and adds the "DTC's QE", $q_{\text{DTC}}[k] \in (-\frac{1}{2}, \frac{1}{2}]$, to $c[k]$. This linearized model of the DTC requantizer is included in the upper branch of Fig. 6.3b.

Polynomial INL of DCD

In the ideal case, where the DCD has perfect linearity, it should be designed to provide a linear gain of τ_{DCD}, which we assume here to be equal to $\frac{\Delta}{M} \cdot T_{\text{out}}$, with units of s/LSB, in order to match the gain of the original injection (lower) branches in Fig. 6.3. Note that, as a design guideline, the value of $\frac{\Delta}{M} \cdot T_{\text{out}}$, where T_{out} equals the minimum output period in the tuning range of the DCO, ought to be the unit time delay provided by one delay cell in the DCD.

When DTC linearity enhancement is not enabled, representative DCDs reported in the literature, e.g., [8, 13, 15, 16, 21, 27, 31, 33, 35, 36, 39–42], have an INL that is concave and is asymmetric with respect to the vertical axis in the middle of the DR; this can be approximated by a third-order polynomial function. Here, we first consider the uncompensated INL reported in [27, Fig. 32.1.2] and [33, Fig. 11] that shows a 12-bit DTC having a concave-shaped INL with a maximum normalized to about +4.4 LSBs in the middle and a worst-case nonlinearity of about -9.0 LSBs at both ends of the DR. A similar DTC nonlinearity was also reported in other literature, e.g., [21, Fig. 11b], which simulates a 10-bit DTC—hence the DR is reduced by a factor of four compared to the 12-bit case—that has a similar concave-shaped INL of +1.1 LSBs in the middle and -2.3 LSBs at the edges of the DR.

In our simulations, as shown in Fig. 6.3a, we have modeled the DCD as a nonlinear memoryless transfer function, $f_{\text{DCD}}^{\mathcal{N}}(\cdot)$, followed by a scaling factor, $\frac{\Delta}{M} \cdot T_{\text{out}}$. We can use $f_{\text{DCD}}^{\text{poly}}(\cdot)$ to represent the constituent transfer characteristic describing a smooth cubic function that best fits and comprises $f_{\text{DCD}}^{\mathcal{N}}(\cdot)$; this is expressed as

$$f_{\text{DCD}}^{\text{poly}}(x) = \alpha_0 + \alpha_1 \cdot x + \alpha_2 \cdot x^2 + \alpha_3 \cdot x^3, \tag{6.5}$$

where the coefficients are

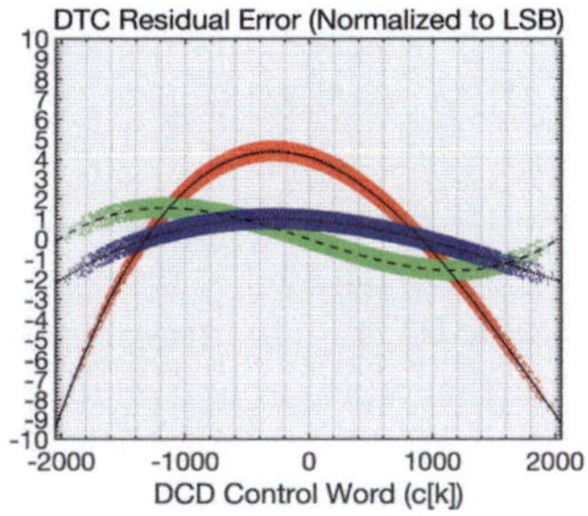

Fig. 6.4 Modeled cubic INLs versus simulated residual errors of state-of-the-art DTCs. Solid curve and red dots: original INL without linearity enhancement (6.6); dashed curve and green dots: wave-shaped INL after linearity enhancement (8.3); dotted curve and blue dots: weaker-concave-shaped INL after linearity enhancement (8.4)

$$\alpha_0 = +4.1,\ \alpha_1 = +0.998,\ \alpha_2 = -3.2 \cdot 10^{-6},\ \alpha_3 = +4.9 \cdot 10^{-10}. \tag{6.6}$$

The polynomial INL of the DCD (normalized to LSB) is the difference between the nonlinear and linear transfer characteristics, namely

$$\text{poly}(x) \triangleq f_{\text{DCD}}^{\text{poly}}(x) - x = \alpha_0 + (\alpha_1 - 1) \cdot x + \alpha_2 \cdot x^2 + \alpha_3 \cdot x^3. \tag{6.7}$$

The modeled cubic INL (normalized to LSB) matching its reported counterpart in [27, 33] is plotted as the black curve in Fig. 6.4. The simulated residual errors are plotted in red. These are located no farther than $\pm 1/2$ LSBs from the INL curve; the deviation is due to the aforementioned DTC's QE. Note that we modeled a 12-bit DTC to cover the variation range of the third-order DDSM (MASH 1-1-1)'s QE. If a MASH 1-1 were used, only 11 bits in the center of the DCD's DR would be exercised.

The polynomial INL and linear paths of the nonlinear DCD are shown in the upper branch of Fig. 6.3b to facilitate noise and spurious analysis.

REM of DCD

Due to PVT variations, the DCD will exhibit static REM once manufactured. Namely, when controlled by the digital code c yielded by the requantizer, different codes will switch on their corresponding delay cells in the DCD, if the access to the delay cells are not shuffled. Each delay cell yields an output with its own static mismatch compared to the desired nominal unit delay; these contribute an overall REM nonlinearity of the delay line.

For example, as reported in [35] that presents the first *constant-slope charging* (CSC) DTC to mitigate its soft INL, the DCD's REM exists when the linearity of the DCD is not improved, namely in the *variable-slope charging* case, see [35, Fig. 7]. On the other hand, when the CSC method is used, the polynomial INL is much attenuated, and yet the REM is not affected; in this case, the residual nonlinearity of the DCD is dominated by its REM, see [35, Figs. 8, 17, 19, 20]. The peak-to-peak variation in the transfer characteristic induced by the REM spans a range of about 0.2% of the FS, e.g., as shown in [35, Fig. 19].

In the upper branch of Fig. 6.3b, in parallel to the polynomial INL and linear paths of the DCD, we use a lookup table REM[·] to represents the static element mismatch associated with the delay line that will be activated by the control code, c. The notation [·] highlights the nature of the modeled REM being a referenced lookup table—rather than a numerical function—namely, a specific value of c excites its corresponding predefined REM value following a one-to-one mapping.

Finally, we denote in Fig. 6.3 the time delay generated by the *Non-Ideal* DTC as $t_{\mathrm{DTC}}^{\mathrm{NI}}[k+1]$. The inexact cancellation of the original $t^{\mathrm{acc}}[k+1]$ yields a residual time error, denoted $t_{\epsilon}[k+1]$, that is injected into the loop.

MATLAB Model of Practical DTC

Listing 6.3 shows the MATLAB behavioral model of a practical DTC. For presentation simplicity, this model exhibits quantization and polynomial integral nonlinearities as introduced in section, but no REM. We present here the model of a non-segmented DTC, as shown in Fig. 6.1a, whose signal flow was illustrated in Fig. 6.3. The MATLAB modeling approach describing different segmented DTCs as shown Fig. 6.1b will be presented in Sect. 6.3. The LMS modeling approach will be introduced in Sect. 6.5.

Listing 6.3 MATLAB model of a non-segmented DTC.

```
function t_DTC = DTC(in,T_out,M,res_DTC,
    DTC_quant_ON,DTC_nonlin_ON,alpha0,
    alpha1,alpha2,alpha3)
  if DTC_quant_ON == 1
      %% Truncation-type D2D quantizer
      c      = floor(in/res_DTC);
  else
      %% No quantization
      c      = in/res_DTC;
  end
  if DTC_nonlin_ON == 1
      %% Nonlinear DCD
      t_DTC = T_out*res_DTC/M * (alpha0
         + alpha1*c + alpha2*c.^2 +
         alpha3*c.^3);
  else
      %% Linear DCD
      t_DTC = T_out*res_DTC/M * c;
  end
end
```

The model is controlled by two logic variables `DTC_quant_ON` and `DTC_nonlin_ON`. When the former is true, the DTC exhibits quantization nonlinearity; if not, the DTC's quantization operation is replaced by a linear division by Δ instead. Similarly, when the latter is true, the DTC's DCD exhibits INL as studied in Sect. 6.2.3; if not, the DCD is assumed to be a linear gain factor of $\frac{\Delta}{M} \cdot T_{\mathrm{out}}$.

Note that, as discussed in Sect. 6.2.3, when the LMS algorithm calibrating the DTC gain has settled, the DTC is uniquely controlled by the digital code representing the AQE of the MMDC to mirror and cancel its injection. Therefore, as the `MMDC` function coded in Listing 3.4, which is invoked in the initialization stage in the main file (in Line 55 of Listing 2.1) before the iteration stage, the `DTC` function can also be executed after the `MMDC` before the main iteration as in Line 57 of Listing 2.1. The resulting cycle-wise time delay, $t_{\mathrm{DTC}}[k]$, is inserted in Line 86 of Listing 2.1 to model the DTC-based correction in the feedback path of the PLL. This helps to improve the readability and simulation speed of the proposed MRDT model.

6.2.4 *Separability of Sources of DTC Nonlinearity and Their Induced Spectral Responses*

In this subsection, the DTC's three systematic nonlinearities will be shown to induce different types of undesired PN phenomena that appear to be independent of each other and can be distinguished in the power spectrum. Here, we analyze at both the DTC-block and PLL-system levels the behaviorally-simulated spectral signatures of a type-II first-order DPLL that is parameterized as per Table 3.3[5]. This corresponds to the BBPLL described in Table 5.2(ii) that has a jitter-minimized system design. Note that the DTC is used to mitigate the QE contribution of the MMDC. In principle, the residual error can be made negligible in comparison to the system jitter associated with integer-N operation and therefore one can reach the integer-N PN/jitter performance [10]. Consequently, the TDC can then be modeled in behavioral simulations as a linear element whose gain, $K_{\mathrm{TDC}} = 1.97 \cdot 10^{12}$ LSB/s, equals the effective steady-state gain of the BPD in the integer-N BBPLL, as reported in Table 5.2(ii); this models TDCs with a wider linear range than a BPD and allows us to illustrate better the transfer mechanism for the DTC-induced PN and spurs to the system output in DPLLs with generic TDCs, without being distracted by the hard-limiting bang-bang distortion caused by the BPD. Unless otherwise stated, these parameters are used everywhere in this section. For completeness, a BBPLL with a BPD will be simulated in Chap. 8 showing the comprehensive strategy to mitigate excess DTC-related PN and spurs.

---

[5] Here, the fractional FCW is $X = 8193$. The coefficient of the proportional path of the DLF $\beta = 47.8$.

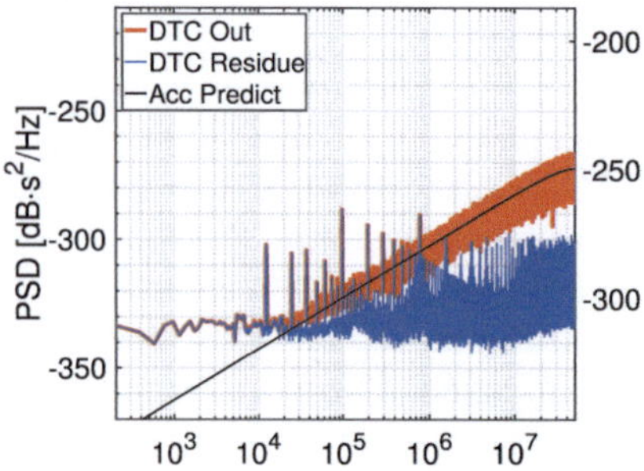

(a) DTC level; polynomial, quantization, and REM nonlinearities simultaneously.

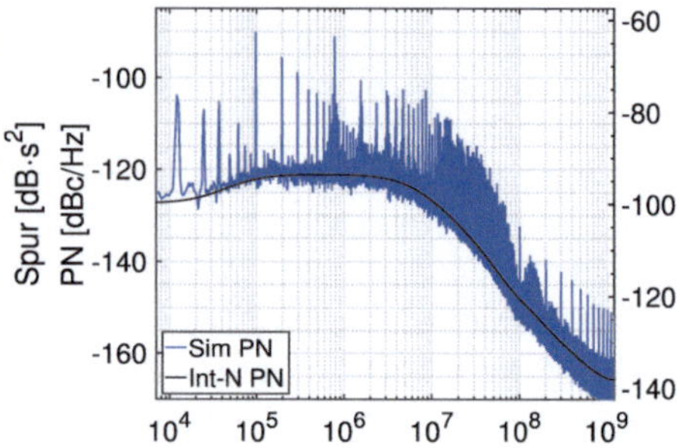

(b) DPLL level; polynomial, quantization, and REM nonlinearities simultaneously.

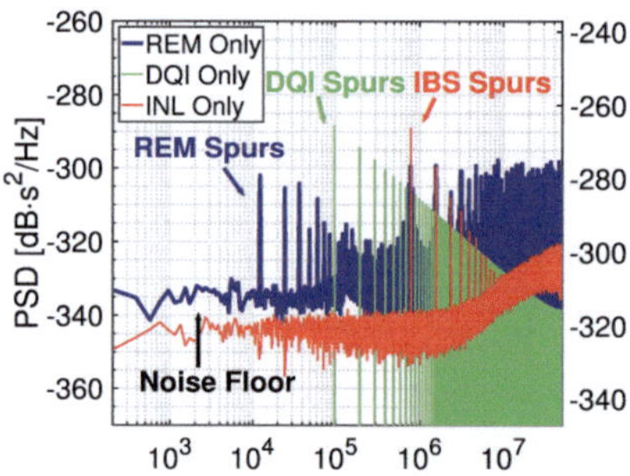

(c) DTC level, one nonlinearity at a time.

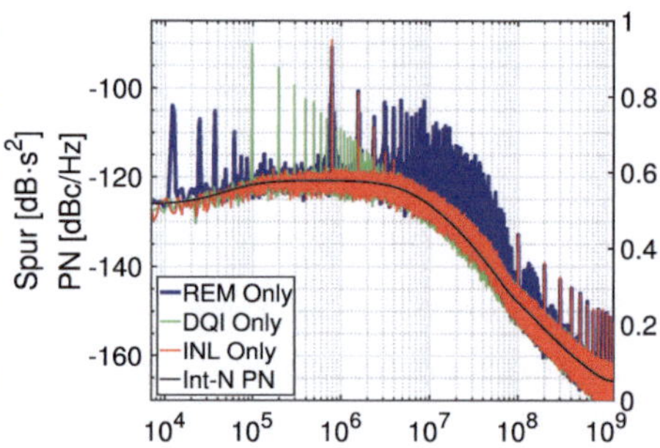

(d) DPLL level, one nonlinearity at a time.

Fig. 6.5 Simulated power spectra at block and system levels of the example fractional-N DPLL (with MASH 1-1 MMDC) with an *undithered DTC* that has different nonidealities

In order to contrast the adverse noise and spurious responses caused by the three nonlinearities of the DTC, we perform simulations considering all the nonidealities simultaneously and also considering only one of the three nonidealities at a time. The results of the simultaneous consideration are shown in the top row of Fig. 6.5 and those of the separate consideration are shown in the bottom row. The block-level and system-level results are shown in the left and right columns respectively.

The black curve in Fig. 6.5a predicts the +20 dB/dec highpass-filtered PSD of the original injected time error related to the MASH 1-1's AQE, i.e., $t^{\mathrm{acc}}[k+1]$ in Fig. 6.3. Figure 6.5a also shows the simulated power spectrum of the nonideal DTC's output, $t_{\mathrm{DTC}}^{\mathrm{NI}}[k+1]$, in orange; and that of the residual error, $t_{\epsilon}[k+1]$, in blue. Here, the original AQE (black) is canceled by the DTC's output (orange). The DTC's output (orange) is the sum of its input (black) and its residual error (blue) due to the modeled nonidealities. The residual power spectrum (blue) constitutes a noise floor together with spurious tones at various frequencies, the spurs of which that are not masked by the black are also present in the DTC's output (orange). Note that the residual spectrum is transferred to the output of the fractional-N DPLL, as shown in Fig. 6.5b, where output spurs exist at the same offset frequencies; in this case, the

DTC-induced noise floor does not have a perceptible impact on the output PN, which recovers the integer-N PN level predicted in Chap. 4.

In Fig. 6.5c, the residual spectrum in red shows the case where the DTC is considered to have infinite resolution but a polynomial INL as per (6.6) and (6.7); the DCD is also assumed to be free of REM. The residue in green shows the case where the DTC has only quantization behavior and is otherwise linear. The residue in blue shows the case where the DTC only exhibits REM in the DCD and is free of quantization and polynomial integral nonlinearities. As expected, the DCD's polynomial INL alone yields (in red) (i) a folded noise floor at low frequencies [19], (ii) IBS spurs at the fundamental and harmonics of $\frac{X}{M} \cdot F_{\text{ref}} = 781$ kHz [18] due to its interaction with the DDSM's AQE, and (iii) a +20 dB/dec noise floor at high frequencies due to inexact cancellation of the QE. Secondly, the linear quantized DTC's residual spectrum in green comprises purely spurious tones at offset frequencies that are different from the IBS frequency and its harmonics; these DQI spurs will be studied further in Sect. 6.3. Furthermore, the residue of the DTC with only REM in blue also shows a folded noise floor and spurs that are at more frequencies that are different from those of the conventional IBS and DQI spurs.

It is interesting to notice that, at the block level, the total residual spectrum, taking into account all of the nonlinearities (blue in Fig. 6.5a), seems to match accurately the sum of the residual spectra in Fig. 6.5c, where the DTC was considered to have only one nonideality at a time (see the magnitudes of their noise floors and the positions and heights of their spurs). The same effect can also be observed in the output spurs in Fig. 6.5b, d.

In order to gain further insight into these mechanisms, we consider the nonideal model shown in Fig. 6.3b which annotates the signal flows throughout the DTC subsystem in every clock cycle. Here, we enumerate five signal paths as highlighted by the colored arrows, namely:

The Dashed Black Path

It shows the linear signal chain that transfers the DTC's input, $q_{\text{MMDC}}^{\text{acc}}|_d[k]$, to the output, $t_{\text{DTC}}^{\text{NI}}[k+1]$. This path has an overall linear gain equal to T_{out}/M in s/LSB. Therefore, it mirrors and compensates the original QE injection in the lower branch. This path guarantees that the original QE can be "canceled exactly" and what remains to affect the loop's performance is the residual error caused by the DTC's nonidealities;

The Red Path

In this path, the linear portion of the DTC requantizer's output passes through the DCD's polynomial INL and is then conveyed to the DTC's output. Hence, it represents the generation of the residual error caused by the DCD's polynomial transfer function only, i.e., the red component in Fig. 6.5c.

The Green Path

The DTC's QE, $q_{\text{DTC}}[k]$, passes through the linear branch of the DCD in this path. It represents the generation of the residual error caused by the DTC's inherent requantization only, i.e., the green component in Fig. 6.5c.

The Blue Path

Here, the digital output of the DTC requantizer exercises the DCD's static REM nonlinearity and is then conveyed to the DTC's output. This represents the generation of the residual error caused by the DCD's REM only, i.e., the blue component in Fig. 6.5c.

The Gray Path

In this path, the DTC's QE, $q_{\text{DTC}}[k] \in (-\frac{1}{2}, \frac{1}{2}]$, passes through the DCD's INL (including both the polynomial and random discontinuous components). However, due to its very limited variation, the DTC's QE can only activate the middlemost two unit delay cells of the N_b^{DTC}-bit DCD. Recall that $N_b^{\text{DTC}} \geq 10$ in our examples and there are in total $2^{N_b^{\text{DTC}}}$ delay cells. Therefore, the DTC's QE is too weak to excite significant nonlinear behavior in the DCD, and hence this path does not yield any perceptible spectral response.

Finally, we observe at the output of the block level a residual power spectrum (shown in blue in Fig. 6.5a) that is a linear combination of the spectrum caused by the DCD's polynomial INL (red in Fig. 6.5c) and REM (blue in Fig. 6.5c) and that caused by the DTC requantizer (green in Fig. 6.5c). The overall block-level residues are transferred and appear in the system's output PN (Fig. 6.5b). Consequently, the nonideal-DTC-based fractional-N DPLL cannot reach the integer-N PN performance.

In Chaps. 7 and 8, we will assess different classes of DTC enhancement techniques to mitigate the categories of DTC's nonidealities and spur families that we have identified.

6.2.5 Conclusion

In this section, we have provided approximate analytical models for practical DTCs, taking into account their quantization behavior and the typical nonlinear transfer characteristics of the DCD. Simulations using these models reveal and distinguish the noise and spurious responses that degrade the system PN and spur performances. In Sect. 6.3.3, we will perform a quantitative prediction of the DTC-induced DQI and IBS spurs. The MATLAB script encoding the analytical prediction model will be provided in Chap. 8.

6.3 DTC Inherent Quantization and DTC-Quantization-Induced (DQI) Spurs

This section provides insights into the DTC's inherent requantization operation introduced in Sect. 6.2. In particular, we reveal the systematic root cause and spectral manifestation of the resulting DQI spurs. The expression for the time-domain waveform is derived analytically.

This chapter is arranged as follows. In Sect. 6.3.2, we catalog the signatures of DQI spurs that can be observed in various DTC configurations. In Sect. 6.3.3, we derive the waveform and mechanism of the DTC's QE and predict the spectral signatures of DQI spurs; their contributions to the system jitter are also discussed quantitatively. Finally, conclusions are drawn in Sect. 6.3.4.

6.3.1 Prior Art

We start by reviewing the prior art in relation to the analysis of the pattern of the DTC's QE and the state of the art of DTC-related design techniques that are used to enhance the DTC's performance. This review motivates our analysis of DQI spurs and the need for mitigation methods.

Some prior art, e.g., [34, Sect. II], has suggested that, if the DTC is assumed to exhibit only a hard quantization nonlinearity, the residual error has a white power spectrum at the DTC-block level. Furthermore, in the case where the MMDC has an order greater than or equal to two (such as MASH 1-1 and MASH 1-1-1, for example), they claimed that the residual error is still free of spurs, regardless of any gain mismatch between the DTC and the original injection of the MMDC's QE. Some others, e.g., [43, Sect. IV], assert that the DTC's QE resulting from its limited resolution is tonal but that these spurs typically occur at *much higher* frequencies than those caused by the INL of the DTC and are effectively filtered by the loop. In this section, we will show that the DTC's QE in fact has a *purely tonal* manifestation and produces output spurs that may lie entirely within the loop bandwidth and at frequencies *lower* than the INL-induced spurs; as these cannot be attenuated by the loop dynamics, they degrade the spectral purity and system jitter. For notational simplicity, we will refer to these DTC-quantization-induced spurs as "DQI spurs". We will derive the closed-form time-domain waveform of the DTC's QE. This will be shown to be determined by the fractional FCW, X, and the DTC's quantization resolution, Δ; with these fixed, the DQI spurs have a well-defined pattern which is independent of the type, order, and modulus (M) of the MMDC. This implies that using large-modulus [44], higher-order [45–47], or stochastic [27, 48–50] MMDCs alone cannot alleviate DQI spurs.

A linear PD, e.g., the PFD used in CP-PLLs [51], transfers to the output spectrum the spurious (or spur-free) pattern presented at its input. For DPLLs using a few-bit TDC, e.g., a BPD, prior art such as [10, 30] claims that, by constraining the DTC's residual error within the SR regime established at the input of the TDC [52–54], the input-referred intrinsic noise within the loop can scramble the DTC-related spur pattern and make the system output free of such spurs. However, as we have analyzed and shown in Chap. 3, the TDC-input-referred noise indeed creates a "pseudo-linear" operation region of the TDC. Because the DTC is typically parameterized to ensure that the DTC's QE lies within this region, which will be explained in Chap. 7, its spectral signature is preserved and transferred to the output exactly by the pseudo-linear detection of the TDC. In other words, the SR mode of operation alone cannot mitigate DQI spurs.

Finally, as analyzed in Sect. 6.2.4 , DQI spurs induced by the *hard quantization nonlinearity* of the DTC can be distinguished from the more widely studied IBS spurs[6] [17, 18, 45, 47] induced by the iteration between $q^{\mathrm{acc}}_{\mathrm{MMDC}}$ and the *soft INL* of the DTC. State-of-the-art DTC linearity enhancement techniques that improve the DTC's INL such as [8, 13, 15, 16, 21, 25, 27, 30, 31, 33, 35, 36, 40–42, 55] and stochastic divider controllers, such as [8, 27, 33, 48–50, 56–58], that exhibit IBS-immunity in the presence of memoryless polynomial nonlinearities, can mitigate the PN and spurs that result from the *soft* INL. However, they do not alter the characteristics of the *hard* DTC requantization, and hence are inadequate in mitigating DQI spurs.

Together, these observations motivate the design of methods that are specifically targeted at DQI spurs and do not belong to the other DTC-enhancement techniques mentioned above. Such methods will be described in Chap. 7.

6.3.2 *Manifestations*

We first report observations about the DQI-spur signatures that arise in various DTC configurations. As per the working principles of the DTC introduced above, the core component within the DTC is a DCD that provides tunable time (phase) delays. The finite-resolution DCD can be implemented in a non-segmented manner, as represented as the "DCD" block in Fig. 6.1a. The "DTC requantizer" marked in magenta performs a *digital-to-digital quantization* that maps the fine-resolution control code, $a[k]$, which we assume has been calibrated by the LMS algorithm [24, 25] based on $q^{\mathrm{acc}}_{\mathrm{MMDC}}|_d[k]$,[7] to a coarser-resolution code, $c[k]$, to control the non-segmented DCD.

---

[6] IBS spurs typically exhibit at $F_{\mathrm{ref}} \cdot X/M$ and its harmonics.

[7] As will be explained in Chap. 7, $q^{\mathrm{acc}}_{\mathrm{MMDC}}|_d[k] \triangleq M \cdot q^{\mathrm{acc}}_{\mathrm{MMDC}}[k]$ is the digital-domain representation of the MMDC's AQE. It is an N_b^{acc}-bit unity-resolution digital code, where $N_b^{\mathrm{acc}} = N_b^{\mathrm{MMDC}} + l - 1$, and N_b^{MMDC} and l are the number of bits and order of the MMDC. The LMS does not change the word length or resolution when transforming $q^{\mathrm{acc}}_{\mathrm{MMDC}}|_d[k]$ to $a[k]$.

Table 6.1 Summary of DQI-spur signatures arising in different DTC configurations

DTC architecture	DCD control method	DQI-spur signature
Non-segmented	Normal requantizer, e.g., [34, Sect. II], [15, 21]	Raw DQI spurs
	EFM1 requantizer, e.g., [12, Fig. 9], [13, Fig. 3], [32, 59]	Shaped DQI spurs
Segmented	Normal coarse & normal fine requantizers, e.g., [30, 33]	Raw DQI spurs
	EFM1 coarse & normal fine requantizers, e.g., [10]	Raw DQI spurs
	EFM1 coarse & EFM1 fine requantizers, e.g., [31]	Shaped DQI spurs

To simplify the hardware consumption and digital control further, the DCD can be partitioned into a cascade of coarse and fine segments, as shown in Fig. 6.1b. In this configuration, both the coarse and fine DCDs are controlled by dedicated requantizers. The "coarse requantizer" at the DTC's control input also takes charge of splitting the digital controls for the coarse and fine stages, namely its quantized output, $c_c[k]$, controls the "coarse DCD", and its QE, $q_c[k]$, representing the desired correction that cannot be provided by the coarse stage, controls the fine stage.

The requantizers that control the DCDs are typically implemented as normal truncating (or rounding) quantizers or as EFM1s, i.e., a first-order DDSM, which interpolates a lower-resolution control word for the DCD.

Thus far, the literature has reported five possible combinations of the DCD implementations and their control methods, which are equally effective in QE cancellation, as will be analyzed in Section 6.5; these are tabulated in Table 6.1. Interestingly, we observe that, despite there being five different possible combinations, the corresponding DQI-spur signatures that arise from them can be categorized into two groups, namely purely tonal raw DQI spurs and highpass-shaped DQI spurs.

These two signatures are illustrated in Fig. 6.6, which shows the simulated power spectra of the time-domain residual errors after cancellation by the linear finite-resolution DTC[8] that performs a Δ-to-1 quantization, i.e., the "DTC's QE", and the output PN. As in Sect. 6.2.4, the behavioral simulation models a BBPLL parameterized as per Table 3.3; the fractional-N operation is enabled by a MASH 1-1 MMDC together with a DTC that has 11-bit equivalent resolution ($\Delta = 2^{10}$), which corresponds to the MMDC and DTC parameterization in [10].

[8] As studied in Sect. 6.2.4, the hard quantization nonlinearity and INL of the DTC result in independent PN/jitter phenomena that can be distinguished in the spectrum and need to be mitigated by different types of techniques. In this example, we focus on the hard nonlinearity; the DTC is assumed to be otherwise linear.

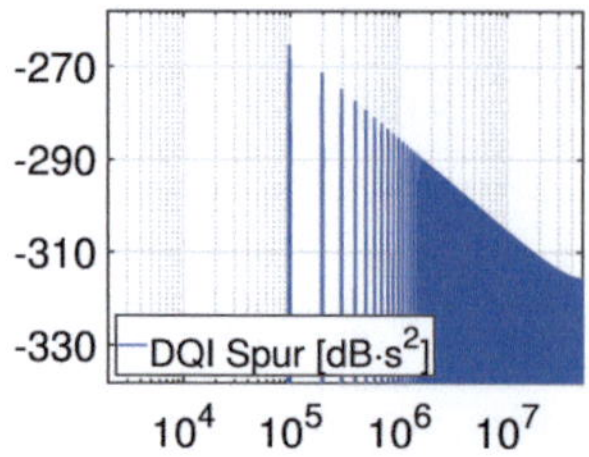

(a) Raw DQI spurs at source, $X = 1$ or $2^{10} + 1$.

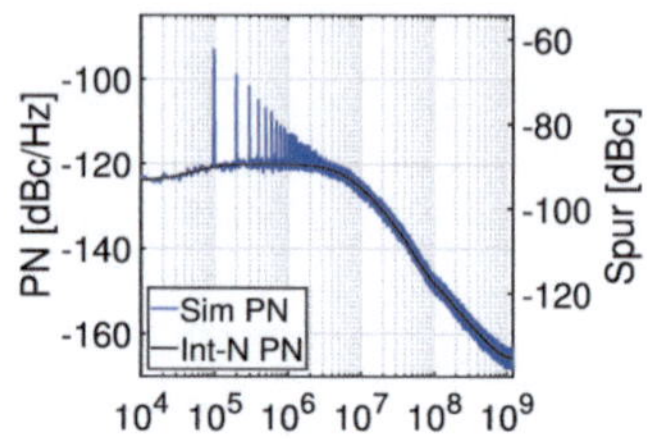

(b) Raw DQI spurs at the output, $X = 1$ or $2^{10} + 1$.

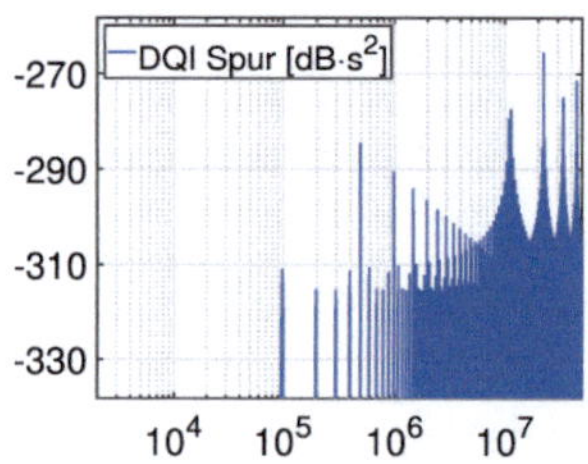

(c) Raw DQI spurs at source, $X = 227$.

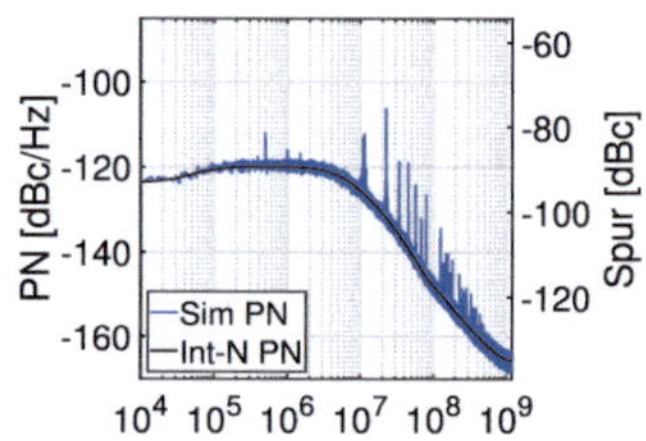

(d) Raw DQI spurs at the output, $X = 227$.

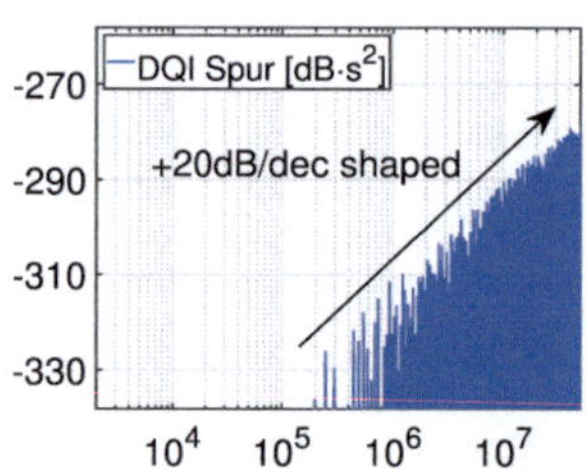

(e) Shaped DQI spurs at source.

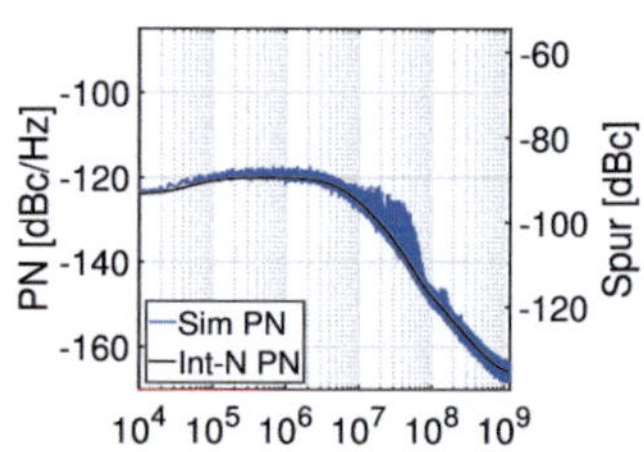

(f) Shaped DQI spurs at the output.

Fig. 6.6 Representative raw (top two rows) and shaped (bottom row) DQI-spur signatures arising from the DTC configurations tabulated in Table 6.1. Left column: power spectra of the DTC's residual error; right column: power spectra of the output PN

MATLAB Model of Segmented DTCs

For the completeness of analysis, we show in Listing 6.4 the simplified MATLAB model of segmented DTCs as shown in Fig. 6.1b. The model is simplified to focus on the segmented control scheme for the coarse and fines DCDs, thus the DCDs *per se* are assumed to be linear (see Lines 34 and 36)—nonetheless, these can be modeled to be nonlinear following the same method presented in Line 11 of Listing 6.3.

Listing 6.4 MATLAB model of segmented DTCs.

```
function t_DTC = DTC_segmented(in,T_out,M,res_DTC_c,res_DTC_f,Segmentation_method,ncycles)
    %% Segmentation_method = 1, 2, or 3 controls the method of DTC segmentation
    %% Local Preallocation
    c_c = zeros(1,ncycles);
    q_c = zeros(1,ncycles);
    c_f = zeros(1,ncycles);
    q_f = zeros(1,ncycles);
    if Segmentation_method == 1
        %% Truncator coarse + Truncator fine
         for k=2:ncycles
            % Coarse quantizer
            [c_c(k),q_c(k)] = adder_modulo_M(in(k),0,res_DTC_c);
            % Fine quantizer
            [c_f(k),~]      = adder_modulo_M(q_c(k),0,res_DTC_f);
         end
    elseif Segmentation_method == 2
        %% EFM1 coarse + Truncator fine
        for k=2:ncycles
            % Coarse quantizer
            [c_c(k),q_c(k)] = adder_modulo_M(in(k),q_c(k-1),res_DTC_c);
            % Fine quantizer
            [c_f(k),~]      = adder_modulo_M(q_c(k)-q_c(k-1),0,res_DTC_f);
        end
    elseif Segmentation_method == 3
        %% EFM1 coarse + EFM1 fine
        for k=2:ncycles
            % Coarse quantizer
            [c_c(k),q_c(k)] = adder_modulo_M(in(k),q_c(k-1),res_DTC_c);
```

```
29                % Fine quantizer
30                [c_f(k),q_f(k)] =
                      adder_modulo_M(q_c(k)-q_c(k
                      -1),q_f(k-1),res_DTC_f);
31            end
32        end
33        %% Coarse DCD
34        t_DTC_c = T_out*res_DTC_c/M * c_c;
35        %% Fine DCD
36        t_DTC_f = T_out*res_DTC_f/M * c_f;
37        %% Total DTC produced delay
38        t_DTC   = t_DTC_c + t_DTC_f;
39    end
```

The segmentation method of this DTC model can be selected via the `Segmentation_method` variable. Namely, when it is set to 1, 2, and 3, the DTC architecture is configured to have (1) truncation-type coarse and fine quantizers, (2) an EFM1 coarse quantizer and a truncating fine quantizer, and (3) EFM1 coarse and fine quantizers. These correspond to the three "segmented" cases in Table 6.1.

Note that, we utilize here the modulo-M adder function in Listing 3.3 to model the quantizers. As expressed in Listing 3.3, when only one of the two inputs is used to drive the adder, the adder becomes a modulo-M truncator acting on the input; when the adder is configured as in Fig. 3.5 and (3.31), it behaves as a digital EFM1 processing its input—like the EFM1s defined in a MASH DDSM, as shown in Listing 3.4.

Here, as annotated in Fig. 6.1b, `c_c` and `c_f` are the digital codes yielded by the quantizers to control the coarse and fine DCDs; `q_c` and `q_f` correspond to the QEs arising from the coarse and fine quantization operations, as derived in (3.34). The QE induced by the coarse stage cannot be corrected therein and should therefore be carried forward to be corrected by the fine stage. Note that, when the coarse quantizer is a truncator (`Segmentation_method` = 1), the QE of the coarse stage, $q_c[k]$, controls the the fine stage (see Line 14); when the coarse quantizer is an EFM1 (`Segmentation_method` = 2 and 3), the QE of the coarse EFM1 that controls the fine stage is $q_c[k] - q_c[k-1]$ (see Lines 22 and 30), as derived in (3.36).

The categorization of the DQI-spur signatures is summarized in Table 6.1. We observe that the spectral signature of the DQI spurs is primarily determined by the method (truncation, Δ-Σ, etc.) employed in the *finest* requantization, marked in magenta in Fig. 6.1, performed by the DTC's control scheme. More specifically, the DQI signature is determined by how the DTC processes the LSBs of its input control code, $a[k]$.

- When the finest requantizer is a normal, e.g., truncating, quantizer, the LSBs of $a[k]$ are simply discarded by the DTC and are injected into the loop as a residual error that cannot be corrected by the DTC. In this case, the DTC's QE incorporates the raw form of the DQI spurs, as shown in Figs. 6.6a (a) and (c). The raw DQI-spur pattern depends on the value of the fractional FCW. With a small FCW, e.g., $X = 1$ in (a), the DQI pattern has a regular -20 dB/dec rolloff envelope in band; with a large value of X, e.g., $X = 227$ in (c), the DQI pattern may still have an in-band component and the spur locations become non-trivial to predict. Note that using large X values does not necessarily push the DQI spurs to high frequencies; for example, when $X = 2^{10} + 1$ in this example, the DQI-spur pattern is identical to the case of $X = 1$, as shown in (a).

- On the other hand, when the finest requantizer is an EFM1, it applies first-order noise shaping to the DTC's QE, which consequently results in the residual error having a characteristic $+20$ dB/dec envelope that does not vary with different X values, as shown in Fig. 6.6(e). Note that using a fine EFM1 requantizer alone does not guarantee that the DQI spurs will be converted into a densely shaped cluster; residual DQI spurs with significant magnitudes can still remain, see e.g., [59, Fig. 10].

System-level results in the right column of Fig. 6.6 show that, despite the presence of the bang-bang detection following the DTC, the DQI-spur pattern is preserved; the DQI spurs are transferred to the PLL's output and affect its spectral purity. This is because the DTC in high-performance PLLs is typically designed to have its QE confined to the "pseudo-linear" operation region at the input of the TDC (see Chap. 7).

This mechanism was explained in Chap. 3. In Fig. 6.6b, d, the individual spurs are clearly distinguishable and the largest of them appear at close-in frequencies and cannot be filtered by the loop *per se*; in Fig. 6.6(f), they appear like a "bump" when plotted on a log frequency scale but, with higher frequency resolution, they can be distinguished.

We note also that the DQI-spur signature—either raw or shaped—is well-defined and independent of the type and order of the MMDC being used, as long as the values of Δ and X are fixed. For example, simulations show that when the MMDC is an EFM1, MASH 1-1, or MASH 1-1-1, or other state-of-the-art stochastic MMDCs such as PDS [27, 33] and ENOP [50, 60] DDSMs, the DQI-spur patterns are identical to those shown in Fig. 6.6.

In summary, regardless of the architectures of the DTC and the MMDC, DQI spurs exist ubiquitously as long as there is a resolution gap between the DTC's control and output; these can significantly degrade the output spectral purity and jitter. This demands specific methods to cancel the DQI spurs, ideally at source.

In the next section, we will derive closed-form time-domain expressions for the raw and shaped DTC's QE signals, which are manifest as DQI spurs in the spectrum. IDQ methods to eliminate DQI spurs at source will be elaborated in Chap. 7.

6.3.3 Analysis

In this subsection, we derive an analytical expression for the DTC's QE; this helps to explain the characteristic shape and architecture independence of the DQI-spur pattern in the cases where the same DTC is used to cancel the AQE of MMDCs with different types, orders, and moduli.

Waveform of Raw DTC's QE

To derive the time-domain waveform of the DTC's QE in the case where the fine DTC requantizer is a normal Δ-to-1 truncator, we review the first principles of the fractional-N synthesis enabled by a generic DDSM-based MMDC, as shown in Fig. 1.5.

To synthesize an output frequency that is a non-integer multiple of the reference frequency, the MMDC with modulus M takes as its input a constant (integer) FCW, X, and outputs (integer) $y[k]$ in the kth reference clock cycle, which, added to N_{int}, provides a randomized instantaneous divide ratio, $N_{\text{div}}[k]$ whose average is $(N_{\text{int}} + X/M)$ [44].

By definition, the QE of the MMDC is $q_{\text{MMDC}}[k] \triangleq y[k] - X/M$, and, hence, the AQE is $q_{\text{MMDC}}^{\text{acc}}[k] \triangleq \sum_{i=1}^{k} q_{\text{MMDC}}[i]$. The digital code of the AQE becomes

$$q_{\text{MMDC}}^{\text{acc}}|_d[k] \triangleq M \cdot q_{\text{MMDC}}^{\text{acc}}[k] = -X \cdot k + M \cdot \sum_{i=1}^{k} y[i]. \tag{6.8}$$

This is an important relationship that is valid for all kinds of divider controllers in principle. It indicates that their AQE comprises two components: (i) a code ramp and (ii) a sum of integers scaled by the MMDC's modulus.

Note that $q_{\text{MMDC}}^{\text{acc}}\big|_d[k]$ is the control input of the DTC which is quantized by the DTC requantizer. Thus, the DTC's QE referred to the input of the DTC requantizer, which is injected into the loop and is manifest as raw DQI spurs, is given by

$$\begin{aligned} q_{\text{in,DTC}}^{\text{raw}}[k] &\triangleq \Delta \cdot \left\lfloor \frac{q_{\text{MMDC}}^{\text{acc}}|_d[k]}{\Delta} \right\rfloor - q_{\text{MMDC}}^{\text{acc}}|_d[k] \\ &\triangleq -\left(q_{\text{MMDC}}^{\text{acc}}|_d[k] \bmod \Delta\right), \end{aligned} \tag{6.9}$$

where $\lfloor \cdot \rfloor$ denotes the floor function.

Note that, in a typical digital implementation, the modulus of the MMDC (M) and the quantization resolution of the DTC (Δ) are powers of two, where $M = 2^m \cdot \Delta$ and $m \in \mathbb{N}^+$. Recall also that the $y[i]$s are integers. When substituting (6.8) into (6.9), the terms associated with the coefficient M vanish after applying mod Δ. Eq. (6.9) simplifies to

$$q_{\text{in,DTC}}^{\text{raw}}[k] = -\left[(-X \cdot k)\, mod \Delta\right], \tag{6.10}$$

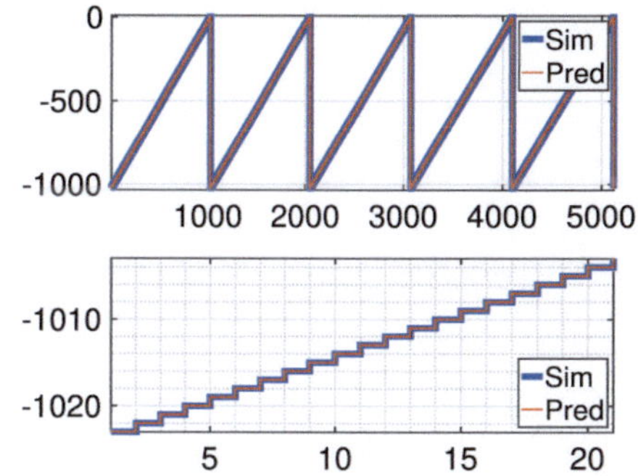

(a) $\Delta = 2^{10}$, $X = 1$ or $2^{10} + 1$.

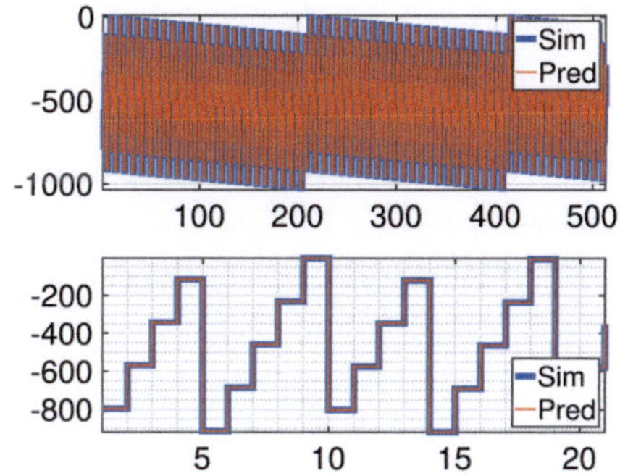

(b) $\Delta = 2^{10}$, $X = 227$ or $2^{10} + 227$.

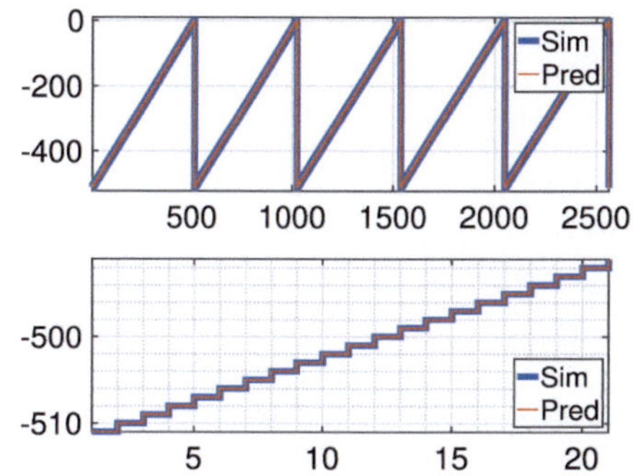

(c) $\Delta = 2^{9}$, $X = 1$ or $2^{10} + 1$.

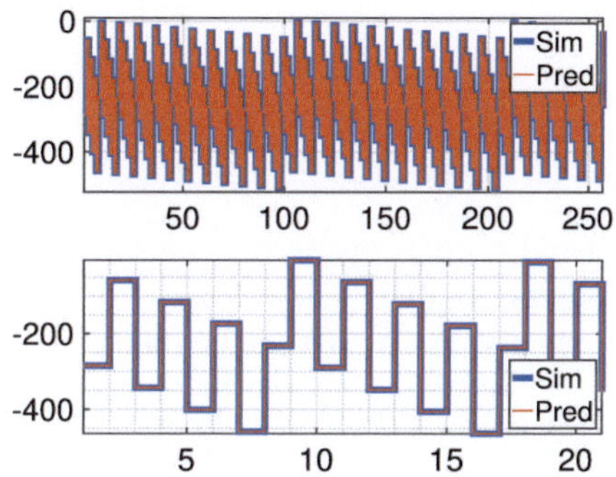

(d) $\Delta = 2^{9}$, $X = 227$ or $2^{10} + 227$.

Fig. 6.7 Simulated waveforms of the raw DTC's QE vs. their prediction using (6.10). Different values of Δ and X are used in the subfigures, in each of which the top strip (zoomed out) shows the periodic pattern and the bottom strip (zoomed in) shows the detailed variation. The match between simulation and prediction is exact

which is the closed-form waveform of the DTC's QE when the fine DTC requantizer is a normal truncator. Fig. 6.7 shows that this prediction exactly matches the simulated DTC's QE.[9]

Notice that Eq. (6.10) has a form that is very similar to the waveform of the input-referred AQE of an first-order DDSM (EFM1) with modulus Δ, as shown in Fig. 3.6, that takes as its input a constant FCW, X. The "sum" signal of the EFM1 can be expressed as

$$s[k] = (x[k] + s[k-1]) \bmod \Delta \tag{6.11a}$$

$$\triangleq x[k] + s[k-1] - \Delta \cdot c[k], \tag{6.11b}$$

[9] These results were obtained with a MASH 1-1 MMDC; the results for MASH 1-1-1, PDS, and ENOP P1 are identical.

where $c[k] = \lfloor (x[k] + s[k-1]) / \Delta \rfloor$ is the "carry" signal, i.e., the overflow of the accumulator. Rearranging (6.11b), the QE of the EFM1 is

$$q_{\text{in,EFM1}}[k] \triangleq \Delta \cdot c[k] - x[k] = s[k-1] - s[k], \tag{6.12}$$

which can be interpreted as the QE of the internal quantizer, $-s[k]$, being first-order highpass shaped. The AQE is

$$q_{\text{in,EFM1}}^{\text{acc}}[k] \triangleq \sum_{i=1}^{k} q_{\text{in,EFM1}}[i] = -s[k] = -\sum_{i=1}^{k} x[i] + \Delta \cdot \sum_{i=1}^{k} c[i]. \tag{6.13}$$

Substituting $x[i] = X$ into (6.13) yields the waveform of the mentioned AQE of an EFM1 as

$$q_{\text{in,EFM1}}^{\text{acc}}[k] = -X \cdot k + \Delta \cdot \sum_{i=1}^{k} c[i] = -\left[(X \cdot k) \mod \Delta\right], \tag{6.14}$$

where Eq. (6.14) is derived by substituting $s[k-1]$ in (6.13) into (6.11a), which is similar to (6.10).

Finally, Eq. (6.10) can be interpreted in the following way:

- The expression represents a monotonic ramping signal with a step equal to $-X$ that is quantized (floored) with a constant quantization resolution Δ. Its similarity to (6.14) indicates that it must have the same spectral signature as the AQE of an EFM1 with constant input X and modulus Δ.
- As shown in Fig. 6.7, the signal described by (6.10) is inherently periodic when X is constant; this explains its purely tonal manifestation (direct-injection spurs [46]) in the spectrum, i.e., raw DQI spurs. The characteristic DQI-spur signatures shown in Figs. 6.6a (a) and (c) can be predicted exactly by substituting the values of X and Δ into (6.10), followed by a spectral transformation.
- The waveform described in (6.10) depends only on X and Δ and is independent of the MMDC's type, order (l), and modulus (M). This explains why the DQI-spur signature is independent of the choice of MMDC, as reported in Sect. 6.3.2.
- With small FCWs, where $X \ll \Delta$, the $q_{\text{in,DTC}}^{\text{raw}}$ waveform has a regular and predictable pattern that is a periodic sawtooth wave with frequency equal to $X/\Delta \cdot F_{\text{ref}}$. This corresponds to the simulated DQI-spur pattern shown in Fig. 6.6a, where the fundamental tone lies at $1/2^{10} \cdot 100$ MHz ≈ 97.7 kHz. In this case , the largest DQI spurs lie entirely within the loop bandwidth; these cannot be filtered by the loop dynamics and will hence contribute to the output jitter. Larger values of FCW can make the waveform and the corresponding DQI-spur pattern irregular and less easy to be described intuitively, such as that shown in Fig. 6.6c.

In particular, note that, contrary to what was asserted in [43], even large FCWs cannot guarantee that the DQI spurs will be sufficiently randomized nor that the spurs are pushed to higher frequencies so that they can be effectively filtered by the loop.

This can be observed in Fig. 6.6a, where the DQI-spur pattern when $X = 2^{10} + 1$ is identical to that when $X = 1$; this is confirmed in Fig. 6.7. To understand this, we further decompose (6.10) by applying the multiplication property of the modulo operation. Namely,

$$q_{\text{in,DTC}}^{\text{raw}}[k] = -\left\{\left[\underbrace{X \bmod \Delta}_{\text{fixed remainder }(R)} \cdot \underbrace{(-k) \bmod \Delta}_{\text{unity-step sawtooth}}\right] \bmod \Delta\right\}, \tag{6.15}$$

the internal expression of which describes a unity-step sawtooth wave that is scaled by the *remainder* of X divided by Δ; the remainder is fixed (time-invariant) when X and Δ are fixed. This indicates that the waveform of the DTC's QE depends only on (i) the DTC's quantization resolution, Δ, and (ii) the remainder, $R = X \bmod \Delta$. In other words, for all valid fractional FCWs, i.e., $X \in \{X \in \mathbb{N} \,|\, 0 < X < M\}$, that are *congruent modulo* Δ, i.e., $X = m \cdot \Delta + R, \forall\, m \in \mathbb{N}$, the DTC's QE waveforms, and hence the DQI-spur patterns, are equivalent. This explains why a larger FCW alone does not automatically mitigate DQI spurs.

Signature of Shaped DQI Spurs

Next, we consider the case shown in Fig. 6.6e, where the fine DTC requantizer is an EFM1. Here, the spectral manifestation of the DTC's QE is a densely spaced cluster of spurs with a +20 dB/dec envelope. This signature is independent of X and the MMDC-related settings. In this subsection, we will prove that this spectral signature is the same as the AQE of a MASH 1-1 MMDC, as shown in Fig. 6.8, with a relatively poor resolution, i.e., with modulus equal to Δ (the quantization resolution of the DTC). In other words, the shaped DQI spurs share the property of another type of what we called "directly injected spurs" in [46] that arise because of the inherent short cycle length of the MASH itself, rather than nonlinear distortion.

Recalling the relationships in (6.13) and (6.14), the sum signal of the input EFM1 of the MASH 1-1 controlled by constant X is

$$s_1[k] = X \cdot k - \Delta \cdot \sum_{i=1}^{k} c_1[i]. \tag{6.16}$$

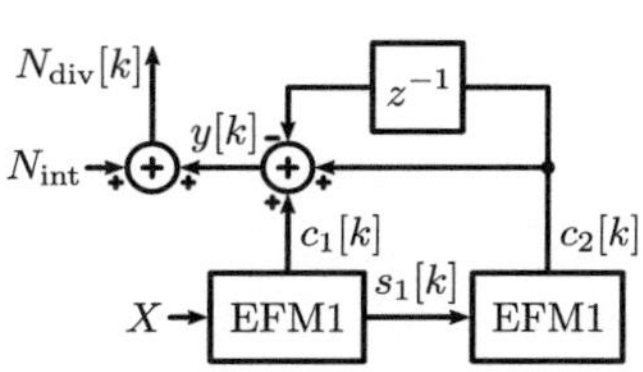

Fig. 6.8 A MASH 1-1 MMDC contains two EFM1s as shown in Fig. 3.6

Note the similarity in the forms of this expression and (6.8) in the sense that their right-hand sides are comprised of (i) an accumulation of X (or $-X$), representing a code ramp with a step having an absolute value of X in the time domain, and (ii) a set of integers scaled by M in (6.8) or Δ in (6.16). As discussed in Sect. 6.3.3, integer multiples of Δ vanish with the modΔ operation. Therefore, quantizing $s_1[k]$ in (6.16) with a quantization resolution equal to Δ yields the same result as quantizing $q^{\mathrm{acc}}_{\mathrm{MMDC}}|_d[k]$ in (6.8) from (6.9) to (6.10) in the sense that the QE of the quantizer, i.e., the remainder of the modulo operation, depends only on the code ramp and is independent of the integer sum.

Thus, when inspecting the spectral signatures of the QE of the internal Δ-to-1 quantizer and, hence, the overall QE of the DTC's EFM1 requantizer, i.e., $-s[k]$ and $q_{\mathrm{in,EFM1}}[k]$ in (6.12), the input of this EFM1 being $q^{\mathrm{acc}}_{\mathrm{MMDC}}|_d[k]$, i.e., the AQE of the MMDC with modulus of M, can be regarded effectively the same as the input being $s_1[k]$, i.e., the sum signal of an EFM1 with modulus of Δ. From this point of view, quantizing $q^{\mathrm{acc}}_{\mathrm{MMDC}}|_d[k]$ by the EFM1 DTC requantizer can be regarded as a cascade of two EFM1s, both with modulus equal to Δ in the MASH 1-1 fashion, where the second stage mimics the DTC requantizer.

Furthermore, substituting $y[k] = c_1[k] + c_2[k] - c_2[k-1]$, $x_1[k] = X$, and $x_2[k] = s_1[k]$ from Fig. 6.8 and (6.11b) into (6.8) yields the AQE of a MASH 1-1 DDSM as

$$q^{\mathrm{acc}}_{\mathrm{MASH\ 1\text{-}1}}|_d[k] = s_2[k-1] - s_2[k],$$

which is the same as (6.12). This indicates that the DTC's QE injected into the loop in this case has the same spectral signature as the AQE of a MASH 1-1 DDSM without LSB-dither [61] and with a relatively poor resolution in the sense that the MASH's modulus equals the value of the DTC's quantization resolution.

As per the definition in Chap. 7, $\Delta = 2^{N_b^{\mathrm{acc}} - N_b^{\mathrm{DTC}}}$, where $N_b^{\mathrm{acc}} = N_b^{\mathrm{MMDC}} + l - 1$ is the equivalent number of bits of the MMDC's AQE, $M = 2^{N_b^{\mathrm{MMDC}}}$ is the modulus of the MMDC, and N_b^{DTC} is the number of bits of the DTC. In the example in Sect. 6.3.2, a MASH 1-1 MMDC with modulus equal to 2^{20} used together with a DTC with 11 equivalent number of bits results in $\Delta = 2^{10}$. In the case of Fig. 6.6e, the shaped DQI-spur pattern conforms to the spectral signature of the AQE of a MASH 1-1 DDSM with modulus equal to 2^{10}, which is exactly the case demonstrated in [46, Fig. 5c]. A MASH 1-1 without LSB-dither and with a relatively small modulus will result in a short cycle length [44] and hence produce direct- injection spurs [46]. This explains the phenomenon of the dense cluster of shaped IBS spurs in Fig. 6.6(e).

Quantitative Discussion

In this subsection, we have derived the waveforms of the raw and shaped DTC's QEs and analyzed their spectral signatures. Recall Fig. 6.3 that shows the cancellation mechanism of the MMDC's QE using a DTC; it considers DTC's nonidealities such as the hard quantization nonlinearity in the DTC requantizer and the DCD's polynomial

INL. The "DTC requantizer" with a quantization resolution of Δ can be linearized by its inherent gain of $1/\Delta$ with its induced QE referred to its output. The linear gain of the "DCD" after calibration by the LMS is assumed to match the original injection of $q_{\mathrm{MMDC}}^{\mathrm{acc}}|_d[k]$.

Recall that the right column in Fig. 6.6 shows the simulated output PN of the example BBPLL taking into account the contribution of the DTC's QE, whilst assuming that the DCD has good linearity, without the "poly" and "REM" marked in Fig. 6.3b. In addition to the integer-N PN level predicted using the method detailed in Chap. 4, raw or shaped DQI spurs arise and dominate the spectral purity. From Sect. 6.3.3, we derive the following conclusions:

- **when the DTC requantizer is a normal Δ -to-1 truncator:** the DQI- spur pattern is that of the AQE of an EFM1 with constant input X and modulus Δ. This means that, with a linear DCD, the quantized DTC cancels the original injection of the AQE of the (normally high-order) MMDC with modulus M that has a range of variation of a few T_{out}'s, e.g., $q_{\mathrm{MMDC}}^{\mathrm{acc}}|_d \in M \cdot \left(-2^{l-2}, 2^{l-2}\right)$ for an lth-order MASH, and hence $t^{\mathrm{acc}} \in T_{\mathrm{out}} \cdot \left(-2^{l-2}, 2^{l-2}\right)$; it instead injects a residual error with a range of variation of $\frac{\Delta}{M} \cdot T_{\mathrm{out}}$, namely $t_{\epsilon}^{\mathrm{raw}} \in T_{\mathrm{out}} \cdot [(1-\Delta)/M, 0]$ (see Fig. 6.7). However, the residual error in this case has the spectral signature of an EFM1's AQE that is purely tonal and is independent of the main MMDC's type, order, and modulus. Consequently, the advantages normally conveyed by using high-order, large-modulus, and novel MMDCs such as sequence randomization, noise shaping, and spur immunity, have no effect on DQI spurs;
- **when the DTC requantizer is an EFM1:** the DQI spurs form a first-order shaped cluster, which is similar to the directly injected spurs associated with the AQE of a MASH 1-1 with a small modulus Δ. Therefore, the DTC's QE induced residual error varies in $t_{\epsilon}^{\mathrm{shaped}} \in T_{\mathrm{out}} \cdot [(1-\Delta)/M, (\Delta-1)/M]$. It is manifest at the output PN spectrum as a dense "bump" of spurs.

Spur Predictions

In order to compare the PN/jitter contributions of the DTC's *hard* quantization nonlinearity and *soft* integral nonlinearity, we also performed a comparative study by modeling the polynomial INL in Fig. 6.3b as per Eq. (6.6), which approximates the measured INL of a commonly used variable-slope charging DTC in Fig. 6.2 that has an asymmetric concave shape and a peak-to-peak range of about 7 LSBs for an 11-bit dynamic range [21, 27, 33]. We perform fractional-N simulations with a MASH 1-1 MMDC that is commonly used in DTC-assisted PLLs. The simulated results are shown in Fig. 6.9.

Figure 6.9a shows the output PN when the DTC requantizer is a simple truncator. We have noted in Section 6.2.4 that the hard and soft nonlinearities of the DTC exhibit separable PN and spurious responses that can be distinguished in the output spectrum. Here, we use the analytical expression of the DTC's QE derived in (6.10) to predict

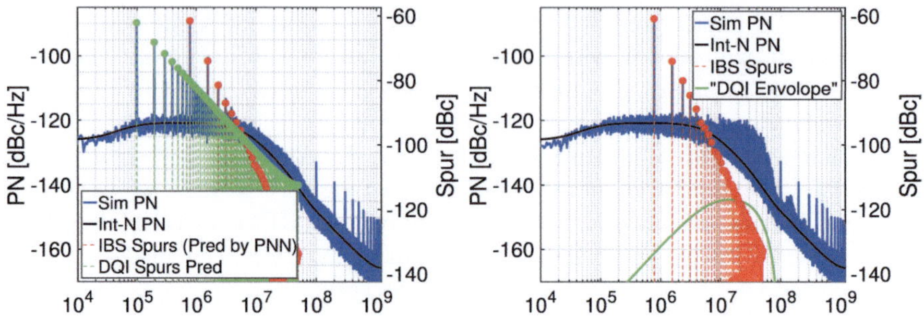

(a) Truncating DTC requantizer.

(b) EFM1 DTC requantizer.

Fig. 6.9 Simulated output PN power spectra of the example fractional-N BBPLL with a MASH 1-1 MMDC and a DTC with a nonlinear DCD controlled by different quantization schemes. $\Delta = 2^{10}$, $X = 8 \cdot \Delta + 1 = 8193$

the DQI spurs,[10] and the periodic nonlinearity noise (PNN) theory[11] [18, 45, 46] to predict the polynomial INL-induced IBS spurs. The matches between the predictions and simulations are exact. We choose the fractional FCW, $X = 8 \cdot \Delta + 1 = 8193$, to illustrate the potential importance of DQI spurs compared to IBS spurs. In this case, according to the analysis of (6.15), the DQI-spur pattern is the same as when $X = 1$, where the tones are at the fundamental and harmonics of $1/\Delta \cdot F_{\text{ref}} \approx 97.7$ kHz. The IBS cluster is located at the (well-known) fundamental and harmonics of $X/M \cdot F_{\text{ref}} \approx 781$ kHz. Notice that, with this value of X, both the DQI and IBS clusters lie within the loop bandwidth and hence cannot be filtered by the loop; the fundamental of the DQI cluster is at an even lower frequency than that of the IBS cluster. This contradicts the assertion made in [43] that the DQI spurs "typically occur at much higher frequencies than those caused by the INL and are effectively filtered by the loop". The "congruent modulo Δ" homogeneity of the DQI-spur pattern derived in (6.15) explains this phenomenon.

For this fraction X/M, t here are more in-band DQI spurs than IBS. Furthermore, i n terms of their amplitudes, the DQI spurs are worse than the IBS; their fundamentals are of similar magnitudes. This can be understood from the fact that the DQI originates from the DTC's QE that is, recalling the quantitative discussion above, a sawtooth wave varying in $T_{\text{out}} \cdot [(1 - \Delta)/M, 0] = T_{\text{out}} \cdot \left[-9.76 \cdot 10^{-4}, 0\right]$; whereas the IBS originates from the PNN, which will be shown in Fig. 8.2a that varies in $T_{\text{out}} \cdot (-3, 6) \cdot 10^{-4}$—their waveforms have similar amplitudes and, hence, result in fundamentals with similar magnitudes.

Figure 6.9b shows the output spectrum when the DTC requantizer is an EFM1. The bump of densely clustered DQI spurs can also be distinguished from the IBS family in this case. Note that Δ-Σ modulation of the DTC's control code does not

---

[10] The DTC's QE in (6.10) is transferred to the output PN via a gain factor equal to $\frac{T_{\text{out}}}{M} \cdot N \cdot 2\pi \cdot F_{\text{ref}}$, where T_{out}/M is the DTC inherent gain as shown in Fig. 6.3, N is the lowpass gain of the TF imposed by the loop, and $2\pi \cdot F_{\text{ref}}$ transfers the time-domain signal to the phase domain.

[11] The PNN analysis framework will be detailed in Chap. 7.

Table 6.2 Simulated RMS output jitters in different operation modes

Case	Mode	Requantizer	DCD	RMS jitter
(a)	Integer-N	N/A	N/A	252 fs
(b)	Fractional-N	Ideal	Nonlinear	301 fs
(c)		Normal	Linear	299 fs
(d)		Normal	Nonlinear	341 fs
(e)		EFM1	Linear	279 fs
(f)		EFM1	Nonlinear	335 fs

affect IBS spurs induced by the distortion resulting from the DCD's soft nonlinearity. To confirm that the nature of the bump is in fact purely spurious, rather than being purely noise-like and defined by the PSD of the accumulated quantization noise of a MASH 1-1 with modulus equal to Δ, we use the fractional-N PN prediction framework detailed in Chap. 4 to show the theoretically predicted envelope of the DTC's QE contribution, assuming that it is purely PN. Notice that the envelope of the bump does not match the PN prediction; its height in fact varies with the choice of resolution frequency. This confirms that the shaped DQI spurs are indeed still purely tonal and cannot be masked by the integer-N PN floor.

Jitter Contributions

Provided that both the DQI and IBS spurs deteriorate the spectral purity significantly, in terms of the output jitters, we have simulated different operation modes of the example BBPLL and recorded the RMS values of the simulated output phase error samples in Table 6.2.

The integer-N (case (a), which exhibits the best jitter performance) and fractional-N modes considering different DTC nonlinearities are recorded. Cases (b) and (c) are when only the DCD's polynomial INL is considered for the DTC (with infinite-resolution), and, in contrast, when a linear DCD and a truncating DTC requantizer are considered. Quantitatively, the hard and soft nonlinearities of the DTC make equal contributions to the output jitter in the case of Fig. 6.9a, which is recorded in case (d) with the worst-case jitter. Fig. 6.9b corresponds to case (f). Cases (e) and (f) show that the first-order Δ-Σ modulation in the requantization helps to highpass shape the DQI spurs, which slightly reduces the fractional-N output jitter, i.e., improving from cases (c) and (d). However, the effect of jitter mitigation is limited due to the fact that the bump cannot be fully masked spectrally. Note also that the EFM1 requantizer may not be able to fully transform the DQI spurs to become a shaped cluster with all the significant components staying out of band [59]; in such cases the residual in-band DQI spurs can still contribute significantly to the system jitter.

Finally, as we will show in Chap. 7, high-order and stochastic divider controllers, as well as the DTC linearity enhancement techniques introduced in Sect. 6.3.1, help to mitigate the spurs and jitter induced by the polynomial INL of the DCD, but have no effect on DQI spurs. With these, the DQI spurs become the significant spur/jitter contributors. This motivates us to design better methods which target DQI mitigation specifically; the proposed IDQ methods to achieve this goal will be described in the next chapter.

6.3.4 Conclusion

In this work, we observed the manifestation of the DQI spurs whose root cause is confirmed to be the resolution mismatch between the AQE of the MMDC and its cancellation subsystem, the DTC, in fractional-*N* frequency synthesizers. We derived an analytical expression for the DTC's QE and analyzed the spectral signature of its induced DQI spurs, when the DTC has a normal, e.g., truncating, or an EFM1 requantizer. As the output DQI spurs, especially those at close-in frequencies, may dominate the spurious response of a DPLL system, they present a further challenge in addition to conventional fractional spurs. Given that conventional DTC linearity enhancement and stochastic design of the MMDC cannot mitigate DQI spurs, we will describe methods to eliminate DQI spurs in Chap. 7.

6.4 Digitally Controlled Delay (DCD) Random Element Mismatch (REM) and Its Induced Spurs and PN

As analyzed in Sect. 6.2.4, the generation mechanism for the spurs and noise induced by the DCD's REM nonlinearity is independent and separable from those of the quantization and polynomial integral nonlinearities. Recall Fig. 6.5 that shows that, even with the other two nonlinearities ultimately mitigated and the DTC exhibiting only the REM nonlinearity, the REM-induced spurs prevail at the DTC-block level (the blue curve in Fig. 6.5(c)), and are transferred to affect the output PN, as shown in blue in (d).

In this section, we will extend the analysis to investigate the REM spurs and PN signatures in the non-segmented and segmented DTCs, as shown in Fig. 6.1; the measured REM variations of representative DTCs reported in the literature will be referred to when modeling the REM in our simulation, in order to facilitate realistic modeling.

6.4.1 REM Spurs in Non-segmented DTC

To model the REM of a representative non-segmented DTC (illustrated in Fig. 6.1a), we refer to Ru et al. [35] that proposes a 10-bit CSC DTC. The CSC method is used to mitigate the smooth concave-shaped INL of the DCD, which is not targeted at the DCD's REM. As shown in [35, Fig. 8], in the best case scenario, the resulting INL of the CSC DCD becomes dominated by only the REM pattern. The INL of the 10-bit DTC is simulated in the span of the FS of 100 ps, corresponding to an LSB resolution equal to 97.7 fs. The random distribution of the REM has a peak-to-peak range of variation of 25 fs or approximately 0.26 LSB. The measured mitigated INLs on [35, page 9] agree well with this degree of mitigation.

From there, we assume that the REM follows a zero-mean Gaussian distribution—central limited, because the REM is more likely to have values concentrated in the vicinity of zero. We specify the standard deviation of the Gaussian distribution so that its 2σ range (encapsulating 95% of probability) equals the peak-to-peak range of the REM variation. To rule out outliers, when configuring the lookup table REM[$\cdot$] in Fig. 6.3, we nullify the randomly generated REM values beyond 2σ. In this way, the modeled range of REM variation is equal to the reported counterpart. In other words, the standard deviation of the REM equals $\sigma = 25$ fs$/97.7$ fs$/4 = 6.4\%$ of the unit delay of the DCD. The modeled REM INL of a representative 10-bit DCD is shown in Fig. 6.10.

For the sake of conciseness, we show in Fig. 6.11 the simulated REM spur pattern *at source*, i.e., in the power spectrum of the DTC's residual error (t_ϵ in Fig. 6.3). Here, we focus on the generation of REM spurs, and hence set the DTC to exhibit only the REM nonlinearity, but no quantization or polynomial integral nonlinearities. These spurs can be transferred exactly to the PLL's output spectrum, which was shown in Fig. 6.5; comparison of their magnitude against those induced by the other nonlinearities was also shown there.

In Fig. 6.11, the DR of the 10-bit DTC is set to cover exactly the FS variation of the AQE of a MASH 1-1 DDSM with modulus $M = 2^{20}$. $T_{\text{out}} = 417$ ps, i.e., 2.4 GHz output frequency, is set. We set the fractional FCW, $X = 27$ and 227, in the two rows, respectively. We can see that the AQE exercising a static REM nonlinearity induces

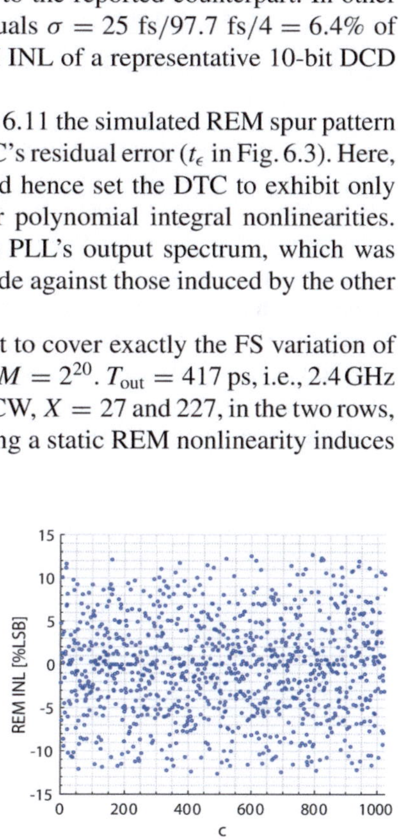

Fig. 6.10 REM INL (normalized to LSB) modeled for a 10-bit DCD. The REM is generated according to a Gaussian distribution, with samples whose absolute values are $\geq 2\sigma$ set to zero. $\sigma = 6.4\%$ LSB as discussed

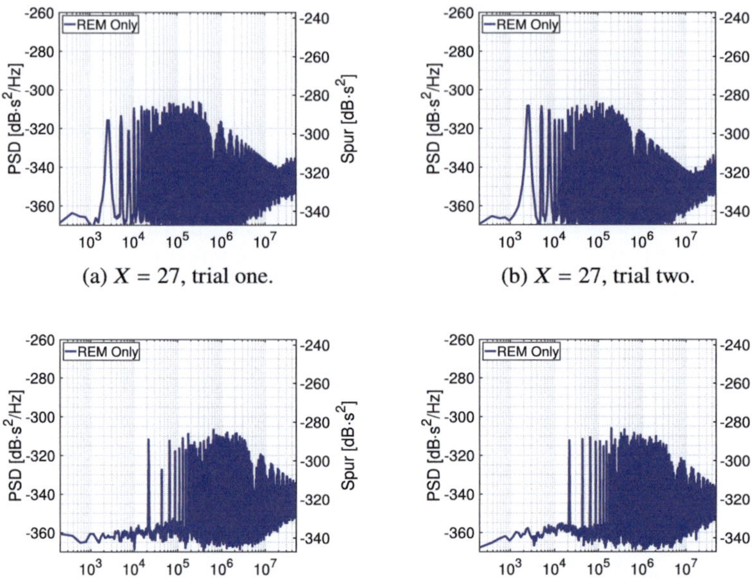

(a) $X = 27$, trial one.

(b) $X = 27$, trial two.

(c) $X = 227$, trial one.

(d) $X = 227$, trial two.

Fig. 6.11 Simulated power spectra of the DTC residual error at the DTC-block level. A 10-bit non-segmented DTC exhibiting only the REM nonlinearity is used to cancel the AQE of a MASH 1-1 DDSM with $M = 2^{20}$. Two different values of fractional FCW are used in the two rows. In each row, the simulation is repeated twice with different randomly generated REMs

a PN floor and a cluster of fractional spurs that do not show a regular and easily predictable pattern. The fundamental of the REM spurs is at $F_{\text{ref}} \cdot X/M$, the same as the fundamental frequency of the conventional IBS spurs, where $F_{\text{ref}} = 100$ MHz in our simulation. This means that a significant amount of REM spurs can appear at in-band frequencies that contribute to the system output PN/jitter which cannot be attenuated by the loop dynamics.

In each row of the figure, we also repeated the simulation by regenerating the modeled REM lookup table. We observe that, with the same intensity, range, and distribution used in REM generation, by just using another realization of the REM nonlinearity, i.e., another randomly generated REM array, in the DCD model, the pattern of REM spurs changes. This is because the REM nonlinearity exercised by the AQE periodically changed. Specifying the intensity of REM variation within the DCD alone can only indicate roughly the level of induced REM spurs (such as close to -280 dB·s^2 in our example), but cannot determine the exact heights and positions of the REM spurs.

6.4.2 REM Spurs in Segmented DTC

In the literature proposing DTCs segmented into a coarse and a fine stage, in the manner illustrated in Fig. 6.1b, the smooth concave-shaped INL of the coarse DTC stage is typically targeted to be mitigated. It is commonly asserted that, "since the fine-DTC covers a small delay range, its non-linearity and jitter are greatly reduced, leading to a negligible impact on PLL performance" as stated in [62] where [21, 63] are also quoted. Therefore, only the INL profile of the coarse DTC is generally reported. To model the REM of the coarse DCD, we refer to the representative profile reported in Rossoni et al. [62].

Here, the reverse-concavity variable-slope (RCVS) DTC is proposed to be used as the coarse DTC; this introduces autocalibrated resistors in series with the capacitors in the capacitor bank of the DCD (in Fig. 6.2), in order to reverse the concavity of the DCD's polynomial INL for its compensation.

In the best case mitigation, i.e., as shown by the blue curve in [62, Fig. 16], the resulting INL does not exhibit a perceptible polynomial component; it is only dominated by the static REM pattern that can also be observed in other partially mitigated INLs. In this case, the peak-to-peak variation of the REM nonlinearity equals 242 fs. The equivalent LSB of the coarse DTC is 1.8 ps.

We use the 2σ Gaussian approximation introduced in the last subsection to derive the standard deviation of the REM, which yields $\sigma = 242\text{ fs}/1.8\text{ ps}/4 = 3.4\%$ of the unity coarse delay. Notably, the REM with $\sigma = 3.4\%$ agrees well with Helal et al. [64] that proposes a DTC consisting of cascaded 1-bit DTC stages with binary weighted delays, which states "Monte-Carlo simulations predict that the DTC's unit element delay mismatch has a standard deviation of 3.1%".

In the DTC block-level simulation, we set the coarse DTC to have 7-bit equivalent resolution and the fine DTC to have 4-bit equivalent resolution to match the design in [62]. For analysis simplicity, we set both the DTCs to be controlled by truncation-type requantizers, as simulation shows that EFM1-type requantizers help to mitigate DQI spurs, but not REM spurs. The behaviorally simulated results are shown in Fig. 6.12.

Similar to the last subsection, in Fig. 6.12, we model a DTC that exhibits only REM nonlinearity, but no quantization or polynomial integral nonlinearity. Two situations are simulated: in both the cases, the coarse DTC is modeled to have $\sigma = 3.4\%$ REM; in (a), we model the fine DTC to be perfectly linear without REM, which approximates the aforementioned observations made in the literature; in (b), the fine DTC is also modeled to have $\sigma = 3.4\%$ REM to approximate the other performance limit.

In both the cases, the DTC's REM results in fractional spurs, the fundamental frequency of which equals $F_{\text{ref}} \cdot X/M = 2.57$ kHz. Notice that, comparing case (b) to (a), including the REM in the fine DTC, which may seem intuitively to randomize the overall REM nonlinearity exercised by the DTC's control code, in fact degrades the REM spurs.

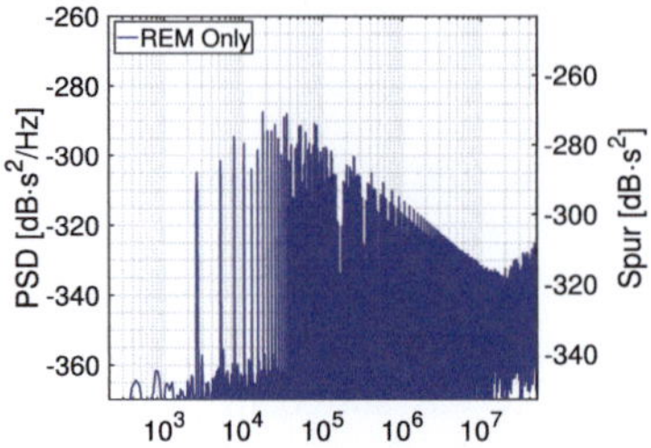

(a) REM only modeled in the coarse DTC.

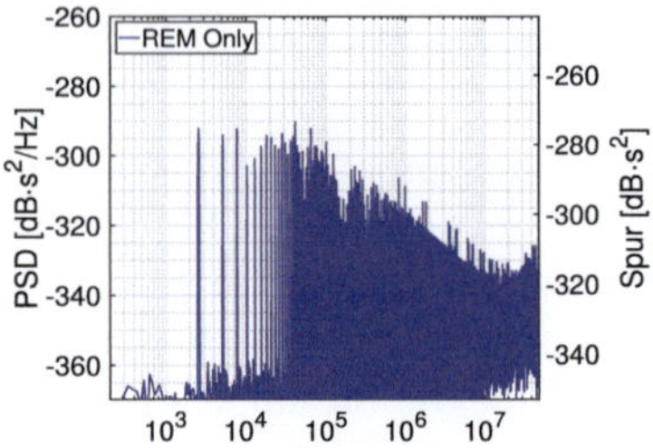

(b) REM modeled in both the coarse and fine DTCs.

Fig. 6.12 Simulated power spectra of the DTC residual error at the DTC-block level. A segmented DTC with 7 coarse bits and 4 fine bits, equivalent to 11 bits, exhibiting only the REM nonlinearity is used to cancel the AQE of a MASH 1-1 DDSM with $M = 2^{20}$. $X = 27$

6.4.3 *Conclusion*

In this section, we have discussed the REM nonlinearity associated with the DCD within the DTC. The modeling method corresponding to the REM profile reported in the literature was introduced. Spectral analysis of the behaviorally simulated DTC residual error shows that the DCD's REM induces a distinct family of fractional spurs, which we call "REM spurs" in this book. A mitigation method specifically targeted at them will be discussed in Chap. 8, where the simulated DTC block and PLL system-level results will show that the REM spurs can be eliminated completely.

6.5 Least-Mean-Square Convergence in Segmented DTCs

To minimize the hardware consumption by the DTC whilst guaranteeing the quality of QE cancellation, Tasca et al. [10] first proposed to use a finite-resolution DTC of at least 10 bits in order to bound the residual error after DTC cancellation to a value lower than the level of the intrinsic integer-N jitter , hence recovering the integer-N PN performance.

In practice, a 10-bit DTC can be implemented using various topologies. For example, the 10-bit DTC can be implemented with a *non-segmented* delay line, which has one capacitor bank that consists of 2^{10} unit capacitors, e.g., as in [15, 21, 32]. In practice, this requires the use of a bus of 2^{10} signals to switch 2^{10} capacitors.

To further simplify the digital control and to reduce the hardware consumption, a *segmented* DTC topology is adopted by many authors, e.g., [10, 13, 30, 31, 33]. It splits the DTC capacitor bank into a cascade of coarse and fine sections, which are typically controlled by an equal number of bits, e.g., 5 coarse bits and 5 fine bits for a total of 10 bits. Therefore, the digital core switches a total of $2^5 + 2^5$ capacitors, of which the coarse and fine banks can be constituted with metal-oxide-metal and metal-oxide-semiconductor varactor-connected capacitors, respectively. The coarse requantizer in a segmented DTC, as shown in Fig. 6.1b, takes charge of splitting the controls for the coarse and fine DCDs.

This section compares and comments on the methods used to implement and control the segmented DTC introduced above. Two methods to split the control signals for the coarse and fine DTC branches in a segmented DTC have been reported in the literature. The first, e.g., proposed in [10], uses an EFM1 as the coarse requantizer in Fig. 6.1b to reduce the resolution and control the coarse DCD; the QE of the EFM 1 then controls the fine DCD [31, 32]. On the other hand, Mahmoud et al. [30] proposed a simplified method that uses a normal rounding (or truncating) quantizer as the coarse requantizer [15, 21, 33]. In spite of its apparent hardware simplicity, interestingly, [30] reports that, in measurement, "the frequency division is not functional for very small division ratios, from $1/2^{10}$ to $21/2^{10}$"; whereas the method in [10] does not incur the same problem as the latter's system works "even in the case of a near-integer fractional channel". This motivates a comparative study of the two DTC segmentation methods to determine whether or not the different control method is responsible for the problems in the case of small fractions in [30].

This section compares and contrasts the two DTC segmentation and control methods [10, 30]. We first show mathematically that the two methods are equivalent from the perspective of canceling the MMDC's QE. Behavioral simulations analyzed in both the time and frequency domains show that the convergence issue of the LMS algorithm calibrating the fine DTC branch, affected by the segmentation method used in [30], is the root cause of the problem in that architecture. Note that, to simplify the analysis by focusing on the control methods for the segmented DTCs, the DCDs in this section are assumed to have good linearity.

6.5.1 DTC Segmentation Methods

The main structures of the segmented DTCs in [10, Fig. 8] and [30, Fig. 3] are similar, as shown in Fig. 6.13b. Here, the DTC is inserted in the feedback path of the DPLL following the MMDC. The MMDCs used in both works have a MASH 1-1 structure. The MASH quantizes X/M: recall that X (integer) is the FCW and $M = 2^{N_b^{\text{MMDC}}}$ is the modulus of the DDSM. A MASH 1-1 DDSM outputs an integer, $y[k] \in [-1, 0, 1, 2]$ at the kth cycle of the reference clock. To compensate for the QE, the DTC is controlled by $q_{\text{MMDC}}^{\text{acc}}\big|_d[k]$ to impose a corresponding cycle-wise time delay to cancel it. $t_{\text{div}_0}[k]$ and $t_{\text{div}}[k]$ in Fig. 6.1 represent the timestamps of the divided signal, before and after DTC cancellation.

In the digital domain, $q_{\text{MMDC}}^{\text{acc}}\big|_d[k] \in \{-M, -M+1, \ldots, M-1, M\}$ in the case of a MASH 1-1 [45], hence being represented by a digital code with resolution equal to one and a number of bits equal to $N_b^{\text{acc}} = N_b^{\text{MMDC}} + 1$. In [30], $N_b^{\text{MMDC}} = 10$, hence $N_b^{\text{acc}} = 11$. The modulus of the MMDC used in [10] is not stated explicitly. For a fair comparison, we assume the same N_b^{MMDC} and N_b^{acc} in its system analysis.

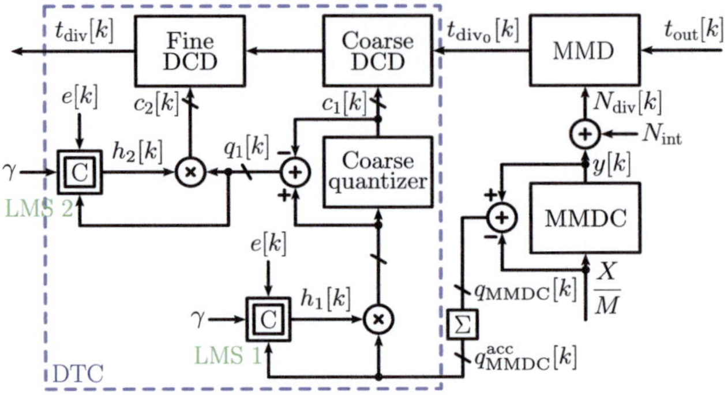

Fig. 6.13 LMS-calibrated segmented DTC, where the DTC's control input and time output have the same resolution

Note that, in principle, an 11-bit AQE can be eliminated completely by an 11-bit DTC if the DCDs are perfectly linear. Readers should not be confused by the segmentation partition of 6 coarse bits and 6 fine bits, which suggests a 12-bit DTC, used in both works. The DTC's DRs are typically doubled for redundancy to accommodate PVT variations. In our analysis, we start by assuming that the DCD is segmented as a coarse bank having 5-bit equivalent resolution and a fine bank having 6-bit equivalent resolution. Therefore, since the overall resolution of the DTC is the same as its input, $q^{\mathrm{acc}}_{\mathrm{MMDC}}\big|_d[k]$, there will be no resolution gap between the DTC's control code and its time output; this means that, after splitting by the coarse requantizer, the resolutions of the fine DCD and its control signal should be the same. Hence, the control to the fine DCD follows a one-to-one mapping and the segmented DTC model in this case can be specialized from Fig. 6.1b to Fig. 6.13, where there is no "fine requantizer". Note that, because there is no resolution gap between the DTC's control and output, no "DQI spurs", which have been described earlier in this chapter, will exist.

As shown in Fig. 6.13, the coarse and fine branches are both auto-calibrated with their individual LMS correlators [10]. The same value $\lambda = 2^{-10}$ is used for both branches, as in [30]. Note that the LMS does not change the resolution of the signal before or after calibration.

Mahmoud et al.'s Method [30]

After the coarse LMS calibration, the DTC proposed in [30] quantizes its input to generate a control code, $c_{\mathrm{c}}[k]$, for the coarse DCD; the QE of this quantization, $q_{\mathrm{c}}[k]$, is then calibrated by the fine LMS to become $c_{\mathrm{f}}[k]$ that controls the fine DCD. Therefore, the "coarse quantizer" block in Fig. 6.13 in this case is a normal *digital-to-digital quantizer* that transfers the 11-bit input signal to its 5-bit output, whose QE is associated with the LSBs of the input.

Assume that the time gains of the DCDs are correctly set to match the gain of the original QE injection path. Ideally, in the steady state, after the LMS algorithms settle, the scaling factors, $h_c[k]$ and $h_f[k]$, will converge to their optimum values, $\hat{h}_c = \hat{h}_f = 1$. In this case, $c_c[k]$ and $c_f[k]$ can be expressed as

$$c_c[k] = \mathbf{Q}_c \left(q_{\text{MMDC}}^{\text{acc}}\big|_d [k] \right), \tag{6.17a}$$

$$c_f[k] = \mathbf{Q}_f \left(q_c[k] \right), \tag{6.17b}$$

where $\mathbf{Q}_c(\cdot)$ represents the coarse quantization operator and $\mathbf{Q}_f(\cdot)$ in (6.17b) is exploited to emphasize that the code controlling the fine DCD is indeed a digital code with a resolution equal to one.

Tasca et al.'s Method [10]

Unlike the basic quantizer used at the DTC's input in [30], its counterpart used in the segmentation proposed in [10] is a first-order *noise-shaping* quantizer, i.e., EFM1, whose behavioral model is shown in Fig. 3.6. The embedded quantizer in the EFM1 is the same Δ_c-to-1 quantizer as that used in the previous case. What is different is that the QE of this internal quantizer, $s_c[k]$, is fed back to be added to the input, $q_{\text{MMDC}}^{\text{acc}}\big|_d [k]$; their sum is quantized by the internal quantizer to yield $c_c[k]$. As in Fig. 3.6, z^{-1} represents the inherent unit delay in the feedback; Δ_c is the resolution of the internal quantizer, which is added when comparing the quantizer's input and output, as the Δ_c-to-1 quantization introduces an inherent gain of $1/\Delta_c$.

Recall from Eq. (6.11b) that, in this case, the transfer relationship of the coarse-stage EFM1 is $s_c[k] = q_{\text{MMDC}}^{\text{acc}}\big|_d [k] + s_c[k-1] - \Delta_c \cdot c_c[k]$. The QE of the coarse stage that (i) cannot be corrected by the stage and (ii) must be carried to control the fine DCD is given by, recalling (6.12), $q_c[k] = s_c[k] - s_c[k-1]$, which, as expected, reflects the ability of the EFM1 to provide first-order noise shaping for the QE.

Note that:

- compared to the previous segmentation whose input to the coarse quantizer is $q_{\text{MMDC}}^{\text{acc}}\big|_d [k]$, the input to the coarse quantizer in this segmentation topology is modified by adding to it the QE of itself, namely $q_{\text{MMDC}}^{\text{acc}}\big|_d [k] + s_c[k-1]$; this implies that, in a practical design, the DR of the coarse branch must be extended by one quantization step on each side to accommodate the added QE. This requirement can be easily satisfied considering the aforementioned redundancy design for combating PVT variations;
- more importantly, considering (6.12), the signal controlling the fine DCD, $q_c[k]$, is the sum of two samples of the coarse quantizer's QE, whose values are mutually independent. This means that, if we represent the unshaped QE of the coarse quantizer as a random variable $s_c \in [-\Delta_c/2, \Delta_c/2]$—which was the input to the fine stage in the previous case—in this segmentation, the input to the fine bank becomes a random variable $q_c \in [-\Delta_c, \Delta_c]$. This indicates that, to ensure that the DTC cancellation functions correctly, the DR of the fine branch must be at

least doubled, i.e., a 7-bit $c_f[k]$ controlling 2^7 fine capacitors rather than 6 fine bits used in the previous case. Redundancy consideration will make the hardware consumption of the fine branch even greater.

In summary, similar to the analysis for Mahmoud's method, we obtain for this segmentation that

$$c_c[k] = \mathbf{Q}_c \left(q_{\mathrm{MMDC}}^{\mathrm{acc}}\big|_d [k] + s_c[k-1] \right), \tag{6.18a}$$

$$c_f[k] = \mathbf{Q}_f \left(s_c[k] - s_c[k-1] \right). \tag{6.18b}$$

Comparing (6.18) with (6.17) describing the segmentation in [30], in which case $q_c[k]$ in (6.17b) is equivalent to $s_c[k]$ in (6.18), we notice that, interestingly, the segmented DTC proposed in [10] effectively "moves" a small component of its input signal, namely $s_c[k-1] \in [-\Delta_c/2, \Delta_c/2]$, from the coarse stage to the fine stage of the DTC during every clock cycle. According to (6.18a), this causes the coarse bank to produce a time delay that deviates from the "without-the-moving case" in (6.17a) by at most a unit coarse delay. Nevertheless, the small component being moved onto the fine stage, if it is correctly designed as mentioned above, can compensate for the deviated time delay by the coarse stage exactly because the fine DCD has sufficient DR and resolution. Therefore, in theory, the two segmentation methods should have the same QE cancellation effect and result in the same time- and frequency-domain patterns, if they function successfully. However, as we will show in the next section, Tasca et al.'s segmented DTC [10] always works, while the other one [30] does not.

6.5.2 Comparative Analysis of Segmented DTCs

In this subsection, we perform behavioral simulations of the example type-II first-order BBPLL ; BBPLL is the DPLL architecture implemented in both [10, 30].

The modeled system operates in fractional-N mode with an LMS-calibrated segmented DTC; the LMS setting and DTC segmentation are introduced in Sect. 6.5.1.

Simulations start with the theoretical worst-case initial time difference $\Delta t[0] = T_{\mathrm{ref}}$, where $\Delta t[k] = t_{\mathrm{ref}}[k] - t_{\mathrm{div}}[k]$ and T_{ref} is the nominal reference period. The gain factors, $h_c[k]$ and $h_f[k]$, trained by the LMS, start at $h_c[0] = h_f[0] = 0$ and will converge to optimal values equal to $\hat{h}_c = \hat{h}_f = 1$, as introduced in Sect. 6.5.1, if the LMS filtering is successful.

The simulated results are shown in Fig. 6.14, where the first row displays 2^{22} simulated samples of the control codes for the coarse and fine capacitor banks, $c_c[k]$ and $c_f[k]$, together with $h_c[k]$ and $h_f[k]$; the second row shows one hundred samples of $\Delta t[k]$ in the locked state; the third row shows the PSD's of the output PNs in locked state.

We make the following observations regarding Fig. 6.14.

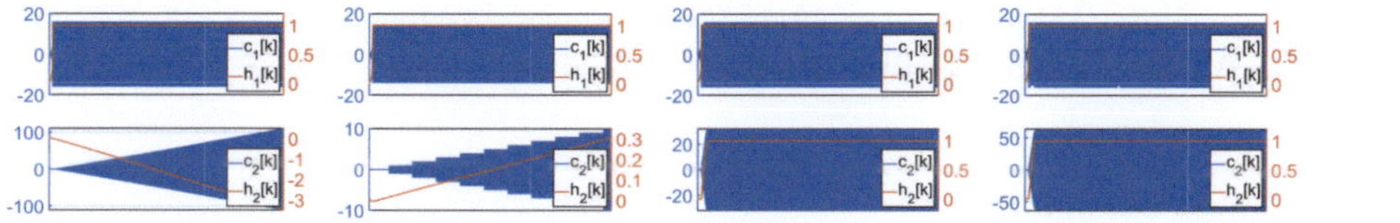

(a) Mahmoud *et al.*, X=15. (b) Mahmoud *et al.*, X=16. (c) Mahmoud *et al.*, X=22. (d) Tasca *et al.*, X=1.

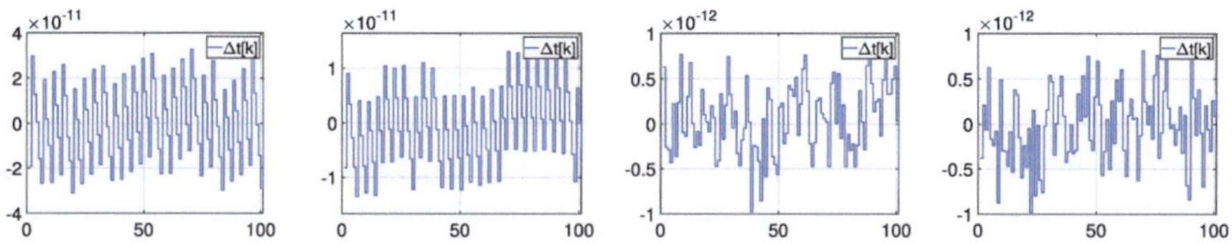

(e) Mahmoud *et al.*, X=15. (f) Mahmoud *et al.*, X=16. (g) Mahmoud *et al.*, X=22. (h) Tasca *et al.*, X=1.

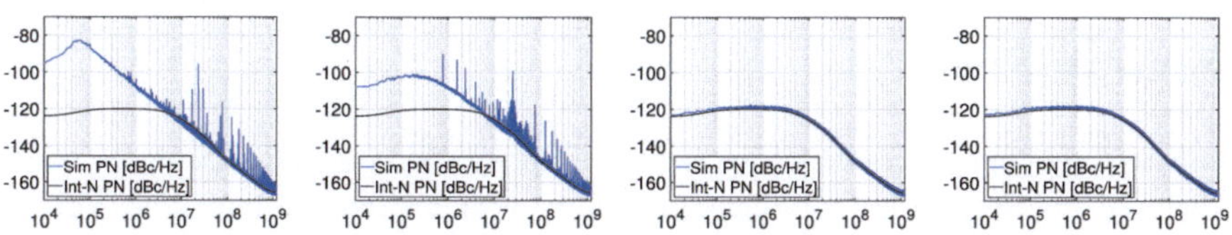

(i) Mahmoud *et al.*, X=15. (j) Mahmoud *et al.*, X=16. (k) Mahmoud *et al.*, X=22. (l) Tasca *et al.*, X=1.

Fig. 6.14 Behavioral simulation results of a fractional-N BBPLL with an LMS-calibrated DTC segmented as per Mahmoud et al. [30] or Tasca et al. [10]. First row: entire simulation with 2^{22} sampled reference cycles showing the convergence (or not) of the LMSs; second row: 100 simulated time differences at the input of the TDC in the locked state; third row: PSD of the output PN

From the First Row

For both DTC segmentation methods, we simulated using "very small" fractional FCWs, namely $X = 15, 16, 22$ for Mahmoud et al.'s topology and $X = 1$ for Tasca et al.'s topology. Simulations show that, for all the cases, the coarse LMS always converges to $\hat{h}_c = 1$ correctly and rapidly. However, in our simulations, when $X \leq 15$, the fine LMS used in the former diverges to negative values; when $X = 16$, the fine LMS starts to converge toward the correct direction, yet the speed of convergence is so slow that it has not settled after 2^{22} simulated reference cycles; when $X \geq 17$, the fine LMS converges with a higher speed. Here, we show the results with $X = 22$—the smallest feasible fractional FCW reported in [30]'s system—as a working example of Mahmoud et al.'s topology; this demonstrates a speed of convergence that is the same as the Tasca et al.'s case; the latter converges properly even when $X = 1$. Notice that, as discussed in Sect. 6.5.1, $c_c[k] \in [-16, +16]$ in both cases, hence it is a 5-bit code; whereas $c_f[k] \in [-32, +32]$ in [30]'s converged cases and $c_f[k] \in [-64, +64]$ in [10]'s case, hence they are 6-bit and 7-bit codes respectively.

From the Second Row

In the nonconverging cases, Fig. 6.14e, f show that their input time differences are dominated by some periodic pattern in the locked state. Their variances are larger than those of the converged cases, where Fig. 6.14e is even worse than Fig. 6.14f; this is due to the fact that the diverged fine LMS induces a large incorrect gain for the fine DTC branch, which introduces more noise power rather than mitigating the errors. Figure 6.14g, h show that, in the converged cases, the DTC helps to eliminate the MMDC's QE in the sense that their TDC-input time differences are dominated by the intrinsic random noises only; they match Table 5.2(ii) in that the measured standard deviation of the integer-N input jitter is 407 fs.

From the Third Row

It is obvious that the nonconverged cases fail to recover the integer-N PN performance predicted using the method introduced in Chap. 4; undesired spurious tones arise at various offset frequencies. On the other hand, the latter two cases (k) and (l), where the LMSs converged, show that the DTC helps to recover the integer-N PN performance in the fractional-N mode, as expected.

Finally, we conclude that the dysfunction of the system in [30], when the fractional FCW is chosen to be very small, is due to the fact that the LMS controlling the fine DTC branch fails to converge. This is because, as discussed in Sect. 6.5.1, the signal training the fine LMS is just the LSBs of the MMDC's AQE, which is not sufficiently exciting. On the other hand, the DTC segmented as per [10] does not suffer from this issue because the training signal for its fine LMS is the first-order high-pass-shaped version of the LSBs, and is able to stimulate the fine LMS effectively, even for very small fractions such as $X = 1$.

6.5.3 Conclusion

In this section, we compared two LMS-calibrated DTC segmentation methods presented in [10, 30]. We showed their equivalence in canceling the MMDC's QE in fractional-N operation. By analyzing the behavioral simulation results, we revealed the reason for the system failure when using small fractional FCWs in [30]; this is due to its DTC segmentation being unable to provide sufficiently excited signals to drive the LMS algorithm for the fine DTC branch to convergence. By contrast, the DTC segmentation proposed in [10] is demonstrated to work successfully even with extremely small FCWs.

6.6 Summary

1. Sect. 6.2 introduces the function, design, and modeling approach of the DTCs. A nonideal model of the DTC is presented, which takes into account the inherent hard quantization nonlinearity and the INL of the DCD. The distinct PN and spurious responses introduced by these are analyzed and distinguished, namely, the DTC's hard quantization induces DQI spurs and the DCD's INL induces excess PN, IBS spurs, and REM spurs; these are separable in the spectrum of the DTC's residual error.
2. Sect. 6.3 provides insights into the generation and transfer mechanisms of DQI spurs at source and at the output PN. The analytical expressions of the DTC's QE and the spectral signatures of the DQI spurs associated with various DTC configuration and control methods have been investigated and compared. The dependence of DQI spurs on the MMDC and DTC-related parameters was revealed. The output-referred DQI spurs were shown to be potentially more of a problem than the conventional IBS spurs in terms of output PN/jitter degradation.
3. Sect. 6.4 shows the distinct fractional spurs and PN induced by the DCD's REM and discusses the REM profiles associated with non-segmented and segmented DCDs.
4. Sect. 6.5 compares two state-of-the-art DTC segmentation and control methods and studies the LMS convergence issue of the DTC-based cancellation. It shows that, in a segmented DTC, when the coarse requantizer is a normal digital-to-digital quantizer and the fractional FCW is small, the LMS auto-calibration fails to converge, which in turn causes inexact cancellation of MMDC's QE; when the coarse requantizer is an EFM1, however, the LMS does converge, which results in the QE being canceled correctly.

References

1. B. Miller, R. Conley, A multiple modulator fractional divider. IEEE Trans. Instrum. Meas. **40**(3), 578–583 (1991)
2. S. Pamarti, L. Jansson, I. Galton, A wideband 2.4-GHz delta-sigma fractional-N PLL with 1-Mb/s in-loop modulation. IEEE J. Solid-State Circuits **39**(1), 49–62 (2004)
3. A. Swaminathan, K.J. Wang, I. Galton, A wide-bandwidth 2.4 GHz ISM band fractional-*N* PLL with adaptive phase noise cancellation. IEEE J. Solid-State Circuits **42**(12), 2639–2650 (2007)
4. C.-M. Hsu, M.Z. Straayer, M.H. Perrott, A low-noise, wide-BW 3.6GHz digital $\Delta\Sigma$ fractional-N frequency synthesizer with a noise-shaping time-to-digital converter and quantization noise cancellation, in *2008 IEEE International Solid-State Circuits Conference - Digest of Technical Papers* (2008), pp. 340–617
5. M. Zanuso, S. Levantino, C. Samori, A.L. Lacaita, A wideband 3.6 GHz digital $\Delta\Sigma$ fractional-N PLL with phase interpolation divider and digital spur cancellation. IEEE J. Solid-State Circuits **46**(3), 627–638 (2011)

6. T. Seong, Y. Lee, S. Yoo, J. Choi, A 320-fs RMS jitter and - 75-dBc reference-spur ring-DCO-based digital PLL using an optimal-threshold TDC. IEEE J. Solid-State Circuits **54**(9), 2501–2512 (2019)
7. L. Avallone, M.P. Kennedy, S. Karman, C. Samori, S. Levantino, Jitter minimization in digital PLLs with mid-rise TDCs. IEEE Trans. Circuits Syst. I Regul. Pap. **67**(3), 743–752 (2020)
8. C. Hwang, H. Park, T. Seong, J. Choi, A 188fs_{rms}-jitter and -243dB-FoM_{jitter} 5.2GHz-ring-DCO-based fractional-N digital PLL with a 1/8 DTC-range-reduction technique using a quadruple-timing-margin phase selector, in *2022 IEEE International Solid- State Circuits Conference (ISSCC)*, vol. 65 (2022), pp. 378–380
9. N.D. Dalt, Linearized analysis of a digital bang-bang PLL and its validity limits applied to jitter transfer and jitter generation. IEEE Trans. Circuits Syst. I Regul. Pap. **55**(11), 3663–3675 (2008)
10. D. Tasca, M. Zanuso, G. Marzin, S. Levantino, C. Samori, A.L. Lacaita, A 2.9–4.0-GHz fractional-N digital PLL with bang-bang phase detector and 560-fs_{rms} integrated jitter at 4.5-mW power. IEEE J. Solid-State Circuits **46**(12), 2745–2758 (2011)
11. S. Levantino, G. Marzin, C. Samori, A.L. Lacaita, A wideband fractional-N PLL with suppressed charge-pump noise and automatic loop filter calibration. IEEE J. Solid-State Circuits **48**(10), 2419–2429 (2013)
12. S. Levantino, C. Samori, Nonlinearity cancellation in digital PLLs (invited paper), in *Proceedings of the IEEE 2013 Custom Integrated Circuits Conference* (2013), pp. 1–8
13. S. Levantino, G. Marzin, C. Samori, An adaptive pre-distortion technique to mitigate the DTC nonlinearity in digital PLLs. IEEE J. Solid-State Circuits **49**(8), 1762–1772 (2014)
14. A. Santiccioli, C. Samori, A. Lacaita, S. Levantino, Power-jitter trade-off analysis in digital-to-time converters. Electron. Lett. **53**(5), 306–308 (2017). https://ietresearch.onlinelibrary.wiley.com/doi/abs/10.1049/el.2016.4577
15. W. Wu, C.-W. Yao, K. Godbole, R. Ni, P.-Y. Chiang, Y. Han, Y. Zuo, A. Verma, I.S.-C. Lu, S.W. Son, T.B. Cho, A 28-nm 75-fs_{rms} analog fractional-*N* sampling PLL with a highly linear DTC incorporating background DTC gain calibration and reference clock duty cycle correction. IEEE J. Solid-State Circuits **54**(5), 1254–1265 (2019)
16. S.M. Dartizio, F. Tesolin, G. Castoro, F. Buccoleri, L. Lanzoni, M. Rossoni, D. Cherniak, L. Bertulessi, C. Samori, A. L. Lacaita, S. Levantino, 4.3 a 76.7fs-Integrated-jitter and -71.9dBc in-band fractional-spur bang-bang digital PLL based on an inverse-constant-slope DTC and FCW subtractive dithering, in *2023 IEEE International Solid-State Circuits Conference (ISSCC)* (2023), pp. 3–5
17. D. Banerjee, *PLL Performance, Simulation and Design* (Dog Ear Publishing, Indianapolis, IN, USA, 2006)
18. Y. Donnelly, M.P. Kennedy, Prediction of phase noise and spurs in a nonlinear fractional-*N* frequency synthesizer. IEEE Trans. Circuits Syst. I Regul. Pap. **66**(11), 4108–4121 (2019)
19. V. Mazzaro, M.P. Kennedy, Folded noise prediction in nonlinear fractional-N frequency synthesizers. IEEE Trans. Circuits Syst. I Regul. Pap. **68**(10), 4038–4048 (2021)
20. T.-K. Kao, C.-F. Liang, H.-H. Chiu, M. Ashburn, A wideband fractional-N ring PLL with fractional-spur suppression using spectrally shaped segmentation, in *2013 IEEE International Solid-State Circuits Conference Digest of Technical Papers* (2013), pp. 416–417
21. W. Wu, C.-W. Yao, C. Guo, P.-Y. Chiang, L. Chen, P.-K. Lau, Z. Bai, S.W. Son, T.B. Cho, A 14-nm ultra-low jitter fractional-N PLL using a DTC range reduction technique and a reconfigurable dual-core VCO. IEEE J. Solid-State Circuits **56**(12), 3756–3767 (2021)
22. J. Jiang, T. Yan, D. Zhou, A.I. Karsilayan, J. Silva-Martinez, A 2.3-3.9 GHz fractional-N frequency synthesizer with charge pump and TDC calibration for reduced reference and fractional spurs, in *2021 IEEE Radio Frequency Integrated Circuits Symposium (RFIC)* (2021), pp. 71–74
23. P.T. Renukaswamy, K. Vaesen, N. Markulic, V. Derudder, D.-W. Park, P. Wambacq, J. Craninckx, 4.1 a 16GHz, 41kHz_{rms} frequency error, background-calibrated, duty-cycled FMCW charge-pump PLL, in *2023 IEEE International Solid-State Circuits Conference (ISSCC)* (2023), pp. 74–76

24. C. Samori, M. Zanuso, S. Levantino, A.L. Lacaita, Multipath adaptive cancellation of divider non-linearity in fractional-N PLLs, in *IEEE International Symposium of Circuits and Systems (ISCAS)*, vol. 2011 (2011), pp. 418–421
25. S. Levantino, L. Grimaldi, C. Samori, Analysis of adaptive pre-distortion in DTC-based digital fractional-N PLLs, in *2016 12th Conference on Ph.D. Research in Microelectronics and Electronics (PRIME)* (2016), pp. 1–4
26. S. Levantino, Calibration techniques in phased-locked loops: theoretical basis and practical applications. IEEE Solid-State Circuits Mag. **16**(3), 37–44 (2024)
27. H. Park, C. Hwang, T. Seong, Y. Lee, J. Choi, 32.1 a 365fs_{rms}-jitter and -63dBc-fractional spur 5.3GHz-ring-DCO-based fractional-N DPLL using a DTC second/third- order nonlinearity cancelation and a probability-density-shaping $\Delta\Sigma$M, in *2021 IEEE International Solid- State Circuits Conference (ISSCC)*, vol. 64 (2021), pp. 442–444
28. Analog Devices, *Microwave Wideband Synthesizer with Integrated VCO*, ADF4372 (2019). https://www.analog.com/media/en/technical-documentation/data-sheets/adf4372.pdf
29. Texas Instruments, *20-GHz wideband RF synthesizer with phase synchronization & JESD204B support*, LMX2595 (2019). https://www.ti.com/lit/gpn/lmx2595
30. A. Mahmoud, P. Andreani, F. Pepe, A 2.8-3.8-GHz low-spur DTC-based DPLL with a class-D DCO in 65-nm CMOS. IEEE Microwave Wirel. Compon. Lett. **27**(11), 1010–1012 (2017)
31. A. Santiccioli, M. Mercandelli, L. Bertulessi, A. Parisi, D. Cherniak, A.L. Lacaita, C. Samori, S. Levantino, A 66-fs-rms jitter 12.8-to-15.2-GHz fractional-N bang-bang PLL with digital frequency-error recovery for fast locking. IEEE J. Solid-State Circuits **55**(12), 3349–3361 (2020)
32. C. Samori, L. Bertulessi, Digital PLLs: the modern timing reference for radar and communication systems, in *ESSCIRC 2021 - IEEE 47th European Solid State Circuits Conference (ESSCIRC)* (2021), pp. 21–27
33. C. Hwang, H. Park, Y. Lee, T. Seong, J. Choi, A low-jitter and low-fractional-spur ring-DCO-based fractional-N digital PLL using a DTC's second-/third-order nonlinearity cancellation and a probability-density-shaping $\Delta\Sigma$M. IEEE J. Solid-State Circuits **57**(9), 2841–2855 (2022)
34. A. Hu, Z. Jin, X. Shan, M. Zhang, D. Liu, X. Zou, Fractional spurs reduction technique using probability density shaping sigma-delta modulator and fractional frequency divider. IEEE Trans. Circuits Syst. II Express Briefs **70**(2), 456–460 (2023)
35. J.Z. Ru, C. Palattella, P. Geraedts, E. Klumperink, B. Nauta, A high-linearity digital-to-time converter technique: Constant-slope charging. IEEE J. Solid-State Circuits **50**(6), 1412–1423 (2015)
36. Y. Liu, H. Gao, H. Xu, P. Lu, N. Yan, A 10-bit 563-fs step constant-slope digital-to-time converter in 40-nm CMOS with nonlinearity cancellation and range extension techniques. IEEE Trans. Circuits Syst. I Regul. Pap. **71**(2), 526–536 (2024)
37. K. Hosseini, B. Fitzgibbon, M.P. Kennedy, Observations concerning the generation of spurious tones in digital delta-sigma modulators followed by a memoryless nonlinearity. IEEE Trans. Circuits Syst. II Express Briefs **58**(11), 714–718 (2011)
38. M.P. Kennedy, V. Mazzaro, D. Mai, Nonlinearity-induced spurs in fractional-N frequency synthesizers. IEEE Trans. Circuits Syst. II Express Briefs **69**(6), 2617–2622 (2022)
39. H. Liu, D. Tang, Z. Sun, W. Deng, H.C. Ngo, K. Okada, A sub-mW fractional-N ADPLL with FOM of -246 dB for IoT applications. IEEE J. Solid-State Circuits **53**(12), 3540–3552 (2018)
40. P. Chen, F. Zhang, Z. Zong, S. Hu, T. Siriburanon, R.B. Staszewski, A 31-μW, 148-fs step, 9-bit capacitor-DAC-based constant-slope digital-to-time converter in 28-nm CMOS. IEEE J. Solid-State Circuits **54**(11), 3075–3085 (2019)
41. M. Rossoni, S.M. Dartizio, F. Tesolin, G. Castoro, R. Dell'Orto, C. Samori, A.L. Lacaita, S. Levantino, 10.1 an 8.75GHz fractional-N digital PLL with a reverse-concavity variable-slope DTC achieving 57.3fs_{rms} integrated jitter and -252.4dB FoM, in *2024 IEEE International Solid-State Circuits Conference (ISSCC)*, vol. 67 (2024), pp. 188–190
42. D. Xu, Z. Liu, Y. Kuai, H. Huang, Y. Zhang, Z. Sun, B. Liu, W. Wang, Y. Xiong, J. Qiu, W. Madany, Y. Zhang, A.A. Fadila, A. Shirane, K. Okada, 10.3 a 7GHz digital PLL with cascaded fractional divider and pseudo-differential DTC achieving -62.1dBc fractional spur and

143.7fs integrated jitter, in *2024 IEEE International Solid-State Circuits Conference (ISSCC)*, vol. 67 (2024), pp. 192–194
43. Y. Hu, W. Tao, R.B. Staszewski, Nonlinearity-induced spur analysis in fractional-N synthesizers with $\Delta\Sigma$ quantization cancellation. IEEE Open J. Solid-State Circuits Soc. 1–1 (2024)
44. K. Hosseini, M.P. Kennedy, Maximum sequence length MASH digital Delta-Sigma modulators. IEEE Trans. Circuits Syst. I Regul. Pap. **54**(12), 2628–2638 (2007)
45. V. Mazzaro, M.P. Kennedy, Spur immunity in MASH-based fractional-N CP-PLLs with polynomial nonlinearities. IEEE Trans. Circuits Syst. I Regul. Pap. **68**(6), 2295–2306 (2021)
46. M.P. Kennedy, X. Lu, X. Wang, Insights into architectural spurs in high-performance fractional-N frequency synthesizers. IEEE Open J. Solid-State Circuits Soc. **4**, 238–251 (2024)
47. X. Lu, M.P. Kennedy, Reconciliation of statistical approaches to predicting nonlinearity-induced spurs in fractional-N frequency synthesizers. IEEE Trans. Circuits Syst. I: Regular Pap. 1–14 (2024)
48. Y. Donnelly, M.P. Kennedy, J. Breslin, S. Tulisi, S. Patil, C. Curtin, S. Brookes, B. Shelly, P. Griffin, M., Keaveney, 4.48-GHz fractional- N frequency synthesizer with spurious-tone suppression via probability mass redistribution. IEEE Solid-State Circuits Lett. **2**(11), 264–267 (2019)
49. T. Seong, Y. Lee, C. Hwang, J. Lee, H. Park, K.J. Lee, J. Choi, A -58dBc-worst-fractional-spur and -234dB-FoM jitter, 5.5GHz ring-DCO-based fractional-N DPLL using a time-invariant-probability modulator, generating a nonlinearity-robust DTC-control word, in *2020 IEEE International Solid- State Circuits Conference - (ISSCC)* (2020), pp. 270–272
50. V. Mazzaro, M.P. Kennedy, A family of $\Delta\Sigma$ modulators with high spur immunity and low folded nonlinearity noise when used in fractional-N frequency synthesizers. IEEE Trans. Circuits Syst. I Regul. Pap. **69**(11), 4499–4509 (2022)
51. M.H. Perrott, M.D. Trott, C.G. Sodini, A modeling approach for Σ-Δ fractional-N frequency synthesizers allowing straightforward noise analysis. IEEE J. Solid-State Circuits **37**(8), 1028–1038 (2002)
52. M. Zanuso, D. Tasca, S. Levantino, A. Donadel, C. Samori, A.L. Lacaita, Noise analysis and minimization in bang-bang digital PLLs. IEEE Trans. Circuits Syst. II Express Briefs **56**(11), 835–839 (2009)
53. M.D. McDonnell, Is electrical noise useful? [point of view]. Proc. IEEE **99**(2), 242–246 (2011)
54. G. Marucci, S. Levantino, P. Maffezzoni, C. Samori, Exploiting stochastic resonance to enhance the performance of digital bang-bang PLLs. IEEE Trans. Circuits Syst. II Express Briefs **60**(10), 632–636 (2013)
55. H. Liu, D. Tang, Z. Sun, W. Deng, H.C. Ngo, K. Okada, A. Matsuzawa, A 0.98mW fractional-N ADPLL using 10b isolated constant-slope DTC with FOM of -246dB for IoT applications in 65nm CMOS, in *2018 IEEE International Solid-State Circuits Conference - (ISSCC)* (2018), pp. 246–248
56. A. Swaminathan, A. Panigada, E. Masry, I. Galton, A digital requantizer with shaped requantization noise that remains well behaved after nonlinear distortion. IEEE Trans. Signal Process. **55**(11), 5382–5394 (2007)
57. E. Familier, C. Venerus, I. Galton, A class of quantizers with DC-free quantization noise and optimal immunity to nonlinearity-induced spurious tones. IEEE Trans. Signal Process. **61**(17), 4270–4283 (2013)
58. E. Familier, I. Galton, Second and third-order noise shaping digital quantizers for low phase noise and nonlinearity-induced spurious tones in fractional-N PLLs. IEEE Trans. Circuits Syst. I Regul. Pap. **63**(6), 836–847 (2016)
59. Z. Zong, P. Chen, R.B. Staszewski, A low-noise fractional-N digital frequency synthesizer with implicit frequency tripling for mm-wave applications. IEEE J. Solid-State Circuits **54**(3), 755–767 (2019)
60. M. P. Kennedy, V. Mazzaro, S. Tulisi, M. Scully, N. McDermott, J. Breslin, 10.4 a 45.5fs-integrated-random-jitter and -75dBc-integer-boundary-spur BiCMOS fractional-N PLL with suppression of fractional, horn, and wandering spurs, in *2024 IEEE International Solid-State Circuits Conference (ISSCC)*, vol. 67 (2024), pp. 194–196

61. S. Pamarti, I. Galton, LSB dithering in MASH Delta-Sigma D/A converters. IEEE Trans. Circuits Syst. I Regul. Pap. **54**(4), 779–790 (2007)
62. M. Rossoni, S.M. Dartizio, F. Tesolin, G. Castoro, R. Dell'Orto, A.L. Lacaita, S. Levantino, A low-jitter fractional-*N* digital PLL adopting a reverse-concavity variable-slope DTC. IEEE J. Solid-State Circuits **60**(6), 2122–2133 (2025)
63. H. Park, C. Hwang, T. Seong, J. Choi, A low-jitter ring-DCO-based fractional-N digital PLL with a 1/8 DTC-range-reduction technique using a quadruple-timing-margin phase selector. IEEE J. Solid-State Circuits **57**(12), 3527–3537 (2022)
64. E. Helal, A.I. Eissa, I. Galton, DTC linearization via mismatch-noise cancellation for digital fractional-N PLLs. IEEE Trans. Circuits Syst. I Regul. Pap. **69**(12), 4993–5006 (2022)

Chapter 7
DTC Enhancement Strategies I: DTC Requantization

7.1 Introduction

The nonlinear behaviors of practical DTCs have been studied and modeled in the previous chapter. In particular, the DTC exhibits three major types of systematic nonlinearity, which cause inexact cancellation and thereby induce excess output PN and spurs, namely

1. the *hard quantization nonlinearity* due to the DTC's limited resolution, resulting in DQI spurs;
2. the *soft integral nonlinearity* associated with the DCD that causes excess PN and IBS [1–3]; and
3. the *random element mismatch nonlinearity* of the DCD due to PVT variations, causing REM spurs.

As analyzed in Chap. 6, DTC-block and DPLL-system-level analyses show that the spectral signatures of the PN and spurs induced by the DTC's nonlinearities are distinguishable in the spectrum; the overall PN and spurious response resembles a linear combination of these.

The separable spectral signatures also suggest the combined use of different classes of DTC enhancement techniques which target different nonlinear sources to mitigate their PN and spurs in order to improve the DTC's overall jitter and spur performance.

In this chapter, we address the DTC's inherent requantization behavior and describe the design methods and enhancement techniques related to it. The enhancement strategies related to the DCD's polynomial and REM INLs will be addressed in the next chapter.

This chapter first presents a novel class of DTC performance enhancement techniques called IDQ methods; these IDQ methods are shown to be able to eliminate DQI spurs completely, with a negligible PN penalty. Consequently, the output

© The Author(s), under exclusive license to Springer Nature Switzerland AG 2026

X. Wang and M. P. Kennedy, *Jitter and Spur Minimization in Fractional-N Digital Frequency Synthesizers*, Analog Circuits and Signal Processing,
https://doi.org/10.1007/978-3-032-27186-0_7

spectral purity will no longer suffer from DQI spurs, and the effect on the system jitter of the dithering noise introduced by IDQ methods themselves will be minimized.

Furthermore, because the DTC has finite resolution and therefore introduces additional quantization noise, quantization resolution becomes a key design parameter that needs to be derived carefully; it determines the residual power of the DTC's QE that is injected into the loop after the DTC-based cancellation. This chapter proposes an enhanced DTC parameterization strategy that considers the noise introduced by different IDQ methods. In theory, DTCs designed following the proposed method help to recover the integer-N PN while using minimum hardware in the DTC.

This chapter is arranged as follows. In Sect. 7.2, the IDQ principle and various implementation methods are elaborated, and the effectiveness in eliminating DQI spurs will be verified. Section 7.3 proposes the method to size the DTC using the minimum amount of hardware in the DTC while ensuring that the DTC's QE is spectrally masked by the intrinsic integer-N PN.

7.2 Input-Dithered Quantization (IDQ) Methods to Mitigate DQI Spurs

In this section, we discuss the working principles of IDQ methods. The methods can be used in place of the EFM1 requantizer by applying a dither with minimum power at the normal requantizer's input in order to realize a dithered DTC quantization and thereby eliminate the DQI spurs at source.

7.2.1 *IDQ Principle and Optimum IDQ Method*

The IDQ methods described in this work were inspired by the SR operation regime in digital PLLs that have a few-bit mid-rise TDC.

Operating in this regime, the intrinsic physical noises, such as the DCO and reference jitters, transferred to the *input of the TDC* randomize the limit-cycle quantization of the TDC and hence eliminate the spurs caused by it [4, 5]. In this case, the input-referred random noise can be regarded as a source of random dither, and the dithered-quantization linearization of a TDC essentially transforms the TDC's hard quantization nonlinearity into an effective linearity; this effect has been studied in Chap. 3. We extend this analysis framework to derive the optimal IDQ method.

The input-dithered DTC requantizer is shown in Fig. 7.1, where a "dither source" is used to generate a dithering signal, $d[k]$, in the digital domain; this is added to the original digital input code applied to the DTC, $a[k]$; the sum is then quantized by the normal DTC requantizer to control the DCD. Note that, in segmented DTCs (Fig. 6.1b), because the LSBs of $a[k]$ are embedded in the QE of the coarse requantizer, i.e., $q_c[k]$, and are passed to and processed by the "fine requantizer," the fine

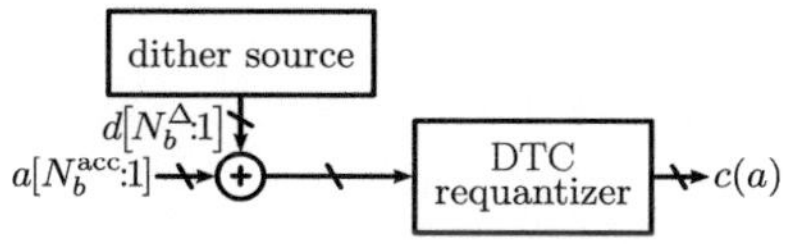

Fig. 7.1 Input-dithered quantization in a IDQ-DTC. Square-bracketed numbers indicate the bits of the digital signals

requantization induces DQI spurs. In such cases, the dither should be applied to the input of the *fine* requantizer.

To maximize the generality of this analysis, we consider an arbitrary deterministic common offset, Θ, applied to the quantization levels of the normal Δ-to-1 DTC requantizer; the *transfer characteristic of the undithered quantization* is

$$c(a) = \alpha, \quad \forall a \in [\alpha \cdot \Delta + \Theta, (\alpha + 1) \cdot \Delta + \Theta - 1], \tag{7.1}$$

where $\alpha \in [0, \pm 1, \pm 2, \ldots, \pm 2^{N_b^{\mathrm{DTC}}-1}]$ represents the quantized output for a specific input, a.[1] Note that this undithered quantizer has an evident quantization nonlinearity.

In terms of the input dither, we start by considering the most fundamental digital dither that can be produced by an N_b^{∇}-bit random number generator (RNG), which has unity resolution and a discrete uniform distribution in $D \sim U[\Phi, \Phi + \nabla - 1]$; similarly to (7.1), we apply an arbitrary deterministic offset, Φ, to the dither.

Our goal here is (i) to derive the values of ∇ that can sufficiently achieve the desired linearization, and (ii) to identify the optimum value of ∇ to guarantee the dithering effect while using the minimum dithering power.

The DTC requantizer is input-dithered by D. Due to the random dither, the output yielded by the dithered quantization is shifted to another unknown level, β, as

$$c(a + d) = \beta, \; \forall (a + d) \in [\beta \cdot \Delta + \Theta, (\beta + 1) \cdot \Delta + \Theta - 1],$$

which represents the *transfer characteristic of the dithered quantization*.

Since the output of the random dithered quantizer is stochastic, it is meaningful to study its effective (or expected) transfer function seen from its input, a, that is

$$c_{\mathbb{E}}(a) = \mathbb{E}\left(c(a + d)\right), \tag{7.2}$$

where $\mathbb{E}(\cdot)$ represents the expectation operator. The *effective transfer characteristic of the IDQ-dithered requantizer* is hence

$$c_{\mathbb{E}}(a) = \sum_{x=\Phi}^{\Phi+\nabla-1} c(a + x) \cdot p_D(x), \tag{7.3}$$

[1] With $\Theta = 0$, Eq. (7.1) describes a truncating quantizer; $\Theta = -\Delta/2$ makes it a rounding quantizer, etc.

where

$$p_D(x) = \begin{cases} \frac{1}{\nabla} & \forall x \in \{\Phi, \Phi+1, \ldots, \Phi+\nabla-1\}, \\ 0 & \text{otherwise}, \end{cases} \tag{7.4}$$

is the probability mass function (PMF) of the dither.

Let us assume that the span of the dithered input $(a+d)$, whose width equals ∇, accommodates n quantization thresholds:$(\beta+m)\cdot\Delta+\Theta, \forall\ m \in \{m \in \mathbb{N}^+ \,|m \le n\}$. Recall that β is a representative unknown integer. Equation (7.3) can be calculated as in

$$c_{\mathbb{E}}(a) = \beta \cdot \sum_{x=\Phi}^{(\beta+1)\cdot\Delta+\Theta-a-1} p_D(x)$$
$$+ (\beta+1)\cdot \sum_{x=(\beta+1)\cdot\Delta+\Theta-a}^{(\beta+2)\cdot\Delta+\Theta-a-1} p_D(x) + \cdots + (\beta+n)\cdot \sum_{x=(\beta+n)\cdot\Delta+\Theta-a}^{\Phi+\nabla-1} p_D(x),$$

substituting (7.4) into which yields

$$c_{\mathbb{E}}(a) = n\cdot\frac{a+\Phi-\Theta}{\nabla} + n\cdot\left(1-\frac{\Delta}{\nabla}\cdot\frac{1+n}{2}\right) + \left(1-\frac{\Delta}{\nabla}\cdot n\right)\cdot\beta. \tag{7.5}$$

Note that not all values of ∇ can result in a linear transfer relationship from a to $c_{\mathbb{E}}(a)$. There is a term in (7.5) related to the unknown integer, β, which, recalling from the above discussion, is unpredictable and depends on the instantaneous values of both the input, a, and random dither, d. Therefore, the effective transfer relationship can be corrupted and fail to become a linear function of a. Residual spurs still exist when using input dithers with arbitrary ranges, ∇.

In order to nullify the uncertain term in (7.5), the condition, $\nabla = n\cdot\Delta$, must be satisfied; this means that the span of the uniform input dither must be an integer multiple of the quantization resolution. If this constraint is satisfied, we get

$$c_{\mathbb{E}}^{n\Delta}(a) = \frac{a+\Phi-\Theta}{\Delta} + \frac{(n-1)}{2}, \tag{7.6}$$

which is a linear function of a.

The optimum (effective and with minimum dithering power) IDQ dither can be achieved when $n = 1$. Equation (7.6) hence becomes

$$c_{\mathbb{E}}^{\Delta}(a) = \frac{a+\Phi-\Theta}{\Delta}, \tag{7.7}$$

which is an affine function of a. This is a key result as it suggests that, no matter what type of uniform quantization, e.g., rounding, truncating, and ceiling, is adopted as the DTC's requantizer and no matter how the dither is biased, as long as the span of the N_b^{Δ}-bit uniform dither matches the quantization resolution ($\Delta = 2^{N_b^{\Delta}}$), the effective transfer function of the dithered quantizer becomes a linear function of its input, a, which in turn eliminates DQI spurs stemming from the quantization nonlinearity. See how the IDQ method transforms the quantization nonlinearity in (7.1) into the effective linearity in (7.7). The coefficient of the linear term, $1/\Delta$, represents the inherent gain of the Δ-to-1 quantizer, as described in Chap. 6.

Listing 7.1 shows the MATLAB behavioral modeling to implement the derived optimum IDQ dither. For a DTC requantizer that is a normal Δ-to-1 truncator, the optimum IDQ dither can be a random integer number following a uniform distribution that has a range of Δ, which is added to the requantizer's input. This can be modeled by replacing Line 4 in Listing 6.3, which describes the original DTC, with Listing 7.1.

Listing 7.1 MATLAB script to enable optimum IDQ dithering.

```
%% Optimum IDQ dither
d = randi([1,res_DTC],1,length(in));
%% Dithered DTC quantization
c = floor((in+d)/res_DTC);
```

We present the following example to illustrate the effects of dither-based linearization using different dithering schemes. In Table 7.1, a in the red row represents the sequence of natural numbers as the input to be quantized. A rounding quantizer, $c(a)$, quantizes a specific a to its nearest multiple of three, i.e., 0, 3, 6, 9, ...; this exhibits a quantization nonlinearity with respect to the progression of a.

Following the derived optimum IDQ method in (7.7), a uniformly distributed dither $d \in \{-1, 0, 1\}$ is used to dither the quantization. The green row, evaluated as per (7.2), shows the expected results of the IDQ quantizer, which restores linearity.

Table 7.1 Example input-to-expected-output relationships using various dithering methods. Red: input; green: optimum IDQ output; blue: other linearized outputs; white: nonlinear outputs

Case	Feature	Value														
a	Input	0	1	2	3	4	5	6	7	8	9	10	11	12	13	14
$c(a)$	No dither	0	0	3	3	3	6	6	6	9	9	9	12	12	12	15
$c_{\mathbb{E}}^{\Delta}(a)$	Optimum IDQ dither	0	1	2	3	4	5	6	7	8	9	10	11	12	13	14
$c_{\mathbb{E}}^{<\Delta}(a)$	Less-power dither	0	3/2	3	3	9/2	6	6	15/2	9	9	21/2	12	12	27/2	15
$c_{\mathbb{E}}^{d,\text{out}}(a)$	Output dither	0	0	3	3	3	6	6	6	9	9	9	12	12	12	15
$c_{\mathbb{E}}^{>\Delta}(a)$	More-power dither	3/4	6/4	9/4	15/4	18/4	21/4	27/4	30/4	33/4	39/4	42/4	45/4	51/4	54/4	57/4
$c_{\mathbb{E}}^{n\Delta}(a)$	$n\Delta$-uniform dither	3/6	9/6	15/6	21/6	27/6	33/6	39/6	45/6	51/6	57/6	63/6	69/6	75/6	81/6	87/6
$c_{\mathbb{E}}^{\text{tri}}(a)$	$\sigma_{D,\min}^2$-triangular dither	0	3/4	9/4	12/4	15/4	21/4	24/4	27/4	33/4	36/4	39/4	45/4	48/4	51/4	57/4
$c_{\mathbb{E}}^{\text{TRI}}(a)$	$2\sigma_{D,\min}^2$-triangular dither	0	1	2	3	4	5	6	7	8	9	10	11	12	13	14

7.2.2 *Suboptimal Dithering Schemes*

To confirm the optimality of the proposed optimum IDQ method, we provide proofs by contradiction in this subsection to show that other dithering schemes are comparatively suboptimal in the sense that they can (i) fail to linearize the transfer characteristic fully and/or (ii) require more dithering power and hence introduce a greater PN penalty.

Dithering with Less Power

We first consider a uniform random dither, whose span is less than the quantization resolution, i.e., $\nabla < \Delta$. Substituting $n = 1$ into (7.5) yields $c_{\mathbb{E}}^{<\Delta}(a) = \frac{a+\Phi-\Theta}{\nabla} + \left(1 - \frac{\Delta}{\nabla}\right) \cdot (1 + \beta)$, where there exists a term related to β that produces uncertainty to the dithered quantization characteristic that depends on a. Hence, dithering with less power than the optimum IDQ cannot fully linearize the quantizer; residual spurs still exist in this case. This case is illustrated in the $c_{\mathbb{E}}^{<\Delta}(a)$ row in Table 7.1, where a unity-resolution input dither $D^{<\Delta} \sim U[0, 1]$ is used rather than $D \sim U[-1, 1]$ in the optimum IDQ case. Clearly, the suboptimum-power input dither fails to linearize the transfer characteristic.

Output Dither

We can check the dithering effect if a random dither is instead added to the *output* of the DTC requantizer, that is $c(a) + d$. Recalling (7.1), it is trivial to work out in this case the *effective transfer characteristic of the output-dithered requantizer* as $c_{\mathbb{E}}^{d,\text{out}}(a) = \mathbb{E}\left(c(a) + d\right) = \alpha + \mathbb{E}\left(d\right), \forall\, a \in [\alpha \cdot \Delta + \Theta, (\alpha + 1) \cdot \Delta + \Theta - 1]$, which is equivalent to merely adding the expected (mean) value of the dither, which is a known constant that is independent of a, to the quantized output of the requantizer. The effective transfer relationship still has an apparent quantization nonlinearity and will consequently produce DQI spurs. Therefore, dithering at the output of the DTC requantizer is ineffective in DQI-spur cancellation. Recall that the SR operation, i.e., the time-domain dither at the output of the DCD, also fails to mitigate the DQI spurs. The $c_{\mathbb{E}}^{d,\text{out}}(a)$ row shows that the quantization nonlinearity in this case is preserved exactly.

Dithering with More Power

As discussed in Sect. 7.2.1, Eq. (7.5) implies that an input dither with an arbitrary power greater than the optimum IDQ can corrupt the linearized operation of the dithered quantizer as well. The $c_{\mathbb{E}}^{>\Delta}(a)$ row in Table 7.1 uses input dither $D^{>\Delta} \sim$

$U[-1,2]$ (wider than the optimum $D \sim U[-1,1]$) to demonstrate this effect. It shows that dithering with more power does not necessarily linearize the transfer function.

The blue row of the $c_{\mathbb{E}}^{n\Delta}(a)$ case in Table 7.1 uses the input dither $D^{n\Delta} \sim U[-2,3]$ which has exactly twice the span as the optimum IDQ case. This recovers the desired linearity for the effective transfer characteristic.

Despite the fact that using $\nabla = n \cdot \Delta$ for the input dither can linearize the DTC requantizer, the cost of doing so is that the unity-resolution uniform dither with a wider span introduces more dithering power, namely $\sigma_{D^{n\Delta}}^2 = \frac{\nabla^2-1}{12}$, which is about n^2 times greater than the minimum dithering power $\sigma_{D,\min}^2 = \frac{\Delta^2-1}{12}$ of the optimum IDQ. Because more (noise-like) dithering power injected into the loop contributes more system jitter, the "$\nabla = n \cdot \Delta$" dither is undesired. The optimum IDQ method derived in this work is still superior.

Central Limited Dithers

In the above discussions, the input dither is assumed to have a uniform distribution. Here, we also check the case when the dither's distribution is nonuniform. As an alternative to uniform dither, other commonly used dithers typically have central-limit-shaped distributions, e.g., triangular and Gaussian. To assess the linearizing effect of such dithers, we use triangularly distributed dithers as the examples and perform the same analysis as in Table 7.1.

For a fair comparison, we apply a discrete triangular dither that has the same power as the optimum IDQ dither ($D \sim U[-1,1]$) for the example quantizer, where $\sigma_{D,\min}^2 = 2/3$. We call this the "$\sigma_{D,\min}^2$-triangular dither" and represent its random variable as D^{tri}, whose PMF is

$$p_{D^{\text{tri}}}(x) = \begin{cases} \frac{1}{4} & \text{for } x = -\frac{2\sqrt{3}}{3}, \\ \frac{1}{2} & \text{for } x = 0, \\ \frac{1}{4} & \text{for } x = \frac{2\sqrt{3}}{3}, \\ 0 & \text{otherwise.} \end{cases}$$

The $c_{\mathbb{E}}^{\text{tri}}(a)$ row in Table 7.1 shows that the triangularly distributed dither that has the same power as the uniformly distributed optimum IDQ dither cannot fully linearize the transfer characteristic. This can be expected intuitively because central-limit-distributed dithers have more chances to produce values that are concentrated near the middle of the span. Thus, compared to the uniformly-dithered case, they are less effective in pushing (randomizing) the dithered signal to other quantization levels.

We notice that the minimum power required for using a triangular dither to achieve the full linearization is twice the $\sigma_{D,\min}^2$; we call this the "$2\sigma_{D,\min}^2$-triangular dither." In this case, the PMF of the random variable, D^{TRI}, becomes

$$p_{D^{\mathrm{TRI}}}(x) = \begin{cases} \frac{1}{9} & \text{for } x = -2, \\ \frac{2}{9} & \text{for } x = -1, \\ \frac{1}{3} & \text{for } x = 0, \\ \frac{2}{9} & \text{for } x = 1, \\ \frac{1}{9} & \text{for } x = 2, \\ 0 & \text{otherwise,} \end{cases}$$

which can be generated by adding two independent and identical optimum IDQ dithers ($D \sim U[-1, 1]$) together. The blue row of the $c_{\mathbb{E}}^{\mathrm{TRI}}(a)$ case in Table 7.1 confirms its linearized effective transfer function.

Notice that the $2\sigma_{D,\min}^2$-triangular dither in our analysis corresponds to the discretized form of the "2-LSB triangular nonsubtractive dither" used in audio signal processing [6–8]. In that field, it is considered the best type of dither, given that the dither signal cannot be removed after the quantizer in a "subtractive" manner, that whitens the total quantization error (equivalent to the "DTC's QE" in our analysis) and makes its power spectrum independent of the input. On the other hand, if the dither could be subtracted after the quantizer, the best solution would be the "1-LSB rectangular subtractive dither" [7]. The discretized version of the dither *per se* corresponds to the optimum IDQ dither derived in this work.

However, due to the resolution limit imposed by the DTC's DCD, the dither in our case cannot be practically subtracted at the output of the DTC's requantizer or DCD. Hence, it is inherently nonsubtractive. Yet, our analysis in (7.7) shows that the optimum IDQ dither (or the discretized "1-LSB rectangular nonsubtractive dither") is sufficient in (i) linearizing the effective transfer characteristic of the DTC requantizer, (ii) eliminating DQI spurs, and (iii) whitening the DTC's residual error, as will be shown in Sect. 7.2.3. Therefore, it is the optimum dither to be used here. Using the above studied $2\sigma_{D,\min}^2$-triangular dither, D^{TRI}, (or the discretized "2-LSB triangular nonsubtractive dither") will introduce twice as much jitter penalty as the optimum IDQ, which, in essence, is redundant and undesirable.

Simulations show that another type of central limited dither with a canonical Gaussian distribution requires even more power than the $2\sigma_{D,\min}^2$-triangular dither to achieve the same effect in terms of DQI-spur mitigation. Note also that to generate a random digital signal that has a specified Gaussian distribution entails much more digital complexity compared to generating a uniformly distributed dither. Therefore, Gaussian-distributed dither is inferior to the optimum uniformly distributed IDQ dither in terms of DQI-spur mitigation.

7.2.3 Verification Results

To verify the sufficiency of the DQI-spur immunity provided by the optimum IDQ, we performed the same block-level analysis in the left column of Fig. 6.6. The simulated power spectrum of the DTC's residual error in Fig. 7.2a shows that, regardless of the

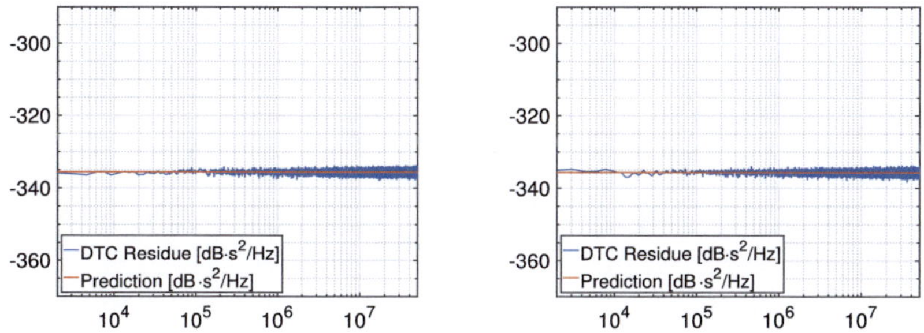

(a) Dithering with an N_b^Δ-bit RNG.

(b) Dithering with a sufficiently random signal from within the system.

Fig. 7.2 Simulated PSDs of the *IDQ-DTC*'s residual error, $t_\epsilon[k]$ in Fig. 6.3. These match their counterparts predicted by (7.8)

values of the FCW, the IDQ-enhanced DTC's residual error is sufficiently whitened by the minimum-power dither. This contrasts with the purely tonal spectra without IDQ (caused only by the DTC's inherent quantization) as shown in Fig. 6.6. The purely white and spur-free block-level spectra confirm the elimination of spurs.

Note that the uniformly distributed optimum IDQ dither (i) has the same power, i.e., $\sigma_{D,\min}^2 = \frac{\Delta^2-1}{12}$, as the original QE of the DTC, and (ii) is independent of the latter. Therefore, recalling that the inherent gain of the DTC requantizer is $1/\Delta$, the power (variance) of IDQ-dithered DTC's QE referred to the DTC quantizer's output, $q_{\text{DTC}}^d[k]$, equals twice the output-referred dithering power, namely $\sigma_{Q^d}^2 = 1/6$.

From Fig. 6.3, the PSD of the whitened IDQ-DTC's residual error in s^2/Hz can be predicted as

$$S_{\Delta t_\epsilon} = \tau_{\text{DCD}}^2 \cdot \frac{\sigma_{Q^d}^2}{F_{\text{ref}}}, \tag{7.8}$$

recalling that τ_{DCD} is the unit time delay provided by the DCD; in this simulation, we set $\tau_{\text{DCD}} = \frac{\Delta}{M} \cdot T_{\text{out}}$. The power of the QE, $\sigma_{Q^d}^2$, is uniformly spread across its noise bandwidth, F_{ref}. The PSD predicted by (7.8) is plotted as the horizontal orange lines in Fig. 7.2; it matches the simulated power spectra exactly.

Notice that a separate RNG is not necessarily required in order to achieve the desired IDQ-based linearization for the DTC requantizer. One can utilize a sufficiently random signal from within the PLL system, which is uncorrelated with the DTC's QE, to function as the dither, in order to save hardware complexity and power consumption. An example of such RNG-free dithering schemes is to use the N_b^Δ most significant bits of the N_b^{MMDC}-bit MASH MMDC's final-stage sum as the IDQ dither, $D \sim U[0, \Delta - 1]$. When seen from the perspective of the DTC's QE, this is sufficiently random and uncorrelated with itself. Together with a normal DTC requantizer, this eliminates DQI spurs. The simulated block-level residual error's spectrum in Fig. 7.2b confirms the effectiveness of this strategy.

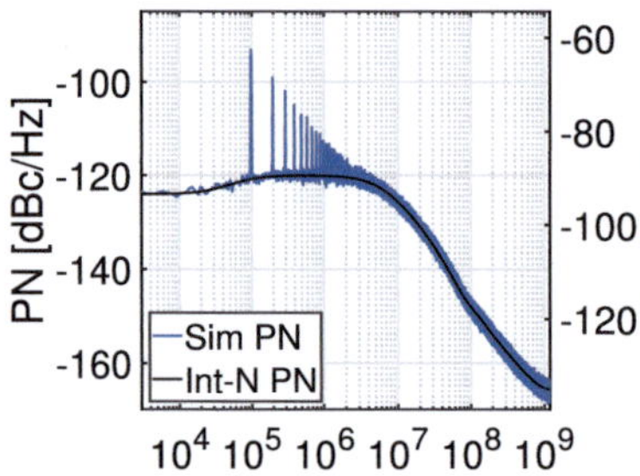

(a) Undithered DTC.

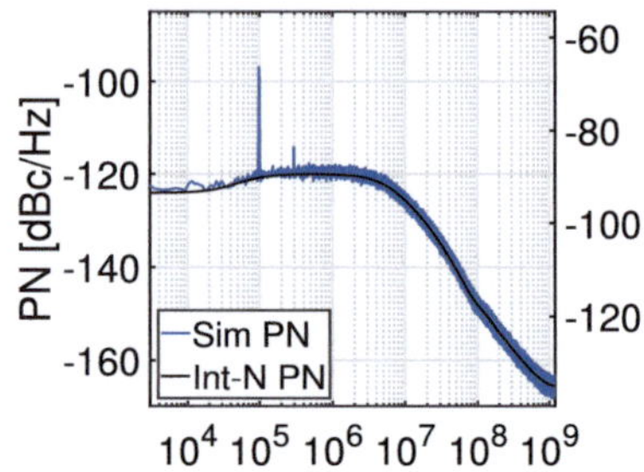

(b) N_b^{Δ} – 1-bit IDQ-DTC.

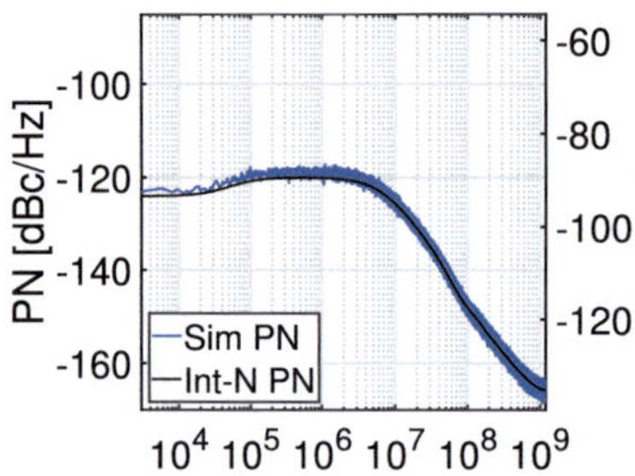

(c) Optimum N_b^{Δ}-bit IDQ-DTC.

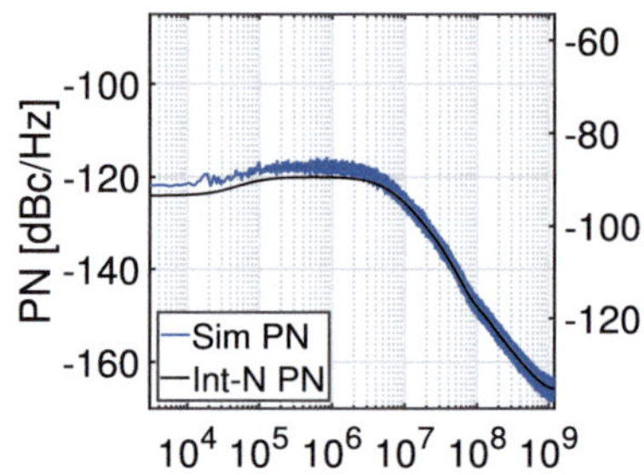

(d) N_b^{Δ} + 1-bit IDQ-DTC.

Fig. 7.3 Simulated power spectra of output PNs of BBPLLs parameterized as per Table 4.1. A *MASH 1–1* MMDC and a *linear quantized DTC with* $\Delta = 2^{10}$ are used to enable fractional-N mode. Different dithered (or undithered) DTCs are assessed

To verify at the PLL-system level the sufficiency and optimality of the derived IDQ methods in eliminating the DQI spurs, we performed behavioral simulations of the example BB PLL parameterized as per Table 3.3, while varying the power of uniformly distributed IDQ dithers used for the finite-resolution DTC. Like the case analyzed in the first two rows of Fig. 6.6, MASH 1–1 MMDCs are used here together with DTCs that have 11-bit equivalent resolution and perform truncating quantization. The DCD is assumed to have good linearity so that we can focus on the effect of DQI mitigation. $X = 1$ is chosen as it exhibits the most evident DQI-spur pattern when the DTC is undithered.

The simulated output PNs shown in Fig. 7.3 illustrate the following cases.

When the DTC Is Not Dithered

Figure 7.3a (the same as Fig. 6.6b), shows that the block-level DQI-spur pattern in Fig. 6.6a is transferred to the BBPLL's output . The worst-case DQI spurs with a magnitude of about –60 dBc degrade the spectral purity of the synthesized signal.

When the DTC Is Input-Dithered with Fewer Bits Than the Optimum IDQ

Here, a uniform input dither that uses one less bit than the optimum IDQ is adopted. Hence, the dither has $N_b^{\Delta} - 1$ bits and a span with width equal to $\Delta/2$. Figure 7.3b shows that using less dithering power helps to mitigate the DQI spurs, especially for the higher harmonics in the DQI-spur pattern. However, it is insufficient to eliminate the DQI spurs completely. The magnitude of the fundamental DQI spur at $\frac{X}{\Delta} \cdot F_{\text{ref}}$ is still significant.

When the DTC Is Dithered Using the Optimum IDQ Method

Figure 7.3c shows that, since the optimum N_b^{Δ}-bit IDQ method eliminates the DQI spurs at source, the PLL output is completely free of DQI spurs. The optimum IDQ utilizes the minimum dithering power in the sense that, while eliminating DQI spurs, the IDQ-DTC-based fractional-N BBPLL recovers its integer-N PN performance almost exactly.

When the DTC Is Dithered with More Bits Than the Optimum IDQ

Figure 7.3d uses one more bit for the input dither than the optimum case, i.e., $N_b^{\Delta} + 1$-bit uniform dither with a span equal to $2 \cdot \Delta$. As analyzed in Sect. 7.2.2, this dither is suboptimal compared to case 3). Although it gets rid of DQI spurs, the excess dithering power contributes more system jitter and PN at close-in frequencies; these are undesired for high-performance frequency synthesizers.

As a final remark, we record the simulated output jitters using the optimum IDQ method (Fig. 7.3c case) in Table 7.2 as a supplement to Table 6.2. Comparing these cases with cases (e) and (f) in Table 6.2, we observe that using an optimum-IDQ-requantizer yields jitter performances that are better than using EFM1 requantizer as a means to mitigate DQI spurs, regardless of the linearity of the DCD. Table 7.2 shows enhanced jitter performances that are the closest among all the simulated cases to the integer-N value (case (a) in Table 6.2). Together, the sufficiency and optimality of the optimum IDQ method are validated.

Finally, as we have analyzed in Chap. 6 the mechanism of DQI-spur generation is independent of the generation of the conventional IBS spurs due to the distortion

Table 7.2 Simulated RMS output jitters with an optimum-IDQ-DTC

Case	Mode	Requantizer	DCD	RMS jitter
(g)	Fractional-N	Optimum IDQ	Linear	272 fs
(h)		Optimum IDQ	Nonlinear	303 fs

of the MMDC's AQE by the INL of the DCD. No matter how well the INL is alleviated (recall that the DCD in Fig. 6.6 is assumed to have perfect linearity), and no matter how well the high-order and/or stochastic MMDC is designed (recall Chap. 6); DQI spurs remain intact without IDQ methods or EFM1 requantization. The IDQ methods proposed in this section function independently and guarantee DQI-spur elimination with/without DTC linearity enhancement and stochastic design of the MMDC. A comprehensive DTC enhancement strategy incorporating IDQ and other classes of techniques that, in principle, can completely cancel various architecture-dependent spurs and thereby enable fractional-spur-free integer-N PN performance in fractional-N frequency synthesizers will be presented in Sect. 8.2.

7.2.4 Conclusion

In this section, based on the analysis in Chap. 6 that conventional DTC linearity enhancement and stochastic design of the MMDC alone cannot mitigate DQI spurs, we presented IDQ methods to linearize the effective transfer characteristic of the DTC requantizer in order to eliminate DQI spurs resulting from the DTC's quantization nonlinearity. Compared to other methods including EFM1 requantization, the N_b^{Δ}-bit uniform IDQ technique that dithers the input to the DTC requantizer is shown to be the optimum solution for DQI-spur mitigation because of its effectiveness in both eliminating DQI spurs and introducing the minimum excess noise. Finally, behavioral simulations at the DTC-block and PLL-system levels confirm the sufficiency and optimality of the derived optimum IDQ method.

7.3 Method to Determine Quantization-Related Parameters of DTC

In PLLs embedding a DTC that has the same resolution as the AQE to be canceled [9, 10], the DTC performs a one-to-one mapping from its control code to its time delay and hence produces no further QE. In practice, a fine-resolution MMDC is required for high-resolution frequency synthesis, which in turn results in a fine-resolution code representing its AQE; the DTC, which is limited by its finite resolution, inevitably introduces additional quantization noise. When investigated as a standalone data converter [11], the resolution and number of bits of the DTC can be determined independently of its external connectivity. In the context of DTC-based cancellation in fractional-N PLLs, the DTC's quantization resolution is indeed a key design parameter that needs to be carefully derived because it determines the residual power of the DTC's QE that is injected into the loop after cancellation.

The DTC in [12] is designed to limit the variation of the DTC's QE to be within $\pm\sigma/2$ of the intrinsic integer-N noise transferred to the TDC's input in order to recover the integer-N noise performance; it also aims to utilize the intrinsic noise to scramble the periodic pattern associated with the DTC's residual error. This parameterization method, which is inspired by the concept of stochastic resonance [4, 5, 13], is widely used in the literature.

In this section, we first demonstrate a limitation of the classical method proposed in [12] to determine the required resolution of the DTC. In particular, we extend the analysis that was presented in previous discussions. As highlighted in those papers, the input-referred intrinsic noise alone is inadequate in randomizing the spurious pattern of the DTC's QE; DQI spurs can still be transferred and affect the spectral purity of the synthesized signal.

Secondly, we describe strategies, which can be adopted for generic (analog and digital) frequency synthesizers, to determine the dynamic range, number of bits, quantization resolution, and unity delay of the DTC that (i) use minimum hardware, (ii) approach the integer-N jitter/PN performance in fractional-N operation, and (iii) incorporate the IDQ method proposed in Sect. 7.2 to achieve an output spectrum that is free of DQI spurs.

The section is arranged as follows. In Sect. 7.3.1, we first review the classic DTC sizing method proposed in [12] and analyze its limitations. Section 7.3.2 extends the integer-N DPLL jitter analysis performed in Chap. 4 to the DTC-enhanced fractional case; this takes into account the system jitter contributed by the DTC's residual error. Section 7.3.3 analyzes the DTC's design criterion and presents an enhanced DTC parameterization strategy that uses minimum DTC hardware to recover the integer-N PN/jitter performance. This strategy also considers the use of IDQ methods in order to eliminate DQI spurs; it results in an output PN that is free of DTC-quantization-related spurs. Finally, conclusions are drawn in Sect. 7.3.4.

7.3.1 Review of the Classical DTC Parameterization Method [12]

The now "classical" DTC parameterization strategy proposed in [12] aims to cancel the MMDC's QE with a finite-resolution DTC. The method is motivated by noticing that, when switching from the integer-N to fractional-N mode, the time jitter in the locked state presented at the TDC's input that is contributed by the modulation operation of the MMDC has a magnitude equal to the MMDC's AQE, $q^{\mathrm{acc}}_{\mathrm{MMDC}}[k]$ in Fig. 1.5, multiplied by the output oscillation period of the DCO, denoted T_{out}.

On the other hand, the input time jitter in the integer-N steady state is intrinsically dominated by the reference and DCO jitters and the TDC's QE, which can be approximated by a Gaussian distribution as [4, 14–16] that has a standard deviation, $\sigma_{\Delta t_{\mathrm{int}}}$. Quantitatively, in radio-frequency synthesizers, the value of $\sigma_{\Delta t_{\mathrm{int}}}$ is typically about one thousand times smaller than that of T_{out} whereas $q^{\mathrm{acc}}_{\mathrm{MMDC}}[k]$ has a variation range equal to 2^{l-1}, where l is the order of the DDSM.

Thus, assuming that the DTC's DCD has sufficiently good linearity—this can be achieved by adopting various DTC linearity enhancement techniques, e.g., adaptive pre-distortion [9, 17, 18], CSC DTC [10, 11, 19, 20], DTC range reduction [21, 22], DTC second/third-order nonlinearity cancellation [23, 24], inverse-constant-slope DTC [25], and others [26, 27]—intuitively, the method parameterizes the DTC to have an effective DR to cover $2^{l-1} \cdot T_{\text{out}}$ and to have equivalent resolution equal to $(9 + l)$ bits, i.e., 10-bit (or 11-bit, 12-bit) DTC for first-order (or second-, third-order) divider controllers.[2] Therefore, the equivalent resolution of the DTC is chosen to be less than the RMS integer-N input jitter, namely

$$\Delta = 2^{l-1} \cdot T_{\text{out}}/2^{9+l} < \sigma_{\Delta t_{\text{int}}}.$$

The entire variation range of the DTC's QE—the residual error after the DTC-based cancellation—is bounded to $(-\Delta/2, \Delta/2)$, which is within the range of $(-\sigma_{\Delta t_{\text{int}}}/2, \sigma_{\Delta t_{\text{int}}}/2)$. If this is achieved, the DTC is claimed to be able to do two things:

1. it reduces the power of the residual error after the DTC to below that of the intrinsic integer-N input jitter, thereby recovering the integer-N PN/jitter performance in the fractional-N operation; and
2. because the residual error is now within the random-noise regime, the latter can destroy any periodic pattern associated with the former, thanks to the stochastic-resonance effect [4, 5, 13]. Therefore, the resulting output spectrum should be free of any DTC-related spurs. This claim is made in [12] and is repeated without question by much of the succeeding literature, e.g., [9, 28].

To verify these claims, we perform behavioral simulations for the example type-II first-order BBPLL with the parameters tabulated in Table 3.3, which describes a jitter-optimized system design. Unless otherwise stated, these parameters are used everywhere in this section.

Five working scenarios are simulated to produce the results shown in Fig. 7.4.

The first row shows the integer-N results, where the divide ratio of the divider, $N_{\text{int}} = 24$, is constant, and hence the divider controller and DTC are not in use. The second row shows the fractional-N mode operation (facilitated by a MASH 1–1 MMDC as in [12]) without DTC-based cancellation. The other three rows show MASH 1–1-based fractional-N operations with a DTC that is modeled with different degrees of nonideality. Namely, the third row shows the case using an "ideal" DTC that is modeled as a linear gain factor that mirrors the original MMDC's QE injection mechanism in order to cancel it exactly. The DTC of the fourth row has

[2] In reported DTC-based fractional-N DPLL implementations, the DTC has a specified resolution; meanwhile, its DR typically covers twice the theoretical range of variation of the MMDC's AQE for redundancy to accommodate PVT variations. In this work, we derive the optimal DTC's resolution and number of bits by considering its "effective input dynamic range" which equals the theoretical variation range of $q^{\text{acc}}_{\text{MMDC}}[k]$. Redundancy considerations do not change the derived resolution but can be included by increasing the number of bits used in the DTC to expand the overall DR.

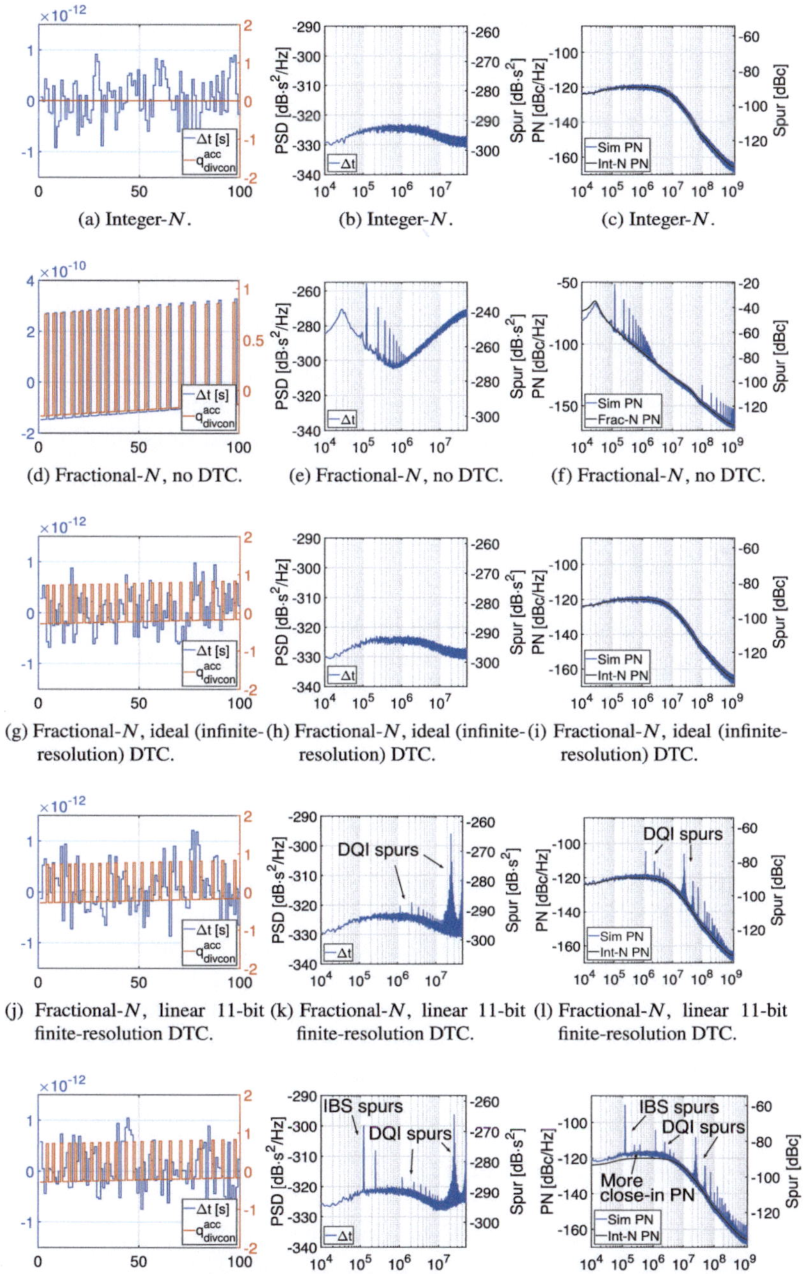

(a) Integer-N. (b) Integer-N. (c) Integer-N.

(d) Fractional-N, no DTC. (e) Fractional-N, no DTC. (f) Fractional-N, no DTC.

(g) Fractional-N, ideal (infinite-resolution) DTC. (h) Fractional-N, ideal (infinite-resolution) DTC. (i) Fractional-N, ideal (infinite-resolution) DTC.

(j) Fractional-N, linear 11-bit finite-resolution DTC. (k) Fractional-N, linear 11-bit finite-resolution DTC. (l) Fractional-N, linear 11-bit finite-resolution DTC.

(m) Fractional-N, nonlinear 11-bit finite-resolution DTC. (n) Fractional-N, nonlinear 11-bit finite-resolution DTC. (o) Fractional-N, nonlinear 11-bit finite-resolution DTC.

Fig. 7.4 Simulated performances in the locked state of the example BBPLL. Here, a MASH 1–1 divider controller with $X = 1277$ is used in fractional-N operation. First column: 100 samples of the input time difference ($\Delta t[k]$), and MMDC's AQE ($q^{acc}_{MMDC}[k]$); second column: power spectrum of $\Delta t[k]$; third column: power spectrum of the output PN finite resolution—11-bit in the MASH 1–1 case introduced above—hence it performs quantization but is otherwise linear. The fifth row's DTC has the same finite resolution as the fourth but its DCD also exhibits a polynomial nonlinearity modeled in Chap. 6 to mimic a more realistic INL of a DCD [21, 23, 24].

Integer-N Mode

In this case, the locked-state jitter/PN performances are dominated by the intrinsic noises only. The simulated RMS input time jitter, $\sigma_{\Delta t_{\mathrm{int}}}$, is 407 fs, as reported in Table 5.2(ii). The power spectra of the input time difference, $\Delta t[k]$, in Fig. 7.4b and the output PN in Fig. 7.4c are smooth, i.e., spur-free; the former can be predicted as described in [29] and the latter is predicted by the black curves in the third row using the method in Chap. 4. The match between theory and simulation in Fig. 7.4c is exact.

Fractional-N Mode Without a DTC

Figure 7.4d matches our prediction in Eq. (4.38) for the waveform of the input time difference as a worst-case performance limit for a fractional-N BBPLL when a DTC is not used to cancel the MMDC's QE. It indicates in this case that the pattern of the TDC's input time difference, $\Delta t[k]$, is no longer like random noise, such as in Fig. 7.4a. Instead, it is dominated by the waveform of the unit-delayed MMDC's AQE, $q_{\mathrm{MMDC}}^{\mathrm{acc}}[k-1]$, scaled by the output period. The magnitude of $\Delta t[k]$ in Fig. 7.4d becomes almost one thousand times greater than its integer-N counterpart, as shown in Fig. 7.4a.

The MMDC's QE overwhelms the input jitter of the DPLL, which in turn drastically reduces the TDC's effective gain and shrinks the loop bandwidth. Therefore, the PN performance is much degraded. The black curve in Fig. 7.4f predicts the degraded output PN as determined by the method provided in Chap. 4. Spurious tones arise at various frequencies. This phenomenon has been analyzed in detail in that chapter, which shows that cancellation of the MMDC's QE, such as DTC-based cancellation, is necessary in fractional-N DPLLs that use mid-rise TDCs.

Fractional-N Mode with an Ideal DTC

The third row shows that an ideal DTC with infinite resolution can in principle eliminate the MMDC's QE entirely before its injection into the TDC. In the time domain, the signal $\Delta t[k]$ becomes purely random and has no correlation with the pattern of $q_{\mathrm{MMDC}}^{\mathrm{acc}}[k]$; the former's variance equals that of the integer-N case. The power spectra shown in Fig. 7.4h and i match their integer-N counterparts, as expected.

Fractional-*N* Mode with a Quantized DTC

When considering the DTC's nonidealities modeled in the fourth and fifth rows, we observe that the real DTC can nearly achieve integer-*N* performance to the extent that, in the time domain, the remaining $\Delta t[k]$ varies randomly and is uncorrelated with $q_{\text{MMDC}}^{\text{acc}}[k]$; its variance is similar to that of the integer-*N* case. In the frequency domain, the floors of the power spectra are close to their integer-*N* counterparts. However, some unwanted features can be observed in the spectra shown in Fig. 7.4k, l, n, and o. When taking the DTC's quantization behavior into account, the undithered DTC induces a distinct family of spurs, which is manifest in the spectra of both the input time difference and the output PN (Fig. 7.4k and l). These are referred to as DQI spurs; their root cause and spectral signatures, as well as a mitigation strategy, were discussed in Chap. 6 and Sect. 7.2.

Fractional-*N* Mode with a Quantized and Nonlinear DTC

When considering also a polynomial nonlinearity associated with the DTC, the latter introduces excess folded noise that increases the close-in output PN floor [3] as well as conventional IBS at the fundamental and harmonics of $\frac{X}{M} \cdot F_{\text{ref}} = 121.784$ kHz [1, 2], where X/M is the fractional part of the nominal divide ratio. Notice that the spectral signatures introduced by the hard quantization and soft integral nonlinearities in Fig. 7.4n and o are mutually independent; this phenomenon was analyzed in Chap. 6.

It is interesting to note that, even when the DTC's DCD is assumed to be perfectly linear and its gain is assumed to match the original QE injection exactly—this was the premise for DTC parameterization in [12] and is the modeled scenario of the fourth row of Fig. 7.4—the output PN suffers from DQI spurs and fails to recover the spur-free integer-*N* performance; in Fig. 7.4l, we have merely considered the inevitable finite-resolution nature of the DTC. This contradicts the aforementioned claim made in [9, 12, 28] that the spectrum will be spur-free due to "stochastic resonance." Furthermore, the DQI spurs at both close-in and out-of-band frequencies in Fig. 7.4o can be regarded as a more challenging spurious response in addition to IBS. The former cannot be mitigated by using conventional DTC linearity enhancement techniques that aim to linearize the DCD's polynomial INL because such techniques do not change the DTC's quantization behavior at all. IDQ methods which can eliminate the DQI spurs have been presented in Sect. 7.2.

In the following subsections, we will propose an enhanced DTC parameterization method that also takes into account the effect of closed-loop filtering; this changes the power of the DTC's residual error when transferring it to the TDC's input, a feature that has been overlooked in the literature. The IDQ-DTC parameterized using this strategy, together with some standard DTC linearity enhancement, can in practice achieve the integer-*N* PN/jitter performance in the fractional-*N* mode; in addition, it is also free of all the aforementioned spurs.

7.3.2 Jitter Analysis for DTC-Enhanced DPLL

In this subsection, we first perform jitter analysis of fractional-N frequency synthesizers in the context of a DPLL. Its generalization to analog PLLs will also be commented on in Sect. 7.3.3. Comprehensive analysis, prediction, and minimization of the jitter and PN of integer-N DPLLs that take into account the most significant integer-N noise sources, i.e., the reference and DCO jitters and TDC's QE, have been presented in Chap. 4.

To extend the jitter analysis technique in Chap. 4 to analyze the system jitter of DTC-enhanced fractional-N type-II DPLLs, we start with the difference equation (4.35), which describes the "timestamps"[3] of the kth rising edges of the divided signal, $\text{div}_0(t)$[4] in Fig. 1.5, when a DTC is not used to cancel the MMDC's QE. The expression is

$$t_{\text{div}_0}[k+1] = t_{\text{div}_0}[k] + \sum_{h=N_{\text{div}}[k]\cdot k+1}^{N_{\text{div}}[k]\cdot(k+1)} \left\{ \underbrace{K_T\left(\beta \cdot e[k] + \psi[k]\right)}_{\text{control via DLF}} + \underbrace{T_{\text{DCO}} + t_{n,\text{DCO}}[h]}_{\text{DCO period with jitter}} \right\}, \tag{7.9}$$

where k and h index the signals operating at the reference and output clock rates.[5] Recall that $N_{\text{div}}[k]$ is the instantaneous divide ratio in the kth cycle, $e[k]$ is output from the TDC, $\psi[k] = \psi[k-1] + \alpha \cdot e[k]$ is the result of the DLF's integral path, $T_{\text{DCO}} = 1/F_{\text{DCO}}$ is the free-running DCO period, and $t_{n,\text{DCO}}[h]$ is the random-walk DCO period jitter associated with the hth cycle. As the DPLL's stability requires that $\alpha \ll \beta$, the integral path of the DLF, $\psi[k]$ in (7.9), can be neglected without sacrificing the accuracy of the jitter transfer analysis [16, 29]; this has been verified and reported in Chap. 4.

The instantaneous divide ratio is equivalent to the nominal divider ratio added to the cycle-wise MMDC's QE, $N_{\text{div}}[k] \triangleq N + q_{\text{MMDC}}[k]$. Equation (7.9) can therefore be rearranged as

$$\begin{aligned} t_{\text{div}_0}[k+1] = t_{\text{div}_0}[k] &+ \beta K_T N_{\text{div}}[k] \cdot e[k] + N \cdot T_{\text{DCO}} \\ &+ q_{\text{MMDC}}[k] \cdot T_{\text{DCO}} + \sum_{h=N_{\text{div}}[k]\cdot k+1}^{N_{\text{div}}[k]\cdot(k+1)} t_{n,\text{DCO}}[h], \end{aligned} \tag{7.10}$$

where $q_{\text{MMDC}}[k] \cdot T_{\text{DCO}}$ represents the contribution by the MMDC's QE, whose magnitude is so significant that it can drastically deteriorate the DPLL's jitter/PN performances. To compensate for this, a DTC is used to impose in the feedback path a corrective time delay that ideally equals

[3] The timestamping notation was explained in detail in Chap. 2.

[4] The subscript "0" indicates the "original" waveform of the signal where DTC-based cancellation is not used.

[5] The multi-rate modeling approach was explained in detail in Chaps. 2 and 4.

$$t_{\text{DTC}}[k+1] = t_{\text{DTC}}[k] - T_{\text{DCO}} \cdot q_{\text{MMDC}}[k] \tag{7.11a}$$
$$= -T_{\text{DCO}} \cdot q_{\text{MMDC}}^{\text{acc}}[k], \tag{7.11b}$$

where the DTC is initialized as $t_{\text{DTC}}[0] = 0$ and

$$q_{\text{MMDC}}^{\text{acc}}[k] = \sum_{n=0}^{k} q_{\text{MMDC}}[n], \tag{7.12}$$

which represents the MMDC's AQE. Equation (7.11b) indicates that the time delay imposed by the DTC at the kth cycle should be controlled by the accumulated version of the MMDC's QE, as shown in Fig. 1.5. Equation (7.11a) added to (7.10) cancels $q_{\text{MMDC}}[k] \cdot T_{\text{DCO}}$ exactly.[6]

However, the performance of a practical DTC is limited by its own nonidealities. On the one hand, in commercial frequency synthesizers achieving high-resolution synthesis, e.g., [30, 31], $q_{\text{MMDC}}^{\text{acc}}[k]$ has very fine, up to 35-bit, equivalent resolution. On the other hand, practical DTCs comprise a finite-resolution DCD with, e.g., 10 to 12 equivalent bits, to reduce the hardware consumption and limit the complexity. This fine-to-coarse mapping from the DTC's control code to the DCD's time delay cannot be achieved without a requantization that we consider to be performed within the DTC. The DTC's behavioral model considering the "DTC requantizer" to control the coarser-resolution "DCD" is shown in Fig. 6.1. The DTC requantizer realizes resolution reduction at the cost of inevitably introducing an extra source of QE in the loop, which we denote as the "DTC's QE," $q_{\text{DTC}}[k]$.

To make a fair comparison with the DTC parameterization method in [12] and to derive guidelines for choosing key parameters, e.g., resolution and number of bits, for the DTC, we assume in this section that the DCD already has good linearity; this can be achieved by e.g., the DTC linearity enhancement techniques mentioned in Sect. 7.3.1. The DTC thermal noise-induced PN [25, 26] has little effect on our derivation, as the method uses no more DTC bits than the classical method, which implies less DTC thermal noise [32]. In recently reported DTC-assisted synthesizers, such noise is already kept to a minimum in order to have negligible impact on the total PN of the system [33]. Hence, we only aim to suppress the DTC's QE effectively when deriving the DTC quantization-related parameters.

It is crucial to note that the classical definitions for the MMDC's QE and its accumulation in (7.11a) and (7.12) implicitly assume that they are fractional quantities with a resolution equal to $1/M$. However, in practical PLL implementations, both $q_{\text{MMDC}}[k]$ and $q_{\text{MMDC}}^{\text{acc}}[k]$ operate in the digital domain and are in fact digital codes with unity resolution. We denote by $q_{\text{MMDC}}|_d[k]$ and $q_{\text{MMDC}}^{\text{acc}}|_d[k]$, as shown in Fig. 6.1, the digital codes of $q_{\text{MMDC}}[k]$ and $q_{\text{MMDC}}^{\text{acc}}[k]$. Note that, $q_{\text{MMDC}}^{\text{acc}}|_d[k] \triangleq M \cdot q_{\text{MMDC}}^{\text{acc}}[k]$, where

[6] Instead of adding $-T_{\text{DCO}} \cdot q_{\text{MMDC}}^{\text{acc}}[k]$ to the feedback path, $T_{\text{DCO}} \cdot q_{\text{MMDC}}^{\text{acc}}[k]$ could be added to the reference path to achieve equivalent QE cancellation.

$$M = 2^{N_b^{\mathrm{MMDC}}} \tag{7.13}$$

is the MMDC's modulus and N_b^{MMDC} is the number of bits used in each stage of the MASH. The digital-domain MMDC's AQE, $q_{\mathrm{MMDC}}^{\mathrm{acc}}\big|_d [k]$, has unity resolution, which depends on the order of DDSM and is represented by a number of bits,

$$N_b^{\mathrm{acc}} = N_b^{\mathrm{MMDC}} + l - 1, \tag{7.14}$$

where l is the order of the DDSM.

The double-framed "C" block in Fig. 6.1 represents a LMS correlator with a step size, λ, that correlates the TDC's output, $e[k]$, and $q_{\mathrm{MMDC}}^{\mathrm{acc}}\big|_d [k]$ to adaptively train a scaling factor, $h[k]$, to calibrate the effective gain of the DTC [18, 34] in the presence of PVT variations. Note that the LMS does not change the word length or resolution when transferring $q_{\mathrm{MMDC}}^{\mathrm{acc}}\big|_d [k]$ to $a[k]$ that controls the DTC.

Here, we study the QE contributed by the DTC requantizer that has a resolution of Δ to map the N_b^{acc}-bit input, $a[k]$, to the N_b^{DTC}-bit code, $c[k]$, to control the N_b^{DTC}-bit DCD. The numbers of bits of $q_{\mathrm{MMDC}}^{\mathrm{acc}}\big|_d [k]$, $a[k]$, and $c[k]$ are highlighted in color in Fig. 6.1. Note that a Δ-to-1 quantizer has an inherent gain equal to $1/\Delta$. Taking the DTC's QE into account, the time delay introduced by a quantized DTC can be updated, from the ideal case in (7.11b), to become

$$t_{\mathrm{DTC}}^{\mathrm{Q}}[k+1] = -\underbrace{\left(\frac{a[k]}{\Delta} + q_{\mathrm{DTC}}[k]\right)}_{c[k]} \cdot \frac{\Delta}{M} \cdot T_{\mathrm{DCO}}, \tag{7.15}$$

where the linear output of the DTC quantizer, $a[k]/\Delta$, is added with the DTC's QE, $q_{\mathrm{DTC}}[k]$, to constitute $c[k]$. $\frac{\Delta}{M} \cdot T_{\mathrm{DCO}}$, where T_{DCO} is the minimum oscillator period of the DCO in its tuning range, is the time delay that should be provided by one LSB delay of the DCD. Recall that $q_{\mathrm{MMDC}}^{\mathrm{acc}}\big|_d [k] \equiv M \cdot q_{\mathrm{MMDC}}^{\mathrm{acc}}[k]$ and $a[k] = q_{\mathrm{MMDC}}^{\mathrm{acc}}\big|_d [k]$ when the LMS algorithm settles. Hence, in the steady state, (7.15) can be rewritten as

$$t_{\mathrm{DTC}}^{\mathrm{Q}}[k+1] = -T_{\mathrm{DCO}} \cdot q_{\mathrm{MMDC}}^{\mathrm{acc}}[k] - \frac{\Delta}{M} \cdot T_{\mathrm{DCO}} \cdot q_{\mathrm{DTC}}[k], \tag{7.16}$$

where the first term represents the linear output, the same as (7.11b), and the second term represents the contribution of the DTC's QE. Recalling (7.12), Eq. (7.16) can be rearranged in the form of a difference equation as

$$\begin{aligned} t_{\mathrm{DTC}}^{\mathrm{Q}}[k+1] = t_{\mathrm{DTC}}^{\mathrm{Q}}[k] &- T_{\mathrm{DCO}} \cdot q_{\mathrm{MMDC}}[k] \\ &- \frac{\Delta}{M} \cdot T_{\mathrm{DCO}} \cdot \left(q_{\mathrm{DTC}}[k] - q_{\mathrm{DTC}}[k-1]\right). \end{aligned} \tag{7.17}$$

Adding (7.17) into (7.10) yields the timestamp of the divided signal that is compensated by a quantized DTC into which we can substitute the following relationships:

- as per Eq. (3.5), the TDC can be linearized with a linear gain, to whose output is added the TDC's QE, as $e[k] = K_{\text{TDC}} \cdot \Delta t[k] + q_{\text{TDC}}[k]$;
- as per Eq. (3.1), $\Delta t[k] = t_{\text{ref}}[k] - t_{\text{div}}[k]$ represents the kth time difference between the timestamps of the reference and divided signals;
- as per Eq. (2.2), the kth period of the TDC that has a synchronous jitter is $T_{\text{ref}}[k] = t_{\text{ref}}[k] - t_{\text{ref}}[k-1] = T_{\text{ref}} + t_{n,\text{ref}}[k] - t_{n,\text{ref}}[k-1]$, where T_{ref} is the nominal reference period; and
- $T_{\text{ref}} = N \cdot T_{\text{DCO}}$ in the locked state.

In this case, the input time difference, similar to the form of (4.3), can be obtained as

$$\begin{aligned}\Delta t[k+1] = (1-X)\cdot\Delta t[k] &+ \left(t_{n,\text{ref}}[k+1] - t_{n,\text{ref}}[k]\right) \\ &- \frac{X}{K_{\text{TDC}}}\cdot q_{\text{TDC}}[k] - \sum_{h=N_{\text{div}}[k]\cdot k+1}^{N_{\text{div}}[k]\cdot(k+1)} t_{n,\text{DCO}}[h] \\ &+ \frac{\Delta}{M}\cdot T_{\text{out}}\cdot\left(q_{\text{DTC}}[k] - q_{\text{DTC}}[k-1]\right),\end{aligned} \tag{7.18}$$

where we use the shorthand notation,[7]

$$X = K_{\text{TDC}} \cdot \beta \cdot K_T \cdot N_{\text{div}}[k]. \tag{7.19}$$

Following the procedure deriving Eq. (4.7), we obtain the variance of the input jitter of a fractional-N DPLL with a finite-resolution DTC as

$$\begin{aligned}\sigma_{\Delta t}^2 = \frac{2}{2-X}\cdot\sigma_{t_{n,\text{ref}}}^2 + \frac{X}{2-X}\cdot\frac{\sigma_{Q_{\text{TDC}}}^2}{K_{\text{TDC}}^2} &+ \frac{N}{X(2-X)}\cdot\sigma_{t_{n,\text{DCO}}}^2 \\ &+ \frac{2}{2-X}\cdot\left(\frac{\Delta}{M}\cdot T_{\text{out}}\right)^2\cdot\sigma_{Q_{\text{DTC}}}^2,\end{aligned} \tag{7.20}$$

where $\sigma_{t_{n,\text{ref}}}$, $\sigma_{Q_{\text{TDC}}}$, $\sigma_{t_{n,\text{DCO}}}$, and $\sigma_{Q_{\text{DTC}}}$ are the RMS values at source of the reference jitter, TDC's QE, DCO random-walk jitter, and DTC's QE.

Note that the TDC's linearized gain and QE power, i.e., K_{TDC} and $\sigma_{Q_{\text{TDC}}}^2$ in (7.19) and (7.20), depend on the TDC's number of bits and resolution and also on the statistical distribution of its input, as discussed in Chap. 3. Here, we represent $K_{\text{TDC}} = Y/\sigma_{\Delta t}$ and $\sigma_{Q_{\text{TDC}}}^2 = Z - Y^2$, as derived in Chap. 3, where the notations of Y and Z are also explained. Notice also that the sufficient condition for this analysis to apply is that the integer-N input jitter has a zero-mean Gaussian distribution. Its extension

[7] The cycle-wise varying ratio, $N_{\text{div}}[k]$, appears in the coefficients of some random noise terms in (7.18), which can over-complicate the subsequent analysis. Recall that $N_{\text{div}}[k] \triangleq N + q_{\text{MMDC}}[k]$, since the variation of the zero-mean QE, $q_{\text{MMDC}}[k]$, is small compared to the nominal divide ratio, N. Here, we approximate $N_{\text{div}}[k]$ in (7.18) and (7.19) by its average value, N, to simplify the analysis.

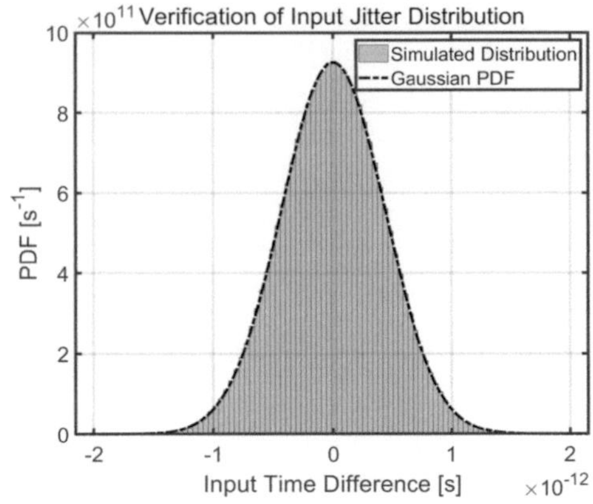

Fig. 7.5 Simulated distribution of the locked-state input jitter, $\Delta t[k]$, in comparison with the Gaussian distribution

to our analysis is valid if the distribution of the fractional-N input jitter, taking into account the DTC's QE, is zero-mean Gaussian as well. To verify this, we plot in Fig. 7.5 the simulated distribution of $\Delta t[k]$ in the locked state that is compared with the PDF of a Gaussian distribution with zero mean and standard deviation equal to $\sigma_{\Delta t}$. Note that the DTC's QE, $q_{\mathrm{DTC}}[k]$, has a zero-mean uniform distribution with $\sigma_{Q_{\mathrm{DTC}}} = 1/\sqrt{12}$, as discussed in Sect. 7.2. Figure 7.5 shows that, regardless of the non-Gaussian shape of the distribution of the DTC's residual error, the input jitter in the DTC-based fractional-N mode always has an approximately zero-mean Gaussian distribution; hence, the extension of the aforementioned expressions to be used here is valid.

Following the procedure to derive Eq. (4.13), the RMS input jitter can be calculated as

$$\sigma_{\Delta t} = \frac{\eta}{2} + \sqrt{\left(\frac{\eta}{2}\right)^2 + \sigma_{t_{n,\mathrm{ref}}}^2 + \left(\frac{\Delta}{M} \cdot T_{\mathrm{out}}\right)^2 \cdot \sigma_{Q_{\mathrm{DTC}}}^2}, \tag{7.21}$$

where η is the same as Eq. (4.14). Equations (7.21) and (4.14) quantify the amounts of jitter contributed by all the noise sources and transferred to the input.

The system jitter induced by other noise sources, i.e., the reference, TDC, and DCO, in (7.20), (7.21) and (4.14) does not arise from the MMDC's modulation and therefore cannot be compensated by the DTC. In the next section, we will derive a design criterion for how much residual error after DTC-based cancellation can be tolerated to be negligible compared to the former; this helps to provide a guideline for DTC parameterization that achieves sufficient QE cancellation with minimum DTC hardware.

Note that an "over-designed" DTC, i.e., one that uses an excessive number of bits and finer resolution, will in general consume larger chip area and power and require higher control complexity. Therefore, over-designing the DTC should always be avoided.

7.3.3 DTC Design

The core component of the DTC is an N_b^{DTC}-bit DCD such as that shown in Fig. 6.2. The DCD comprises an even number of inverter stages. A resistor and a capacitor bank forming an RC network with a tunable time constant is inserted in the middle to provide a cycle-wise controllable time delay to emulate the desired one, as derived in (7.11).[8]

DTC Design Criterion

As discussed in Sect. 7.3.2, in high-resolution frequency synthesis, the Δ-to-1 DTC requantizer maps the DTC's fine-resolution (N_b^{acc}-bit) control code to control the (N_b^{DTC}-bit) DCD with a coarser resolution in order to reduce the hardware consumption. The key parameter that determines the power of the DTC's residual error related to its QE is hence the input resolution of the DTC quantizer, Δ.

In practice, to ensure effective mitigation of the MMDC's QE, one should choose as a design criterion an appropriate value for Δ that is large enough to minimize the hardware requirement for the DTC. It should also be small enough so that the noise power contributed by the DTC's QE is limited below that of the integer-N input jitter, thereby recovering the integer-N PN/jitter performance.

Enhanced DTC Parameterization

It is important to note that, when comparing the power of the DTC's residual error to that of the intrinsic system jitter at the input of the TDC, the former's waveform is not exactly the original waveform of the residual error as in the third line of (7.18). Instead, it is also filtered by the loop; hence, the closed-loop transfer mechanism must be taken into account when describing the jitter power contributed by the DTC's residue. This was overlooked in the original DTC parameterization method proposed in [12].

The jitter transfer mechanism of a DTC-based fractional-N DPLL considering this closed-loop filtering effect was studied in Sect. 7.3.2. The variance of the input jitter in this case was expressed in (7.20), which is the key relationship that we will leverage in this section to derive the optimum DTC design.

Notice that Eq. (7.20) can be trivially transformed to describe the case of a DTC-based analog PLL, which, in turn, can be used to derive the optimum DTC parameterization in an equivalent analog loop. Specifically, recall that X in (7.19) represents the loop gain of the PLL; $\sigma^2_{Q_{\mathrm{TDC}}}/K^2_{\mathrm{TDC}}$ in (7.20) represents the power of the input-

[8] The DTC could be implemented in other ways, e.g., using constant-slope DTCs [10, 11, 19, 20]. The exact implementation method for the DTC's DCD is not important in this section, where the highlight is the strategy to size and parameterize the DTC considering the properties of the loop and MMDC.

referred TDC's QE, which is equivalent to the power of the input-referred physical noise of the charge pump. The other terms carry the same physical meanings.

Without loss of generality, we use DPLL terminology in the following derivation. It is interesting to observe that the first line of (7.20) precisely matches the accurate prediction for the power of the input jitter in the integer-N case, $\sigma^2_{\Delta t_{\mathrm{int}}}$, provided in (4.7) ; in the fractional-N expression, this magnitude is independent of the DTC in the sense that, no matter how well the DTC is designed, the jitter power contributed by the integer-N intrinsic noises represented by the first line of (7.20) does not change. On the other hand, the quantity in the second line, representing the contribution from the DTC's QE, is added to $\sigma^2_{\Delta t_{\mathrm{int}}}$ to constitute the fractional-N jitter power. Apart from the original power of the DTC's QE at source, $\left(\frac{\Delta}{M} \cdot T_{\mathrm{out}} \cdot \sigma_{Q_{\mathrm{DTC}}}\right)^2$, its coefficient, $\frac{2}{2-X}$, describes the gain imposed by the closed loop when transferring it to the TDC's input. For a stable loop, $0 < X < 2$ is always satisfied [29]. Therefore, $\frac{2}{2-X} \in (1, +\infty)$ is an amplifying factor. For example, in the optimized BBPLL design (see Table 5.2), $X \approx 0.33$, hence $\frac{2}{2-X} \approx 1.20$; in the optimized 3-bit-TDC DPLL, $X \approx 0.43$, giving $\frac{2}{2-X} \approx 1.27$.

Parameterization for Undithered DTC: in order to limit the power of the DTC's residual error (the second line of (7.20)) to a level below that of the integer-N jitter (the first line), the following inequality must be satisfied,

$$\frac{2}{2-X} \cdot \left(\frac{\Delta}{M} \cdot T_{\mathrm{out}}\right)^2 \cdot \sigma^2_{Q_{\mathrm{DTC}}} < \sigma^2_{\Delta t_{\mathrm{int}}}. \tag{7.22}$$

This can be rearranged to yield a useful inequality for the choice of the DTC's input resolution, Δ, as the key design parameter, as in

$$\Delta < \sqrt{\frac{2-X}{2}} \cdot \frac{\sigma_{\Delta t_{\mathrm{int}}}}{T_{\mathrm{out}}} \cdot \frac{1}{\sigma_{Q_{\mathrm{DTC}}}} \cdot M, \tag{7.23}$$

where the four terms on the right-hand side have their own physical meanings and are different in different DPLL-system designs. Namely,

- $\sqrt{(2-X)/2}$ represents the effect of the loop dynamics on the choice of Δ: the greater is the value of X, the smaller (finer) Δ should be chosen for the DTC;
- $\sigma_{\Delta t_{\mathrm{int}}}/T_{\mathrm{out}}$ is the ratio of the integer-N RMS input jitter to the output period. As per [12], the ratio is approximately 1/1000 in radio-frequency synthesizers;
- $\sigma_{Q_{\mathrm{DTC}}}$ is the standard deviation of the DTC's QE, whose random variable, Q_{DTC}, has a uniform distribution in $Q_{\mathrm{DTC}} \sim U(-\frac{1}{2}, \frac{1}{2})$ as studied in Sect. 7.2. Therefore,

$$\frac{1}{\sigma_{Q_{\mathrm{DTC}}}} = \sqrt{12}, \tag{7.24}$$

  for an undithered DTC quantizer;
- $M = 2^{N_b^{\mathrm{MMDC}}}$ is the modulus of the MMDC.

The effective input DR of the N_b^{DTC}-bit DTC quantizer should, in principle, match the variation range of its input as

$$2^{N_b^{\mathrm{acc}}} = 2^{N_b^{\mathrm{DTC}}} \cdot \Delta,$$

substituting which into (7.23), together with (7.13) and (7.14), we obtain an inequality, which does not contain Δ, for the choice of the number of bits in the DTC, N_b^{DTC}, namely

$$N_b^{\mathrm{DTC}} > (l-1) - \log_2\left(\sqrt{\frac{2-X}{2}} \cdot \frac{\sigma_{\Delta t_{\mathrm{int}}}}{T_{\mathrm{out}}} \cdot \frac{1}{\sigma_{Q_{\mathrm{DTC}}}}\right). \tag{7.25}$$

Recall that l is the order of the MASH MMDC.

To evaluate the DTC parameterization, we substitute into (7.23) the values associated with the example BBPLL parameterized as per Table 3.3, namely $\sqrt{(2-X)/2} = \sqrt{1/1.20}$, $\sigma_{\Delta t_{\mathrm{int}}}/T_{\mathrm{out}} = 407\ \mathrm{fs}/417\ \mathrm{ps}$, (7.24), and $M = 2^{20}$, and we get that the input resolution of $\Delta < 3236$ can satisfy the relationship (7.23). In digital design, the input resolution of a digital-to-digital quantizer is usually specified as a power of two. Therefore, $\Delta = 2^{11}$ is good enough to suppress the DTC QE's contribution to be below the integer-N jitter level. Δ values smaller than this imply finer resolution and a larger number of bits to be used in the DTC in order to sustain the same DR—this is undesired. Note that, as (7.23) is not related to l, the derived parameterization of Δ does not change with the order of the MMDC.

Interestingly, Eq. (7.25) has l on the right; hence the choice of N_b^{DTC} depends on the MMDC's order. Evaluating (7.25) with the above values, we obtain that $N_b^{\mathrm{DTC}} > (l-1) + 8.34$, together with $\Delta = 2^{11}$, can satisfy the design criterion . This indicates that a 9-bit (or 10-bit, 11-bit) DTC is good enough when a first-order (or MASH 1–1, 1–1–1) DDSM is used as the MMDC.

Compared to the parameterization proposed in [12] and described in Sect. 7.3.1, we can see that the classical method uses a DTC that has twice as many bits as this parameterization strategy, and the resolution of each unit delay cell in their case needs to be twice as fine. Quantitatively, this is mainly due to the fact that the classical method aims to bound the variation of the DTC's residual error within the range of $(-\sigma_{\Delta t_{\mathrm{int}}}/2, \sigma_{\Delta t_{\mathrm{int}}}/2)$; hence the resultant DTC's resolution is too fine. However, in principle, if one wanted to limit the power of the residual error to be lower than that of the integer-N jitter, their RMS values, rather than variation ranges, should be compared.

As a final remark, one may notice that the physical definition of the input resolution of the DTC quantizer requires $\Delta \geq 2^0 = 1$, but the value of the right-hand side of (7.23) could be less than unity. For instance, if the stages within the DDSM MMDC have $N_b^{\mathrm{MMDC}} = 8$, with the remaining factors unchanged, Eq. (7.23) will yield an aphysical $\Delta < 0.79$. Similar "$\Delta < 1$" relationships can be obtained for $N_b^{\mathrm{MMDC}} < 8$. In these cases, the DTC parameterization studied in this subsection is not needed. When the DTC's input is so coarse, one can choose N_b^{DTC} to match N_b^{acc}, the number

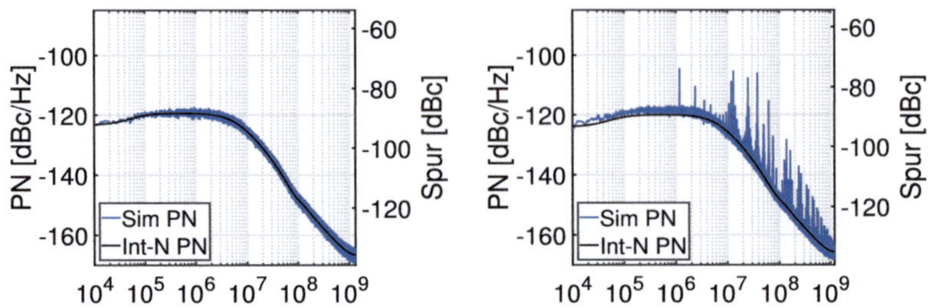

(a) $N_b^{\mathrm{MMDC}} = 9$, $N_b^{\mathrm{acc}} = 10$, $N_b^{\mathrm{DTC}} = 10$, $\Delta = 1$.

(b) $N_b^{\mathrm{MMDC}} = 20$, $N_b^{\mathrm{acc}} = 21$, $N_b^{\mathrm{DTC}} = 10$, $\Delta = 2^{11}$.

Fig. 7.6 Simulated power spectra of the output PN of the fractional-N BBPLL with a MASH 1–1 MMDC ($X = 1277$) and an *undithered finite-resolution linear DTC*. Different MMDC modulus and DTC are analyzed. In case (a), there is no DTC QE. Case (b) shows spurs due to DTC quantization

of bits of the DTC's input. Recall that $N_b^{\mathrm{acc}} = N_b^{\mathrm{MMDC}} + (l-1)$ as per (7.14). The unit delay provided by the DTC delay line can also be chosen so that the DTC can cancel the MMDC's QE exactly; hence the DTC will, in principle, generate no QE to affect the DPLL.

The critical condition of the example system is when $N_b^{\mathrm{MMDC}} = 9$. As per the enhanced parameterization strategy, $\Delta = 1$ and $N_b^{\mathrm{DTC}} = (l+8)$ are needed. Therefore, the DTC quantizer performs a one-to-one mapping and $N_b^{\mathrm{acc}} = N_b^{\mathrm{DTC}}$. There is no resolution gap between the DTC's input and output, and hence there are no DQI spurs. As shown in Fig. 7.6a, when the input and output resolutions of the DTC match, the integer-N PN can be recovered completely with a finite-resolution DTC; no DQI spur exists, even though the DTC is not dithered.

Parameterization for the IDQ-DTC: in more realistic applications, a higher-resolution DDSM ($N_b^{\mathrm{MMDC}} \geq 10$) is typically used as the MMDC. In these cases, because of the resolution gap between the control and output of the finite-resolution DTC, the issue of DQI spurs prevails. Figure 7.6b shows the adverse PN response using the enhanced parameterization method discussed in the previous case. Recall from the fourth and fifth rows of Fig. 7.4 that, even using the DTC proposed in [12] that has extremely fine resolution, the DQI spurs still exist. Clearly, the "stochastic resonance" at the TDC's input fails to randomize the periodicity associated with the DTC's QE, even though the latter's variation has been much limited; the reason for this has been explained in Chap. 3.

The IDQ method described in Sect. 7.2 adds dither at the input of the DTC quantizer in order to eliminate DQI spurs. The dither injects more noise power into the loop. Therefore, when considering DTC parameterization, we need to take into account the injected dithering power when evaluating (7.23) and (7.25). The optimum IDQ method in Sect. 7.2 injects at the DTC's input a random number that has a uniform distribution, $U\left[-\frac{\Delta}{2}, \frac{\Delta}{2} - 1\right]$; this results in an excess noise power equal

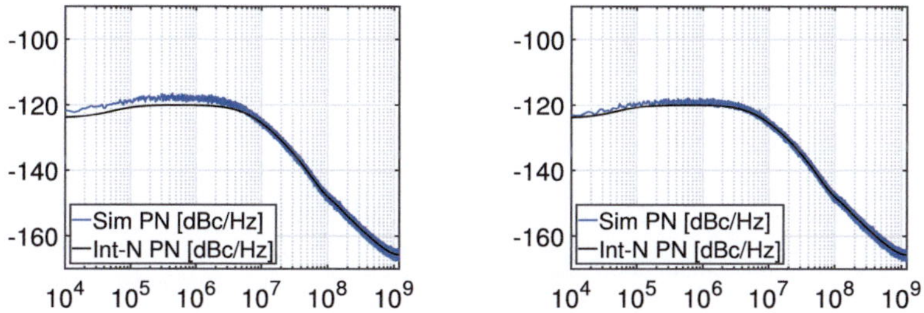

(a) $N_b^{\text{MMDC}} = 20$, $N_b^{\text{acc}} = 21$, $N_b^{\text{DTC}} = 10$, $\Delta = 2^{11}$.

(b) $N_b^{\text{MMDC}} = 20$, $N_b^{\text{acc}} = 21$, $N_b^{\text{DTC}} = 11$, $\Delta = 2^{10}$.

Fig. 7.7 Simulated power spectra of the output PN of the fractional-N BBPLL with a MASH 1–1 MMDC ($X = 1277$) and a *finite-resolution linear IDQ-DTC*. Different DTC parameterizations are analyzed. Case (a) has excess folded noise due to dithering. Adding a bit to the DTC in case (b) reduces the excess noise. Because of the IDQ dither, both cases are free of DQI spurs

to $1/12$ added to $\sigma^2_{Q_{\text{DTC}}}$ in (7.22).[9] Therefore, the value of (7.24) needs to be updated to $\sqrt{6}$ instead of $\sqrt{12}$ when being substituted into (7.23) and (7.25). In this case, $\Delta < 2288$ and $N_b^{\text{DTC}} > (l - 1) + 8.84$ must be satisfied. Notice that these suggest that the previous parameterization of $\Delta = 2^{11}$ and $N_b^{\text{DTC}} = (l + 8)$ can still cause the jitter power contributed by the DTC's QE to be lower than that of the intrinsic integer-N jitter. However, these parameters will push the value of the former to close to that of the latter, hence elevating the output PN level compared to the integer-N case; this is shown in Fig. 7.7a. In addition, considering PVT variations, the parameters backed off to $\Delta = 2^{10}$ and $N_b^{\text{DTC}} = (l + 9)$, together with the aforementioned IDQ method, give a DTC parameterization that can successfully recover the integer-N PN performance and is free of DQI spurs, as shown in Fig. 7.7b.

Finally, we summarize the enhanced DTC parameterization strategy in Table 7.3 for generic DPLL systems. Table 7.4 is customized from Table 7.3, which calculates the numerical values for the example BBPLL used in this paper. Clearly, the parameterization depends on the modulus and order of the MMDC. For other DPLL systems that have different characteristics expressed on the right-hand side of (7.23), Table 7.3 needs to be customized accordingly. Regardless of the parameters associated with the DTC quantizer, the unit delay (or nominal time gain) provided by the DTC's DCD in Fig. 6.1 should equal $\frac{\Delta}{M} \cdot T_{\text{out}}$, as per (7.15).

[9] Note that EFM1 requantization in place of the IDQ-dithered requantization effectively dithers the input to the requantizer with the same power. Therefore, the derivation here is also applicable to DTCs with an EFM1 requantizer.

Table 7.3 Enhanced DTC parameterization (general case)

R.H.S. of (7.23)	IDQ	Δ	N_b^{DTC}	DCD unit delay
≤ 1	Not needed	1	(7.14): $N_b^{\mathrm{DTC}} = N_b^{\mathrm{MMDC}} + l - 1$	$\frac{\Delta}{M} \cdot T_{\mathrm{DCO}}$
> 1	Needed	(7.23): $\Delta < \sqrt{\frac{2-X}{2}} \cdot \frac{\sigma_{\Delta t_{\mathrm{int}}}}{T_{\mathrm{out}}} \cdot \frac{1}{\sigma_{Q_{\mathrm{DTC}}}} \cdot M$	(7.25): $N_b^{\mathrm{DTC}} > l - 1 - \log_2\left(\sqrt{\frac{2-X}{2}} \cdot \frac{\sigma_{\Delta t_{\mathrm{int}}}}{T_{\mathrm{out}}} \cdot \frac{1}{\sigma_{Q_{\mathrm{DTC}}}}\right)$	

Table 7.4 DTC parameterization for the example BBPLL in Table 3.3

N_b^{MMDC}	IDQ	Δ	N_b^{DTC}
≤ 8	Not needed	1	$N_b^{\mathrm{MMDC}} + (l - 1)$
9	Not needed	1	$l + 8$
≥ 10	Needed	$2^{N_b^{\mathrm{MMDC}}-9}$ as per (7.23)	$l + 8$ as per (7.25)
		$2^{N_b^{\mathrm{MMDC}}-10}$ (backed off)	$l + 9$ (backed off)

7.3.4 *Conclusion*

In this section, we reviewed the classical parameterization strategy in [12] for the DTC used in fractional-N PLLs to cancel the QE arising from the DDSM MMDC. Its limitations in terms of hardware consumption and generation of DQI spurs are revealed. An enhanced DTC parameterization method is proposed, which yields optimum quantization-related parameters such as the effective DR, quantization resolution, number of bits, and unit delay. A DTC designed in this way uses a minimum amount of hardware to suppress the DTC-quantization-induced jitter power below the integer-N counterpart in order to recover as closely as possible the integer-N PN/jitter performance. The parameterization strategy includes the use of IDQ methods to eliminate DQI spurs. Behavioral simulations at the DPLL-system level have confirmed our theoretical analysis.

7.4 Summary

1. In Sect. 7.2, the optimum IDQ method that can be used together with the DTC to eliminate DQI spurs was derived. Various implementation methods of it involving different hardware complexities have been presented.
2. In Sect. 7.3, a DTC sizing method is described, which uses a minimum amount of hardware to suppress the DTC-quantization-induced jitter power below the integer-N counterpart in order to recover efficiently the integer-N PN/jitter performance. The sizing strategy includes the use of IDQ methods to eliminate DQI spurs.

References

1. D. Banerjee, *PLL Performance, Simulation and Design* (Dog Ear Publishing, Indianapolis, IN, USA, 2006)
2. Y. Donnelly, M.P. Kennedy, Prediction of phase noise and spurs in a nonlinear fractional-N frequency synthesizer. IEEE Trans. Circuits Syst. I Regul. Pap. **66**(11), 4108–4121 (2019)
3. V. Mazzaro, M.P. Kennedy, Folded noise prediction in nonlinear fractional-N frequency synthesizers. IEEE Trans. Circuits Syst. I Regul. Pap. **68**(10), 4038–4048 (2021)
4. M. Zanuso, D. Tasca, S. Levantino, A. Donadel, C. Samori, A.L. Lacaita, Noise analysis and minimization in bang-bang digital PLLs. IEEE Trans. Circuits Syst. II Express Briefs **56**(11), 835–839 (2009)
5. G. Marucci, S. Levantino, P. Maffezzoni, C. Samori, Exploiting stochastic resonance to enhance the performance of digital bang-bang PLLs. IEEE Trans. Circuits Syst. II Express Briefs **60**(10), 632–636 (2013)
6. S. Lipshitz, R. Wannamaker, J. Vanderkooy, Quantization and dither: a theoretical survey. J. Audio Eng. Soc. **40**(05), 355–374 (1992)
7. S.J. Orfanidis, *Introduction to Signal Processing* (Sophocles J. Orfanidis, 2010). [Online]. https://www.ece.rutgers.edu/~orfanidi/intro2sp/
8. A.A. Eielsen, A.J. Fleming, Improving digital-to-analog converter linearity by large high-frequency dithering. IEEE Trans. Circuits Syst. I Regul. Pap. **64**(6), 1409–1420 (2017)
9. A. Mahmoud, P. Andreani, F. Pepe, A 2.8–3.8-GHz low-spur DTC-based DPLL with a class-D DCO in 65-nm CMOS. IEEE Microwave Wirel. Compon. Lett. **27**(11), 1010–1012 (2017)
10. H. Liu, D. Tang, Z. Sun, W. Deng, H.C. Ngo, K. Okada, A. Matsuzawa, A 0.98mW fractional-N ADPLL using 10b isolated constant-slope DTC with FOM of –246dB for IoT applications in 65nm CMOS, in *2018 IEEE International Solid-State Circuits Conference—(ISSCC)*, pp. 246–248 (2018)
11. P. Chen, F. Zhang, Z. Zong, S. Hu, T. Siriburanon, R.B. Staszewski, A 31-μW, 148-fs step, 9-bit capacitor-DAC-based constant-slope digital-to-time converter in 28-nm CMOS. IEEE J. Solid-State Circuits **54**(11), 3075–3085 (2019)
12. D. Tasca, M. Zanuso, G. Marzin, S. Levantino, C. Samori, A.L. Lacaita, A 2.9–4.0-GHz fractional-N digital PLL with bang-bang phase detector and 560-fs_{rms} integrated jitter at 4.5-mW power. IEEE J. Solid-State Circuits **46**(12), 2745–2758 (2011)
13. M.D. McDonnell, Is electrical noise useful? [point of view]. Proc. IEEE **99**(2), 242–246 (2011)
14. S. Levantino, G. Marzin, C. Samori, A.L. Lacaita, A wideband fractional-N PLL with suppressed charge-pump noise and automatic loop filter calibration. IEEE J. Solid-State Circuits **48**(10), 2419–2429 (2013)
15. H. Xu, A.A. Abidi, Design methodology for phase-locked loops using binary (bang-bang) phase detectors. IEEE Trans. Circuits Syst. I Regul. Pap. **64**(7), 1637–1650 (2017)
16. L. Avallone, M.P. Kennedy, S. Karman, C. Samori, S. Levantino, Jitter minimization in digital PLLs with mid-rise TDCs. IEEE Trans. Circuits Syst. I Regul. Pap. **67**(3), 743–752 (2020)
17. S. Levantino, G. Marzin, C. Samori, An adaptive pre-distortion technique to mitigate the DTC nonlinearity in digital PLLs. IEEE J. Solid-State Circuits **49**(8), 1762–1772 (2014)
18. S. Levantino, L. Grimaldi, C. Samori, Analysis of adaptive pre-distortion in DTC-based digital fractional-N PLLs, in *2016 12th Conference on Ph.D. Research in Microelectronics and Electronics (PRIME)*, pp. 1–4 (2016)
19. J.Z. Ru, C. Palattella, P. Geraedts, E. Klumperink, B. Nauta, A high-linearity digital-to-time converter technique: constant-slope charging. IEEE J. Solid-State Circuits **50**(6), 1412–1423 (2015)
20. Y. Liu, H. Gao, H. Xu, P. Lu, N. Yan, A 10-bit 563-fs step constant-slope digital-to-time converter in 40-nm CMOS with nonlinearity cancellation and range extension techniques. IEEE Trans. Circuits Syst. I Regul. Pap. **71**(2), 526–536 (2024)
21. W. Wu, C.-W. Yao, C. Guo, P.-Y. Chiang, L. Chen, P.-K. Lau, Z. Bai, S.W. Son, T.B. Cho, A 14-nm ultra-low jitter fractional-N PLL using a DTC range reduction technique and a reconfigurable dual-core VCO. IEEE J. Solid-State Circuits **56**(12), 3756–3767 (2021)

22. C. Hwang, H. Park, T. Seong, J. Choi, A 188fs$_{rms}$-jitter and –243dB-FoM$_{jitter}$ 5.2GHz-ring-DCO-based fractional-N digital PLL with a 1/8 DTC-range-reduction technique using a quadruple-timing-margin phase selector, in *2022 IEEE International Solid- State Circuits Conference (ISSCC)*, vol. 65, pp. 378–380 (2022)
23. H. Park, C. Hwang, T. Seong, Y. Lee, J. Choi, 32.1 a 365fs$_{rms}$-jitter and –63dBc-fractional spur 5.3GHz-ring-DCO-based fractional-N DPLL using a DTC second/third- order nonlinearity cancelation and a probability-density-shaping $\Delta\Sigma$M, in *2021 IEEE International Solid- State Circuits Conference (ISSCC)*, vol. 64, pp. 442–444 (2021)
24. C. Hwang, H. Park, Y. Lee, T. Seong, J. Choi, A low-jitter and low-fractional-spur ring-DCO-based fractional-N digital PLL using a DTC's second-/third-order nonlinearity cancellation and a probability-density-shaping $\Delta\Sigma$M. IEEE J. Solid-State Circuits **57**(9), 2841–2855 (2022)
25. S.M. Dartizio, F. Tesolin, G. Castoro, F. Buccoleri, L. Lanzoni, M. Rossoni, D. Cherniak, L. Bertulessi, C. Samori, A.L. Lacaita, S. Levantino, 4.3 a 76.7fs-lntegrated-jitter and –71.9dBc in-band fractional-spur bang-bang digital PLL based on an inverse-constant-slope DTC and FCW subtractive dithering, in *2023 IEEE International Solid-State Circuits Conference (ISSCC)*, pp. 3–5 (2023)
26. W. Wu, C.-W. Yao, K. Godbole, R. Ni, P.-Y. Chiang, Y. Han, Y. Zuo, A. Verma, I.S.-C. Lu, S.W. Son, T.B. Cho, A 28-nm 75-fs$_{rms}$ analog fractional-*N* sampling PLL with a highly linear DTC incorporating background DTC gain calibration and reference clock duty cycle correction. IEEE J. Solid-State Circuits **54**(5), 1254–1265 (2019)
27. A. Santiccioli, M. Mercandelli, L. Bertulessi, A. Parisi, D. Cherniak, A.L. Lacaita, C. Samori, S. Levantino, A 66-fs-rms jitter 12.8-to-15.2-GHz fractional-N bang–bang PLL with digital frequency-error recovery for fast locking. IEEE J. Solid-State Circuits **55**(12), 3349–3361 (2020)
28. A. Hu, Z. Jin, X. Shan, M. Zhang, D. Liu, X. Zou, Fractional spurs reduction technique using probability density shaping sigma-delta modulator and fractional frequency divider. IEEE Trans. Circuits Syst. II Express Briefs **70**(2), 456–460 (2023)
29. L. Avallone, M. Mercandelli, A. Santiccioli, M.P. Kennedy, S. Levantino, C. Samori, A comprehensive phase noise analysis of bang-bang digital PLLs. IEEE Trans. Circuits Syst. I Regul. Pap. **68**(7), 2775–2786 (2021)
30. Texas Instruments, 20-GHz wideband RF synthesizer with phase synchronization & JESD204B support, LMX2595, 2019, rev. C. [Online]. https://www.ti.com/lit/gpn/lmx2595
31. Analog Devices, Microwave Wideband Synthesizer with Integrated VCO, ADF4372, 2019, rev. 0. [Online]. https://www.analog.com/media/en/technical-documentation/data-sheets/adf4372.pdf
32. A. Santiccioli, C. Samori, A. Lacaita, S. Levantino, Power-jitter trade-off analysis in digital-to-time converters. Electron. Lett. **53**(5), 306–308 (2017). [Online]. https://ietresearch.onlinelibrary.wiley.com/doi/abs/10.1049/el.2016.4577
33. M. Rossoni, S.M. Dartizio, F. Tesolin, G. Castoro, R. Dell'Orto, C. Samori, A.L. Lacaita, S. Levantino, 10.1 an 8.75GHz fractional-N digital PLL with a reverse-concavity variable-slope DTC achieving 57.3fs$_{rms}$ integrated jitter and –252.4dB FoM, in *2024 IEEE International Solid-State Circuits Conference (ISSCC)*, vol. 67, pp. 188–190 (2024)
34. C. Samori, M. Zanuso, S. Levantino, A.L. Lacaita, Multipath adaptive cancellation of divider non-linearity in fractional-N PLLs, in *2011 IEEE International Symposium of Circuits and Systems (ISCAS)*, pp. 418–421 (2011)

Chapter 8
DTC Enhancement Strategies II: Nonlinear DCD

8.1 Introduction

The analysis in Chap. 6 explains that the three sources of DTC nonlinearity, i.e., the inherent quantization nonlinearity and the smooth polynomial and REM nonlinearities associated with the DCD, interacting with the AQE of the MMDC induce fractional spurs and PN that can be distinguished in the spectrum. This suggests using a combination of different types of DTC enhancement techniques to mitigate them. In Chap. 7, the sizing method and the IDQ method addressing the quantization-related behavior and the DQI spurs of the DTC were described. In this chapter, we focus on enhancements in relation to the DCD's nonlinearity.

The chapter analyzes and compares the effects of the following types of DTC enhancement techniques:

1. *conventional DTC-linearity-enhancement techniques*, e.g., in [1–16], which aim to linearize the polynomial INL of the DTC's DCD;
2. *statistical methods*, e.g., in [10–12, 17–22], which are used in the MMDC to mitigate IBS spurs resulting from interaction between its AQE and a known polynomial nonlinearity; and
3. *dynamic element matching* (DEM) [23] that can be used to mitigate the REM spurs.

These methods are shown to be applicable together with the IDQ methods, described in the last chapter, which can eliminate DQI spurs at source. We will present a comprehensive DTC design strategy using these classes which minimizes the PN and spurs in order to approach the fractional-spur-free integer-N PN limit in the case of the fractional-N frequency synthesizer.

The chapter is arranged as follows. Section 8.2 addresses the smooth polynomial INL associated with the DCD, where different DCD linearity enhancement techniques are analyzed and compared. Stochastic DDSMs used as the MMDC are demonstrated to be able to gain spur immunity in the presence of the polynomial

© The Author(s), under exclusive license to Springer Nature Switzerland AG 2026

X. Wang and M. P. Kennedy, *Jitter and Spur Minimization in Fractional-N Digital Frequency Synthesizers*, Analog Circuits and Signal Processing,
https://doi.org/10.1007/978-3-032-27186-0_8

INL of the DCD. Section 8.3 addresses the REM nonlinearity of the DCD; we show that using a DEM method that shuffles the control code accessing the DCD capacitor bank can eliminate REM spurs at source. Finally, Sect. 8.4 proposes a comprehensive strategy combining different classes of DTC enhancement techniques in order to mitigate excess PN and spurs that arise in practical DTCs. The enhanced DTC enables the fractional-N DPLL to approach the fractional-spur-free integer-N PN performance limit with minimum additional hardware.

8.2 Analysis and Mitigation of Fractional Spurs Induced by Polynomial Integral Nonlinearity of DCD

This section addresses performance enhancement in relation to the DCD's polynomial nonlinearity. In Sect. 8.2.1, we present models that describe the residual nonlinearity of the DCD when conventional DTC linearity enhancement techniques are used; the effectiveness of PN/spur mitigation using these techniques will also be studied. Section 8.2.2 demonstrates that statistical methods, i.e., so-called stochastic divider controllers, can be used to maximize IBS-spur immunity in the presence of a typical nonlinearity associated with the DTC's DCD.

8.2.1 DCD Linearity Enhancement Techniques

Spurs induced by the DCD's nonlinearity interacting with the MMDC's AQE have been reported in the literature. To date, various design techniques have been proposed to linearize the transfer characteristic of the DCD, in other words, to suppress the DCD's polynomial INL, such as that expressed in (6.7). These include, e.g., adaptive pre-distortion [1, 3, 4], CSC [2, 7, 14], isolated CSC [24], memory-effect suppression [8], DTC range reduction [8, 9, 12, 25], second-/third-order nonlinearity cancellation [10, 11], inverse-constant-slope DTC [13], RCVS DTC [15], and pseudo-differential DTC [16], among others.

This subsection assesses the effectiveness of IBS mitigation using such DTC linearity enhancement techniques that aim to suppress the DCD's polynomial INL. We will primarily compare the cases where MASH 1–1 and 1–1–1 DDSMs are used as the MMDC, and their QE is canceled by DTCs with polynomial INLs that (i) have or (ii) have not been linearized. Before making this comparison, we first perform the same simulations as shown in Fig. 6.5 for the case where a first-order DDSM, i.e., EFM1, is used as the MMDC, in order to show why an EFM1 MMDC is not a good choice in DTC-based fractional-N PLLs, fundamentally because of nonidealities in the DTC.

Canceling EFM1's QE with the Nonideal DTC

We note that, in some reported DTC-based DPLLs, e.g., in [5, 24], an EFM1 is chosen to serve as the MMDC in order to facilitate fractional-N frequency synthesis. The motivation for this choice can be understood because an EFM1 yields an AQE that has a narrower range of variation than higher-order DDSMs, e.g., the span of an EFM1's AQE is half of that of a MASH 1–1. Accordingly, the DTC can be designed to have a smaller DR to accommodate the AQE when canceling it; this reduces the DTC's hardware demand and slightly reduces the thermal noise that it introduces [13, 26].

However, an EFM1 MMDC is notorious for its AQE being rich in spurious tones because, when its input X is constant, it outputs a deterministic sawtooth waveform with a period equal to $\frac{X}{M} \cdot F_{\text{ref}}$ [27, 28]. For this reason, an EFM1 is never used as the MMDC in high-performance analog PLLs, where DTC-based QE cancellation is not used [20, 29, 30].

To assess the spectral response of EFM1-driven DPLLs with DTC-based cancellation, we perform the same simulations as shown in Fig. 6.5, replacing the MASH 1–1 with an EFM1. The results are shown in the first row of Fig. 8.1. Comparing that row with Fig. 6.5c and d, we observe that, with the DTC modeled to have the same polynomial INL as per (6.6) and the same quantization resolution, using an EFM1 DDSM in place of a MASH 1–1 causes the heights of the IBS spurs to increase by about 6 dB.

What makes it even worse is that, in this case, the IBS spur and its harmonics roll off at -20 dB/dec, in contrast to being -40 dB/dec shaped in the MASH 1–1 case. Furthermore, the IBS-spur family severely degrades the spectral purity; as these spurs are all located in-band, and therefore cannot be attenuated by the loop dynamics, they contribute strongly to the integrated jitter.

To make matters even worse, one can notice that the modeled polynomial INL in (6.6) with $\alpha_1 = +0.998$ deviates by only 0.2% from the ideal unity linear gain. If one considers a slightly higher gain mismatch of 1%, i.e., $\alpha_1 = +0.99$, which may occur due to PVT variations, using an EFM1 will further degrade the spurious response by increasing the envelope of the IBS-spur family by about 10 dB; this can be observed by comparing the second row with the first row of Fig. 8.1. This is because a larger gain mismatch will directly result in a higher magnitude of the residual error that is proportional to the EFM1's AQE, and is inherently IBS-spur-rich.

For completeness, the third row of Fig. 8.1 shows the results where a 1% gain mismatch is assumed for the MASH 1–1 case. By comparing with Fig. 6.5, we observe that the increase of gain mismatch does not affect the spectral signature or the magnitudes of the IBS spurs. This can be explained by the PNN theory [31–34]; in particular, the IBS pattern caused by a polynomial-distorted AQE of a MASH 1–1 is independent of the coefficient of the linear term in the nonlinearity. It is also worth noticing that, in all the cases in Figs. 6.5 and 8.1, the signatures of the DQI spurs are identical because they are related to the DTC's quantization characteristic and are not correlated with the INL or the order of the MASH (see relevant analysis in Chap. 6).

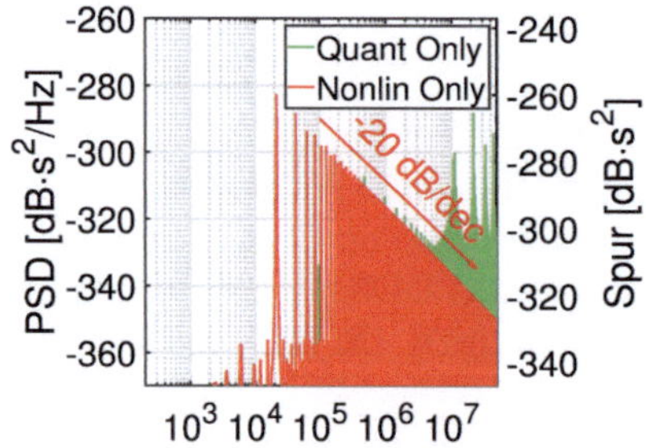

(a) DTC level, EFM1, DCD's INL with 0.2% gain mismatch.

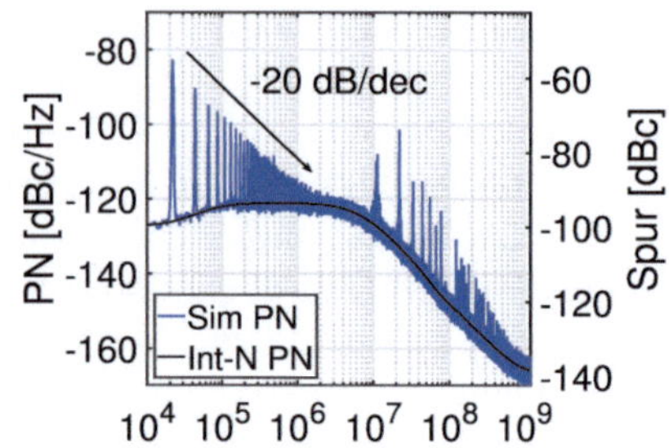

(b) DPLL level, EFM1, DCD's INL with 0.2% gain mismatch.

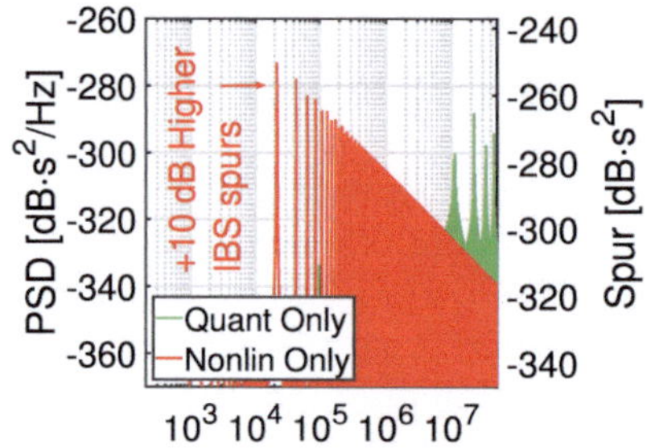

(c) DTC level, EFM1, DCD's INL with 1% gain mismatch.

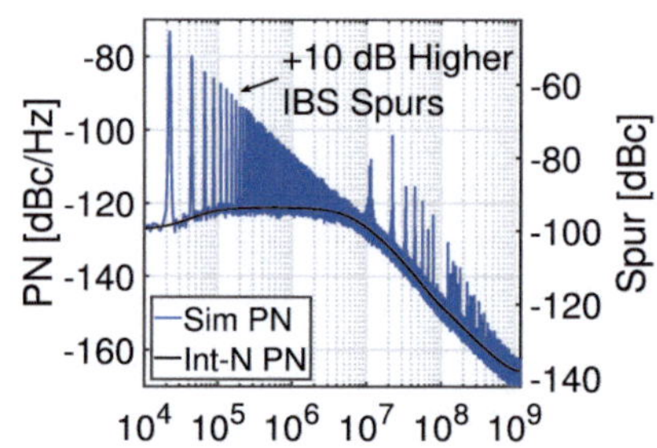

(d) DPLL level, EFM1, DCD's INL with 1% gain mismatch.

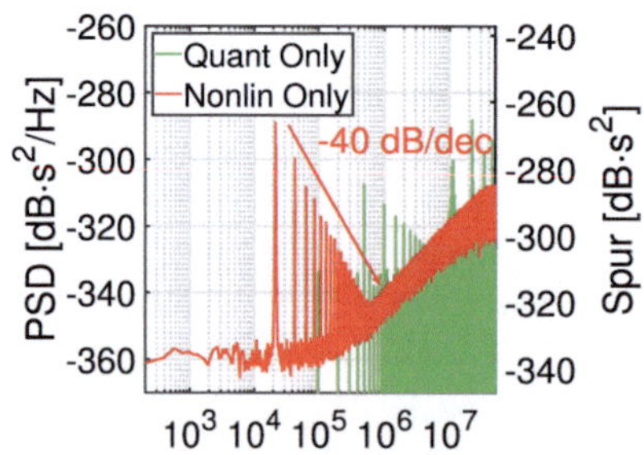

(e) DTC level, MASH 1-1, DCD's INL with 1% gain mismatch.

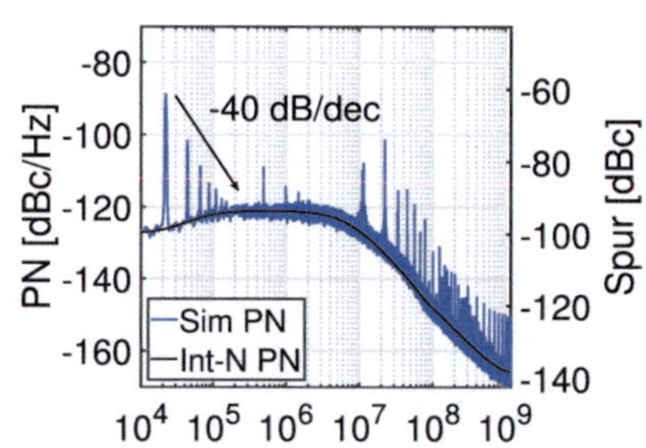

(f) DPLL level, MASH 1-1, DCD's INL with 1% gain mismatch.

Fig. 8.1 Simulated power spectra at block and system levels of the example fractional-N DPLL (*with EFM1 or MASH 1–1 MMDC*) with an *undithered finite-resolution DTC* that has *different polynomial INLs*

We conclude from the analysis above that, even when the AQE of an EFM1 MMDC experiences the midmost DCD's polynomial INL—that has a more moderate amplitude than the full-range INL—counterintuitively, the EFM1 in fact produces much worse IBS spurs compared to the higher-order DDSM case. This contradicts the "trade-off" in terms of spur mitigation asserted in, e.g., [35], between the extra spur immunity provided by using a higher-order DDSM and the "worsened" DTC's polynomial INL it thereby exercises. This is due to the different statistical properties resulting from different types of DDSMs that can only be explained via nonlinear analysis [31, 32, 34]. Therefore, using an EFM1 instead of a higher-order MMDC should not be considered a practical means to reduce the DTC range with the aim of a better spur/PN performance. Instead, sophisticated DTC range reduction techniques, such as [8, 25], should be used together with higher-order MMDCs to achieve the desired mitigation of the DCD's polynomial INL, IBS spurs, and the DTC's physical noise.

Analysis for Higher-Order MMDCs

It is crucial to note that, prior to DTC linearity enhancement, all the original DCD's polynomial INLs reported in the published literature have a similar concave shape, as modeled by the solid black curve in Fig. 6.4; after enhancement, the DCD always exhibits a residual INL, which can have one of the following qualitatively different signatures.

1. "Wave-shaped" residual INL: in [10, 11, 15, 16, 24], etc., the suppressed DCD's INL has a "wave shape" like that shown as the dashed curve in Fig. 6.4.
2. Weaker concave-shaped residual INL: in [2, 7–9, 12, 14], etc., the DCD's nonlinearity becomes weaker, yet the residual INL still has a concave shape, as shown in the dotted curve in Fig. 6.4.
3. Irregular-wave-shaped residual INL: in [1, 13], the shape of the reported residual INL is more irregular than the previous cases, but can still be approximated by a cubic polynomial function overlaid with random jumps.

The shapes of such residual INLs can be intuitively understood as due to the mismatch between the characteristic curve introduced by INL-mitigation techniques to match and cancel the original DCD's INL and the curve of the original INL *per se*. In this subsection, we will describe in detail the models of these representative residual INLs and comment on the effectiveness of DTC linearity enhancement techniques.

We notice that, except for the EFM1-based fractional-N DTC-DPLLs introduced in Sect. 8.2.1, a MASH 1–1 MMDC is typically favored in high-performance DTC-DPLLs, e.g., [1, 4, 6, 8, 9, 13, 15, 16, 21]. This is because, compared to a first-order DDSM, the MASH 1–1's AQE is more randomized and can, under certain circumstances, be made spur-free *per se* [36, 37]. Compared to higher-order DDSMs, the MASH 1–1's AQE has a smaller range of variation, which allows the DTC to have a smaller DR to accommodate its variation, hence saving DTC hardware. For the completeness of analysis, in this subsection we compare the cases where MASH 1–1 and 1–1–1 are used as the MMDC; the DTCs in both cases are modeled to have the same equivalent resolution (finest unit delay) and to cover the variation range of the MMDC's AQE.

The PNN analysis framework [31, 32, 34] is used in this subsection (i) to predict precisely the IBS spurs arising from the MMDC's AQE passing through the modeled DCD's polynomial INL and (ii) to help to understand which type of DTC linearity enhancement mitigates IBS spurs more effectively when different orders of MMDC are used. PNN is a zero-mean time-domain waveform that is calculated as

$$\mathrm{PNN}[k] = \sum_i f^{\mathcal{N}}\left(\tau_i[k]\right) \cdot P\left(\tau_i[k]\right). \tag{8.1}$$

It describes the time-dependent expected value of the DDSM's AQE that lies on a number of parallel tracks (τ_i) indexed by i and is distorted by a nonlinearity ($f^{\mathcal{N}}(\cdot)$); it also takes into account the probability distribution of the AQE whose PDF is denoted as $P(\cdot)$. A non-constant PNN manifests itself in the spectrum as IBS spurs. Note that, in [31, 32], the PNN is derived based on the PDF of the classic definition of the DDSM's AQE, i.e., $q^{\mathrm{acc}}_{\mathrm{MMDC}}[k]$ in (7.12) (see [32, Eqs. (15)–(17)]), rather than its digital-domain version, $q^{\mathrm{acc}}_{\mathrm{MMDC}}\big|_d[k]$ in (6.3). Therefore, recalling the red path in Fig. 6.3b generating the DCD's polynomial INL-induced spurs, that is

$$t_\epsilon^{\mathrm{poly}}[k+1] = \frac{\Delta}{M} \cdot T_{\mathrm{out}} \cdot \mathrm{poly}\left(\frac{a[k]}{\Delta}\right),$$

and substituting into it the relationships in (6.3) and (6.4), the red path can be transformed and expressed as

$$t_\epsilon^{\mathrm{PNN}}[k+1] = T_{\mathrm{out}} \cdot \mathrm{poly}_{\mathrm{PNN}}\left(q^{\mathrm{acc}}_{\mathrm{MMDC}}[k]\right), \tag{8.2}$$

where

$$\mathrm{poly}_{\mathrm{PNN}}(x) = \frac{\Delta}{M} \cdot \alpha_0 + (\alpha_1 - 1) \cdot x + \frac{M}{\Delta} \cdot \alpha_2 \cdot x^2 + \left(\frac{M}{\Delta}\right)^2 \cdot \alpha_3 \cdot x^3,$$

represents the nonlinear function used in our PNN analysis, i.e., $f^{\mathcal{N}}(\cdot)$ in (8.1). Recall that $\alpha_0, \alpha_1, \alpha_2, \alpha_3$, e.g., as modeled in (6.6), are the coefficients of the DCD's nonlinear transfer characteristic $f^{\mathrm{poly}}_{\mathrm{DCD}}(\cdot)$ in (6.5).

As a final remark, to show the effect of DTC linearity enhancement clearly, we consider in this subsection DTCs that have a polynomial nonlinear DCD but do not perform quantization. The DTC requantizer is replaced by a linear gain of $1/\Delta$. As shown in Fig. 6.3b, only the red path generates noise and spurs in this case. In fact, DTC linearity enhancement by suppressing the DCD's INL has no effect on mitigating the DTC's inherent quantization and DQI spurs. This is demonstrated by the green plots in Fig. 6.5c and d, which show that, even if the DCD has perfect linearity, the DQI spurs are intact and distinguishable from INL-induced responses.

Original polynomial INL: the unmitigated polynomial INL of the original DCD is modeled as per (6.6) and (6.7). Distorted by this nonlinearity, the MMDC's AQEs yield PNNs (PNN[k] in (8.1)) as shown in the first row of Fig. 8.2. The second row shows the power spectra of the residual errors induced by the nonlinear DTCs ($t_\epsilon[k+1]$ in Fig. 6.3), and also the power spectra resulting from the PNN prediction ($T_{\text{out}}\cdot$

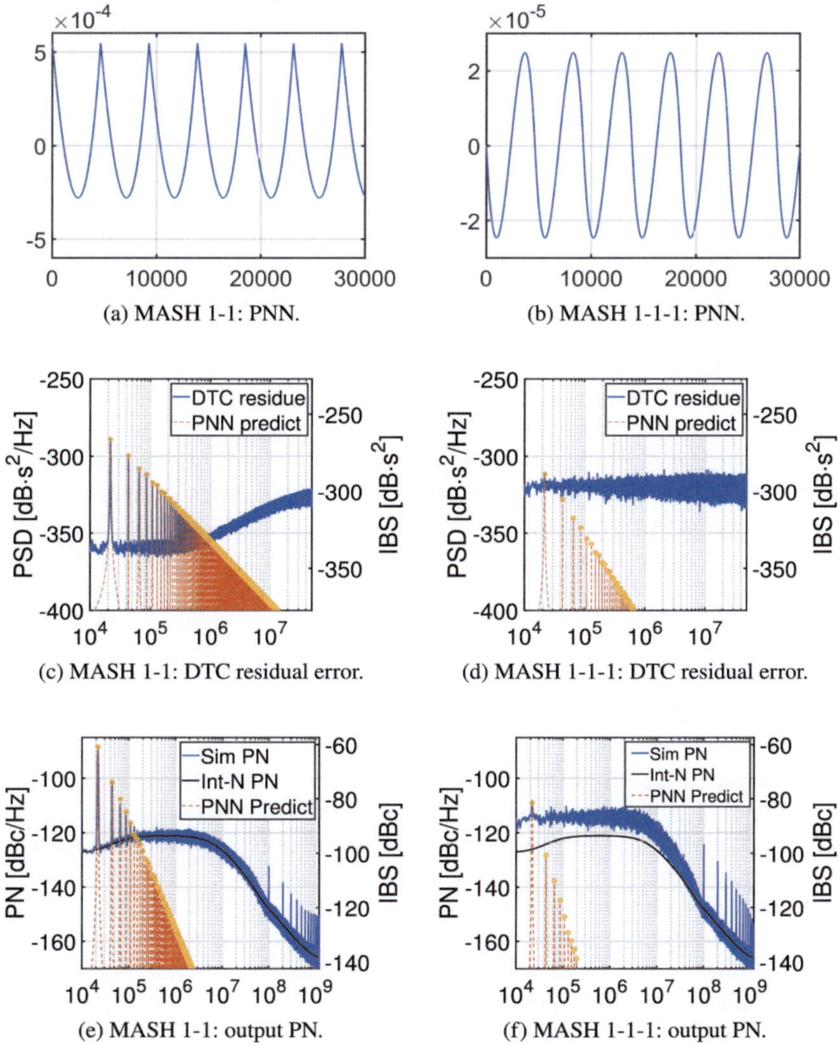

(a) MASH 1-1: PNN. (b) MASH 1-1-1: PNN.

(c) MASH 1-1: DTC residual error. (d) MASH 1-1-1: DTC residual error.

(e) MASH 1-1: output PN. (f) MASH 1-1-1: output PN.

Fig. 8.2 Analysis of the PN and spurs induced by the *infinite-resolution nonlinear DTC* with the *unmitigated DCD's polynomial INL* modeled by (6.5) and (6.6)

PNN[k] as per (8.2)). The third row shows the simulated output PN spectra, where the PNN-predicted output-referred IBS spurs at close-in frequencies are transferred from (8.2) and scaled by $N \cdot 2\pi \cdot F_{\text{ref}}$, where N is the lowpass gain of the noise transfer function imposed by the loop. The scaling by $2\pi \cdot F_{\text{ref}}$ transfers the time-domain signal to the phase domain. The left and right columns of Fig. 8.2 correspond to the cases where MASH 1–1 and 1–1–1 are used as the MMDC. Notice that the simulated IBS spurs match their counterparts predicted by the PNN *exactly* at both DTC-block and DPLL-system levels.

One can see that, in this case, a MASH 1–1–1 MMDC yields a PNN waveform that has a much smaller magnitude than that using a MASH 1–1. This is because, as per the PNN theory [32], MASH 1–1–1 provides inherent spur immunity against polynomial nonlinearities up to second order. In other words, the DC, linear, and quadratic terms in the DCD's INL produce constant components in the PNN; only the cubic term yields a time-varying PNN. On the other hand, a MASH 1–1 DDSM is not spur-immune, even to quadratic distortion [32]. Recall from Fig. 6.4 and (6.6), that the original DCD's INL has a strong quadratic component, which yields the biggest proportion of PNN magnitude in the case of a MASH 1–1, but not MASH 1–1–1.

Consequently, in the frequency domain, spurs in the nonlinearity-induced IBS family in the MASH 1–1–1 case have much smaller (≈ 20 dB less) magnitudes compared to the MASH 1–1 case. Hence, in terms of IBS mitigation for the DCD before linearization, using a MASH 1–1–1 is preferable to a MASH 1–1.

However, the problem of using a MASH 1–1–1 MMDC with the original DCD is that, as shown in Fig. 8.2d and f, since the MASH 1–1–1's AQE exercises the full DR of the modeled DCD, and the polynomial INL (solid black curve in Fig. 6.4) gets worse approaching both edges, it produces a higher folded PN floor compared to the MASH 1–1 case; the latter drives only the central portion of the DR which has a more moderate INL, hence producing less PN. In Fig. 8.2f, the close-in PN is elevated from the integer-N level due to the MASH 1–1–1 induced folded noise. In Fig. 8.2e, the INL-induced PN is masked in the spectrum by the intrinsic integer-N noise level.

In Fig. 8.2e and f, "reference spurs" at the fundamental and harmonics of $F_{\text{ref}} =$ 100 MHz are amplified due to the DCD's INL.

Wave-Shaped Residual INL: after linearity enhancement, the residual INL may have a wave-shaped curve (like that shown dashed in Fig. 6.4). Apart from the inevitable residual mismatch between the linearization provided by any such technique and the original INL, this phenomenon is particularly prominent in cases where the linearization technique, such as the pseudo-differential DTC in [16], is designed to cancel the even-order terms of the DCD's INL. As discussed above, such terms typically dominate the concave shape of the original INL. Hence, we update the DCD's nonlinear model from (6.6) to

$$\alpha_0 = 0,\ \alpha_1 = +0.998,\ \alpha_2 = 0,\ \alpha_3 = +4.9 \cdot 10^{-10}, \tag{8.3}$$

which is plotted as the dashed curve in Fig. 6.4.

The same simulation in Fig. 8.2 is performed for this case, as shown in Fig. 8.3. In terms of INL-induced PNs, the mitigated INL yields much less folded PN compared to the original INL in both the MASH 1–1 and 1–1–1 cases, which recover the integer-N output PN level even for the MASH 1–1–1 case.

In terms of IBS spurs, one can see that canceling the even-order terms of the DCD's polynomial INL suppresses the PNN's amplitude for the MASH 1–1 case; this results in IBS mitigation by $\approx$ 15 dB compared to the original case (compare (c)

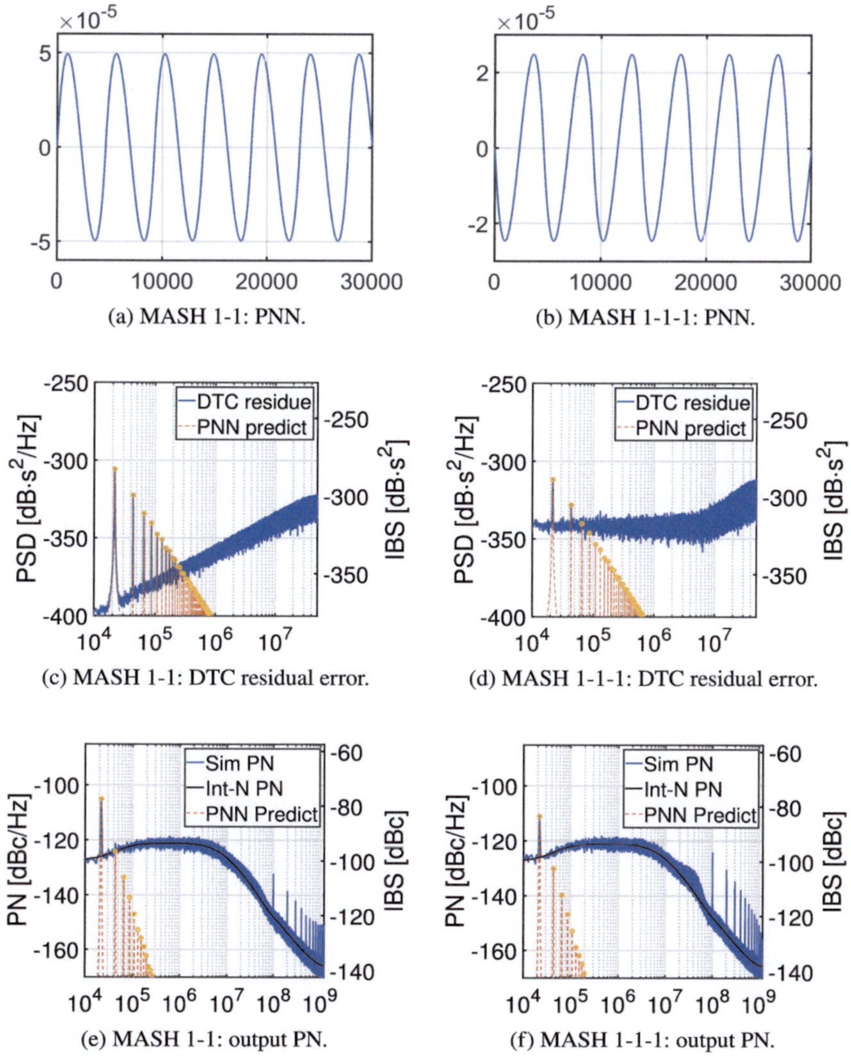

(a) MASH 1-1: PNN. (b) MASH 1-1-1: PNN.

(c) MASH 1-1: DTC residual error. (d) MASH 1-1-1: DTC residual error.

(e) MASH 1-1: output PN. (f) MASH 1-1-1: output PN.

Fig. 8.3 Analysis of the PN and spurs induced by the *infinite-resolution nonlinear DTC* with the *wave-shaped DCD's residual INL* modeled in (8.3) and (e) in Figs. 8.2 and 8.3). On the other hand, for the MASH 1–1–1 case, because the PNN is only contributed by the cubic term that is not affected by this type of linearity enhancement, the PNNs in Fig. 8.3b and 8.2b are identical. As expected, this type of DTC linearization does not alleviate the IBS spur family at all (see (d) and (f) in Figs. 8.2 and 8.3).

Notice that, comparing Fig. 8.3e and f, the MASH 1–1–1 case recovers almost the same PN level as the MASH 1–1 case. Despite the fact that it exhibits a PN "bump" outside its bandwidth, a MASH 1–1–1 is still superior to a MASH 1–1, in terms of IBS magnitudes.

Weaker Concave-Shaped Residual INL: as introduced above, the residual INL after some linearization techniques can still feature a concave shape, like that shown dotted in Fig. 6.4, whose peak-to-peak range is reduced by about three to four times, e.g., in [2]. For a fair comparison, we modeled the residual INL in this case to have the same peak-to-peak range as the previous wave-shaped INL; the coefficients in (6.7) are reduced by a common factor. The updated coefficients are

$$\alpha_0 = +0.95,\ \alpha_1 = +0.9995,\ \alpha_2 = -7.4 \cdot 10^{-7},\ \alpha_3 = +1.1 \cdot 10^{-10}. \tag{8.4}$$

The resultant INL is plotted as the dotted black curve in Fig. 6.4.

The simulation results are shown in Fig. 8.4. In this case, both MASH 1–1 and 1–1–1 approach the integer-N PN performance. In terms of IBS spurs, as the coefficients of the polynomial INL in this case are scaled down by the same proportion from the original INL, the PNNs of both MASH 1–1 and 1–1–1 have the same shape and reduced amplitude compared to their counterparts in the original INL analysis (compare (a) and (b) in Figs. 8.2 and 8.4). Consequently, the IBS spurs in both MASH cases are reduced by the same proportion ($\approx$ 11 dB) (compare (c)–(f) in Figs. 8.2 and 8.4).

Finally, we conclude this section with the following observations.

1. Despite it exercising the INL less than higher-order modulators, using an EFM1 MMDC yields the worst spectral response among the analyzed cases when considering the DCD's INL.
2. In our simulations, if a MASH 1–1 MMDC is used, the PN induced by the DCD's polynomial INL is not comparable to the intrinsic integer-N PN; DCD linearity enhancement mainly helps to mitigate IBS spurs and its PN mitigation effect is imperceptible.
3. In terms of the IBS-spur-mitigation effect, MASH 1–1–1 is superior to MASH 1–1 in all the modeled cases.
4. The original excess PN induced by the MASH 1–1–1 (in Fig. 8.2f) can be much reduced by linearity enhancement, after which the output PN approaches the integer-N performance (see (f) in Figs. 8.3 and 8.4).

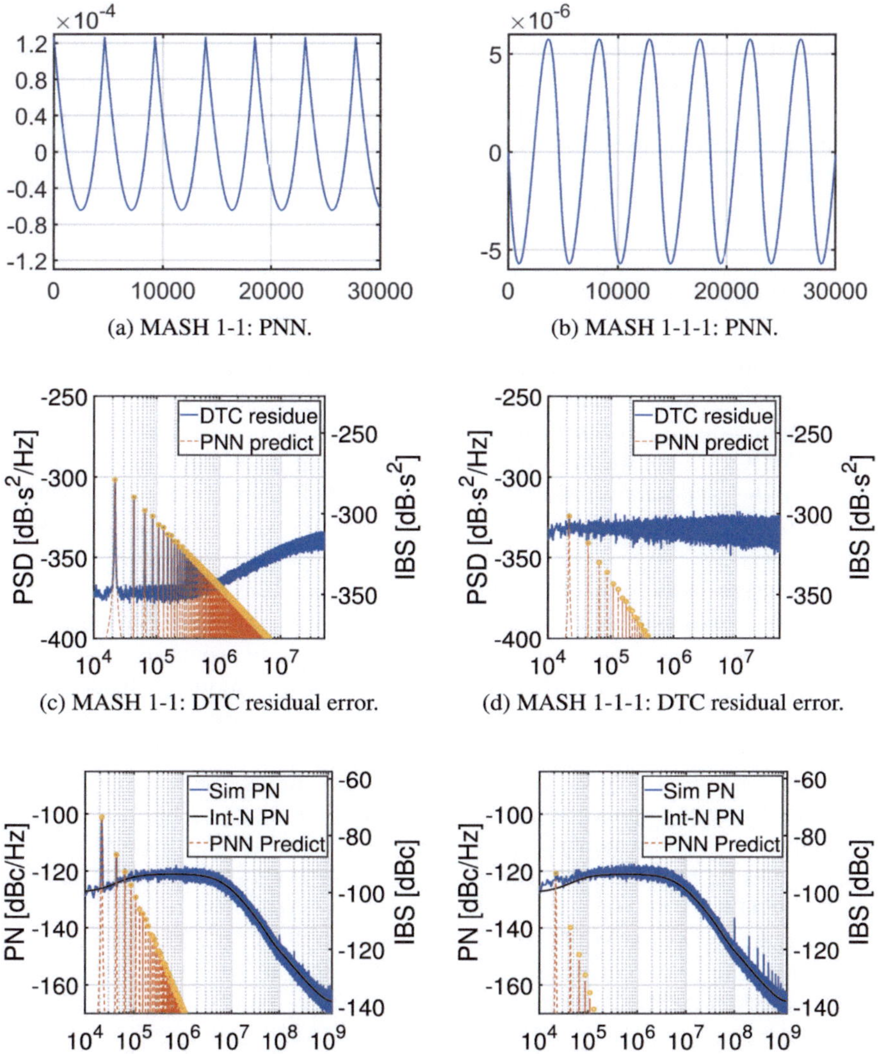

(a) MASH 1-1: PNN.

(b) MASH 1-1-1: PNN.

(c) MASH 1-1: DTC residual error.

(d) MASH 1-1-1: DTC residual error.

(e) MASH 1-1: output PN.

(f) MASH 1-1-1: output PN.

Fig. 8.4 Analysis of the PN and spurs induced by the *infinite-resolution nonlinear DTC* with the *weaker concave-shaped DCD's residual INL* modeled in (8.4)

5. Linearizing the DCD *per se* can, in some cases, mitigate IBS spurs but cannot eliminate them completely. IBS spurs due to the residual INL can still dominate the fractional-spur performance at close-in frequencies. As long as the residual INL can be approximated by a polynomial of up to third order, IBS spurs persist. This motivates the use of stochastic divider controllers that will be discussed in the next subsection in order to gain IBS-spur immunity against soft nonlinearities.

8.2.2 Stochastic Divider Controllers

"Statistical methods," e.g., successive requantization [17–19], PMR [20], time-invariant-probability modulator [21], PDS $\Delta\Sigma$M [10–12], and ENOP DDSM [22], have been proposed in the literature as alternatives to conventional MASH DDSM-based MMDCs. These methods reshape the statistical distribution of the AQE so that, when the proposed modulator is used as the MMDC in the fractional-N PLL, its AQE interacting with certain types of nonlinearities within the system does not produce spurs.

As discussed in the previous subsection, the INLs of the DCD before and after linearization exhibit nonlinear characteristics that can be approximated by third-order polynomial functions; their induced IBS spurs dominate the spurious performance at close-in frequencies, even when the DCD's linearity is improved. Therefore, in this section, we demonstrate the use of an ENOP DDSM as an example in order to eliminate these polynomial-INL-induced spurs.

The family of ENOP MMDCs proposed in [22] exhibits immunity against IBS, horn, and wandering spurs [38] when their AQEs are distorted by a memoryless polynomial nonlinearity. In particular, the ENOP P1 provides spur-immunity against nonlinearities up to third order. Hence, it is suitable for use as the MMDC in the DTC-based DPLL.

Listing 8.1 shows the MATLAB model of an ENOP P1 DDSM which can used as the MMDC in place of a conventional MASH DDSM, as shown in Listing 3.4. To enable the ENOP MMDC in simulation, Line 55 of the top file, Listing 2.1, should be changed from invoking the `MASH` to the `ENOP_P1` function.

Listing 8.1 MATLAB behavioral model of an ENOP P1 DDSM.

```
function [N_div,q_acc] = ENOP_P1(X,M,N_int,ncycles)
    %% Local preallocation
    c1 = zeros(1,ncycles);
    s1 = zeros(1,ncycles);
    c2 = zeros(1,ncycles);
    s2 = zeros(1,ncycles);
    q_acc = zeros(1,ncycles);
    N_div = N_int * ones(1,ncycles);
    %% ENOP P1 generation
    for k=4:ncycles
        [c1(k),s1(k)] = adder_modulo_M(X,s1(k-1),M);
```

```
        [c2(k),s2(k)] = adder_modulo_M(s1(k),s2(k-1) + s2(k-2) -
            s2(k-4),M);
        N_div(k) = N_int + c1(k) + c2(k) - c2(k-1);
        q_acc(k) = - s2(k) + s2(k-1) + s2(k-2) - s2(k-4);
    end
end
```

In brief, the advantage of ENOP P1 is that its AQE has the same variation range as that of a MASH 1–1–1; at the same time, it can provide IBS-spur immunity that is the same as a MASH 1–1–1–1–as per PNN theory, a MASH 1–1–1–1 is immune to IBS spurs resulting from interaction with polynomial nonlinearities up to third order [32]. The DTC-block and DPLL-system-level results shown in Fig. 8.5 confirm this property. Compared with (d) and (f) in Figs. 8.3 and 8.4, which shows the residual spectra of linearity-enhanced DTCs with a MASH 1–1–1 MMDC, Fig. 8.5 demonstrates that the ENOP P1 achieves complete IBS-spur elimination and, since the DCD's INL has also been suppressed, the output PN reaches the theoretical integer-N limit (shown in black) as well.

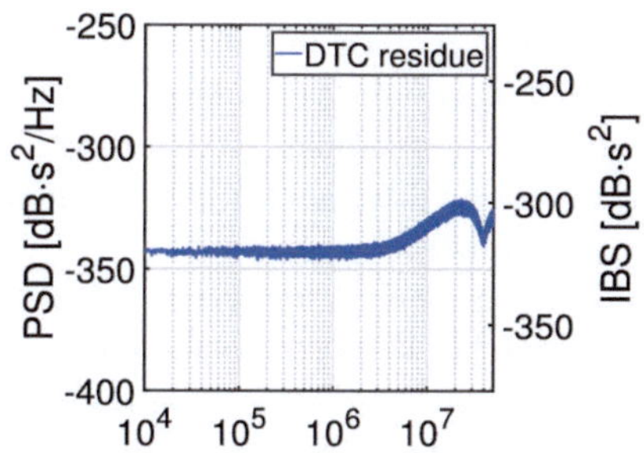

(a) Wave-shaped residual INL: DTC residual error.

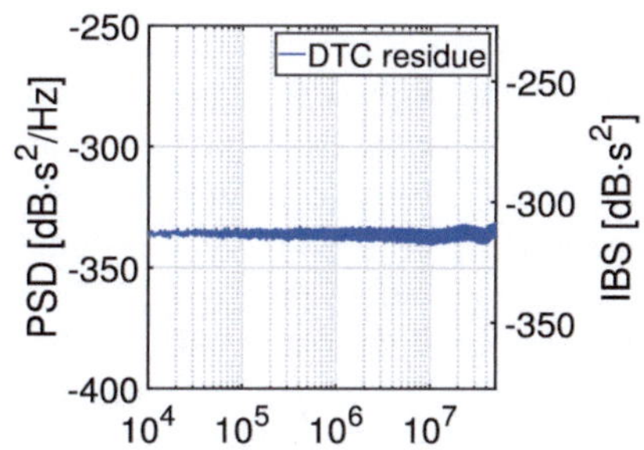

(b) Weaker concave-shaped residual INL: DTC residual error.

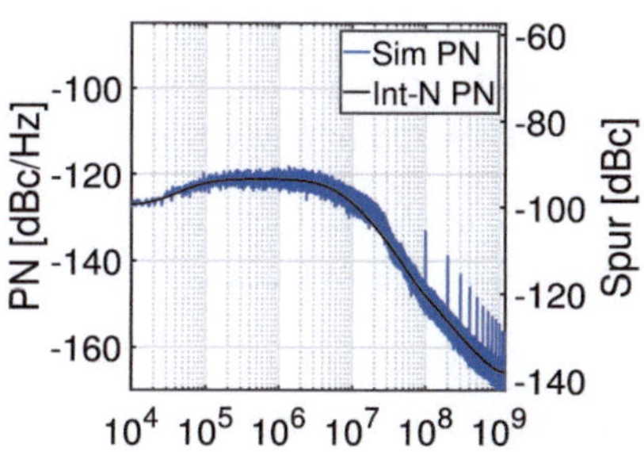

(c) Wave-shaped residual INL: output PN.

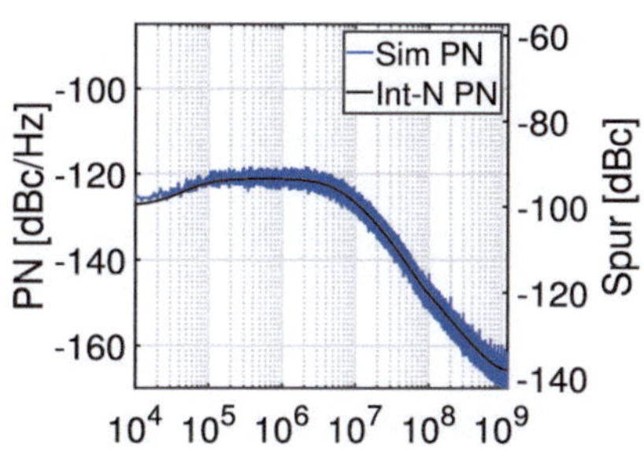

(d) Weaker concave-shaped residual INL: output PN.

Fig. 8.5 Analysis of the spectrum using the *infinite-resolution nonlinear DTC* with the *residual polynomial INLs* modeled in Sect. 8.2.1 and an *ENOP P1 MMDC*

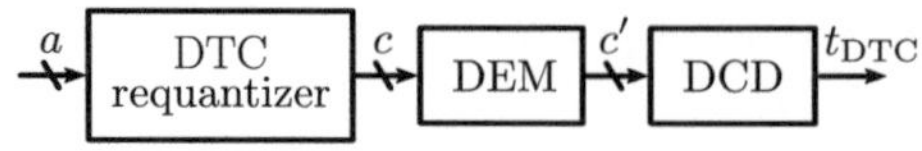

Fig. 8.6 DEM block controlled by the digital code, c, generated by the DTC requantizer. The DEM updates c' that shuffles the access to the delay cells in the DCD

8.3 Dynamic Element Matching to Eliminate REM Nonlinearity-Induced Spurs

In this section, we analyze the effectiveness of the DEM method used in [23] to mitigate a DCD's REM-induced spurs. Notice that, Helal *et al.* described in [23] a DTC architecture that cascades 1-bit DTC stages with binary-weighted delays. The element mismatch of this can be inherently worse than conventional DTCs with a DCD, as shown in Fig. 6.2 whose delay is digitally controlled by switching a unitary-weighted capacitor bank.

From the analysis in Chap. 6, we concluded that the generation mechanism of the REM spurs is related to the MMDC's AQE's periodic exercise of the static, albeit seemly random, REM pattern of the DCD. The DEM method therefore can be used to shuffle the accessed delay cells within the DCD, as shown in Fig. 8.6. In this way, when the same control code yielded by the DTC requantizer accesses the DCD repetitively, it always activates different delay cells with different static REM; hence, the periodicity of the experienced nonlinearity is broken.

Simulations show that the DEM method works equally effectively for non-segmented and segmented DTCs with modeled REM characteristics, as discussed in Chap. 6. Fig. 8.7 shows the simulated result. The first row shows that, when the DTC is modeled to exhibit only REM nonlinearity and the control of the DCD's delay cells is not shuffled, REM spurs appear in the spectrum of the DTC's residual error in (a), which are transferred to the output PN as shown in (b). Unfortunately, the highest amplitude REM spurs appear at in-band frequencies and can therefore contribute significantly to the overall jitter. The second row shows what happens when the DEM method is used. DEM eliminates REM spurs at source and turns the DTC's residual error into white noise, as shown in (c); the output PN in (d) in this case approaches the integer-N level as the DEM-produced white noise is negligible compared to the integer-N noise sources.

8.4 Comprehensive Strategy to Mitigate Excess PN and Spurs in DTC-Enhanced Synthesizers

In this section, we present a comprehensive DTC design strategy using different classes of DTC enhancement techniques which target different nonlinear sources to mitigate their PN and spurs in order to approach the fractional-spur-free integer-N

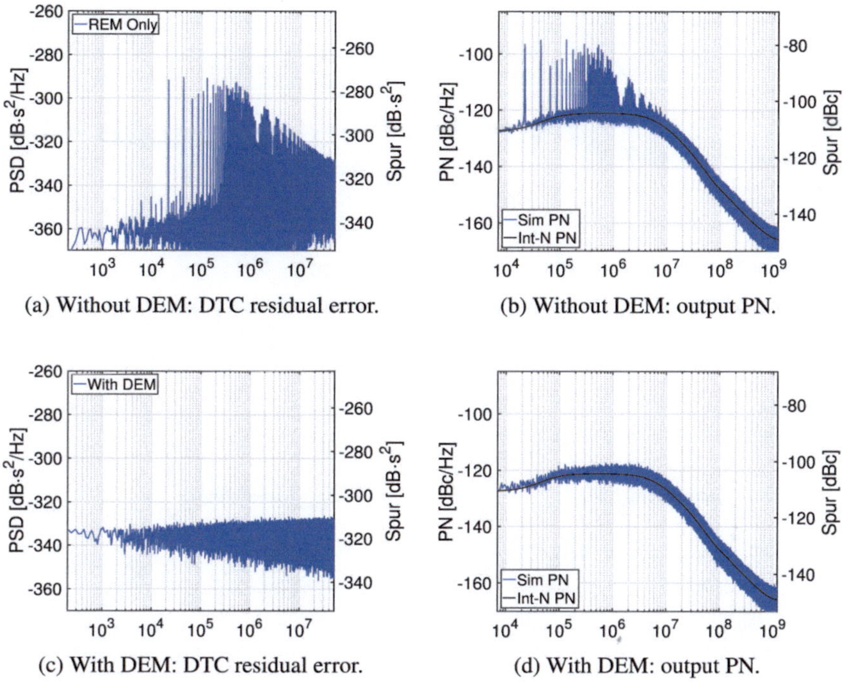

Fig. 8.7 Analysis of the spectrum using the *infinite-resolution DTC with a DCD that has only REM nonlinearity* and a *MASH 1–1 MMDC*

PN limit in the case of the fractional-N frequency synthesizer. Section 8.4.1 shows that using the IDQ method can eliminate DQI spurs caused by mismatch between the resolutions of the MMDC's QE and the DTC; this can be applied together with the DCD linearity enhancement, stochastic MMDC, and DEM methods described in this chapter to improve the DTC's performance. Section 8.4.2 presents a comprehensive qualitative DTC enhancement strategy that provides insight into why and how to combine all the methods studied in this work; this helps to reach the fractional-spur-free integer-N PN performance in fractional-N PLLs.

8.4.1 Reconciliation of DCD Enhancement with IDQ Method

The DTC requantizer in Sects. 8.2.1 and 8.2.2 was assumed to have infinite resolution, and hence it does not perform quantization. Here, we reconsider the DTC's quantization operation and redo the simulations shown in Fig. 8.5. The DTC requantizer now has 11-bit equivalent resolution with $\Delta = 2^{10}$, as per the parameterization method described in [25]. Figure 8.8 shows that conventional DCD linearity enhancement, statistical methods, and DEM method do not help to alleviate DQI spurs that arise

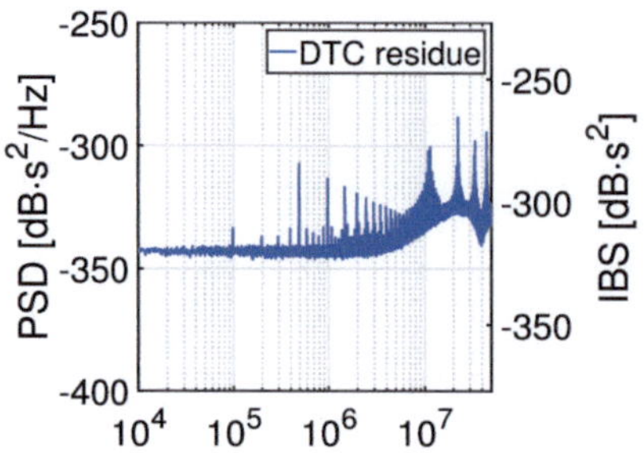

(a) Wave-shaped residual INL: DTC residual error.

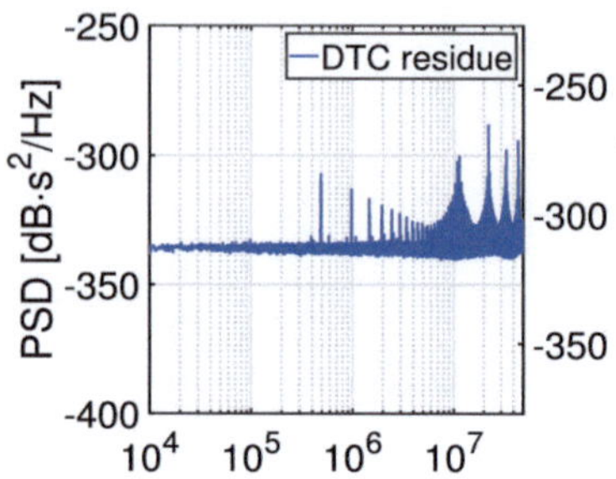

(b) Weaker concave-shaped residual INL: DTC residual error.

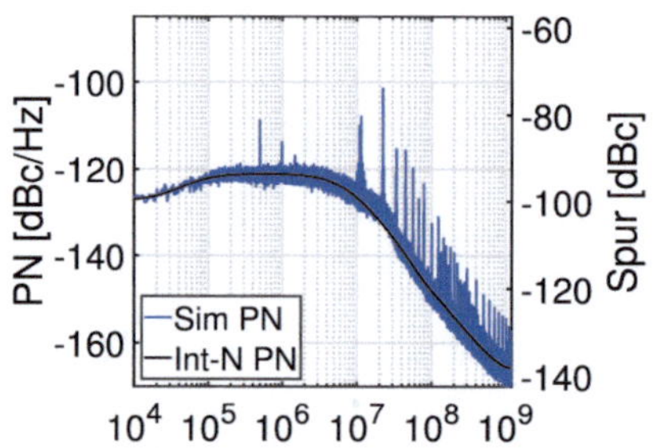

(c) Wave-shaped residual INL: output PN.

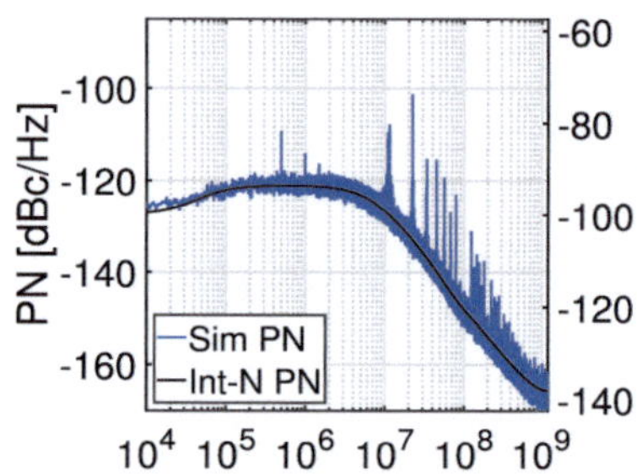

(d) Weaker concave-shaped residual INL: output PN.

Fig. 8.8 Analysis of the spectrum using the *finite-resolution nonlinear DTC* with a *DEM-shuffled DCD with residual INLs* modeled in Sect. 8.2.1 and an *ENOP P1 MMDC*

from the resolution gap between the DTC's control and output. DQI spurs prevail in this case; these are identical to the green plots in Fig. 6.5, where the DCD is considered linear and the only nonlinearity modeled in the DTC is the hard quantization nonlinearity.

The phenomenon of DQI spurs can be intuitively understood as, when using a DTC with limited (N_b^{DTC}-bit) resolution to match and cancel the AQE with finer resolution, only the N_b^{DTC} most significant bits of the error can be canceled in the best-case scenario; the LSBs of the AQE will still be injected into the loop as the residual error after DTC-based cancellation. Although the full-range MMDC's AQE can be made spur-free, its LSBs inevitably produce spurs.

The DQI-spur pattern is not related to the order of the MMDC or the shape of DCD's INL. However, the positions and magnitudes of the DQI spurs are related to the fractional part of the FCW, X, and the requantization resolution, Δ; detailed analysis for the DQI spurs has been provided in Chap. 6. Contrary to the assertion made in [35] that DQI spurs occur at out-of-band frequencies that can be filtered by the loop, our analysis shows that, with certain X values and a large mismatch ratio Δ, the dominant components of the DQI spurs will be present at in-band frequencies;

these can contribute significantly to the system jitter. The IDQ methods proposed in Chap. 7 can be used within the DTC; they linearize the transfer characteristic of the DTC requantizer by transforming it from a hard quantization nonlinearity to an effective linearity.[1]

Figure 8.9 adopts the optimum IDQ method derived in Chap. 7 and uses it to improve the performance of the DTCs modeled in Fig. 8.8. The DTC is also parameterized as per the sizing method described in Chap. 7 so that the (noise-like) dithering power of the IDQ is suppressed to be insignificant compared to the integer-N jitter power. The first two rows of Fig. 8.9 show that using the IDQ method eliminates DQI spurs completely; the recovered integer-N PN level remains intact in this case. The third row shows the simulated output PN of a DPLL using a bang-bang TDC. These confirm that, because all the identified fractional spurs have been eliminated at source and the DTC residual error is limited within the linearized region of bang-bang detection, defined in Chap. 3, the output PN can approach the fractional-spur-free integer-N performance even in DPLLs with quantized TDCs.

Notice that, by using a combination of DCD linearity enhancement, ENOP, IDQ, and DEM methods simultaneously, the output PN performance can be improved from the worst-case performance in Fig. 6.5b to Fig. 8.9c–d, which best approximate the integer-N PN performance and are also free of fractional spurs.

8.4.2 Comprehensive DTC Enhancement Strategy

Simulations and analysis in this section suggest that, in order to recover the fractional-spur-free integer-N PN in fractional-N PLLs, state-of-the-art DCD linearity enhancement techniques should always be used to linearize the transfer function of the DCD as much as possible. For the inevitable residual smooth INL of the DCD, which produces a residual family of IBS spurs, modulators based on statistical methods such as ENOP can be used for their elimination. The DEM method can be used to shuffle the access to the DCD's delay cells to gain spur immunity again the DCD's REM nonlinearity. It is crucial to recall that, even if the DCD is perfectly linear, as long as the DTC has limited resolution and produces time delays that are coarser than the MMDC's AQE, DQI spurs will be present. Therefore, the IDQ technique should be used to eliminate DQI spurs in cases where the resolution of the DTC's input and output are not the same.

[1] As analyzed in Sect. 6.3.3, using an EFM1 requantizer helps to shape the DQI spurs but cannot eliminate them entirely, its induced output jitter in Table 6.2 is inferior to that using the optimum IDQ method in Table 7.2. Therefore, in this section, we use the latter as a means to mitigate DQI spurs. The effect of using an EFM1 requantizer was shown in Fig. 6.9.

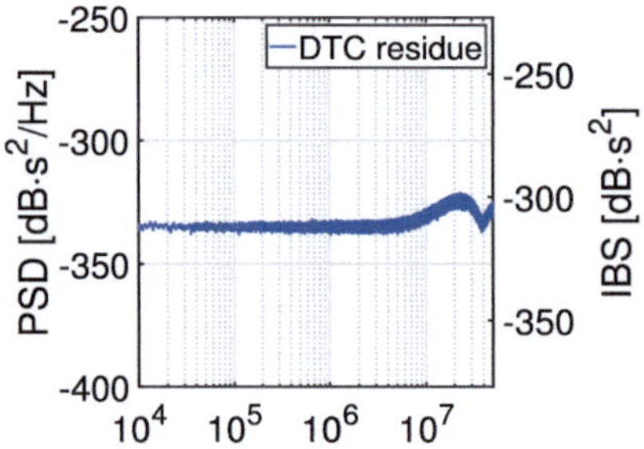

(a) Wave-shaped residual INL: DTC residual error.

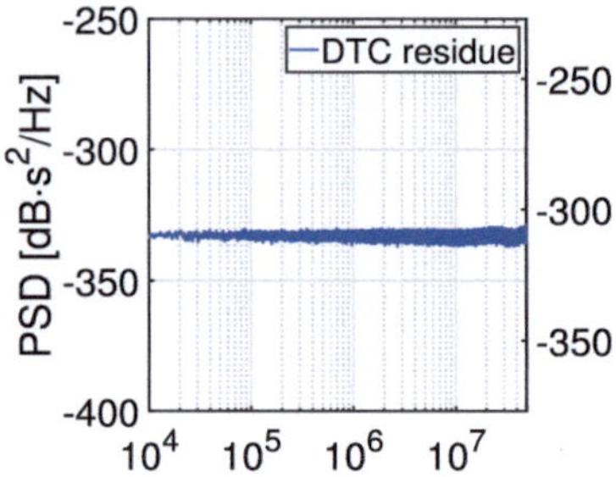

(b) Weaker concave-shaped residual INL: DTC residual error.

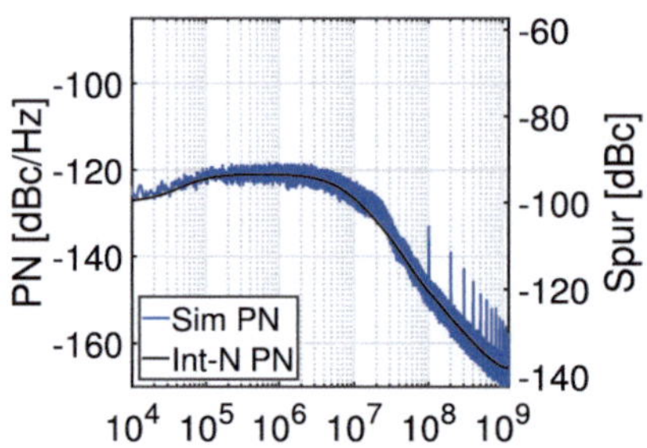

(c) Wave-shaped residual INL: output PN.

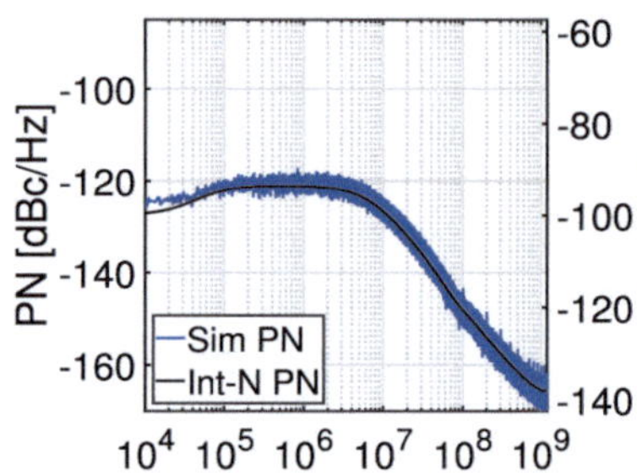

(d) Weaker concave-shaped residual INL: output PN.

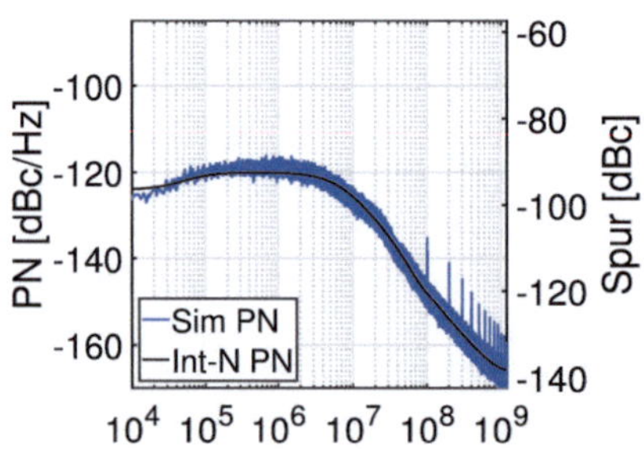

(e) DTC with wave-shaped residual INL & BPD: output PN.

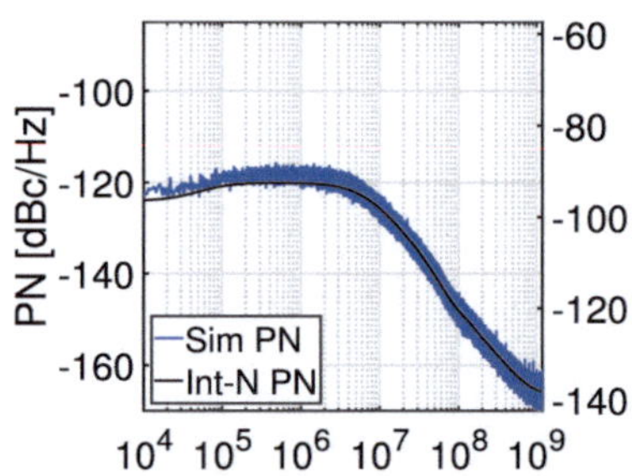

(f) DTC with weaker concave-shaped residual INL & BPD: output PN.

Fig. 8.9 Analysis of the spectrum using the *finite-resolution nonlinear IDQ-DTC* with a *DEM-shuffled DCD with residual polynomial INLs* modeled in Sect. 8.2.1 and an *ENOP P1 MMDC*. The third row shows the simulated PNs of BBPLLs using a BPD

MATLAB Script for DTC-Related Analysis

Listing 8.2 shows the MATLAB script that compares the predicted DTC-induced DQI spurs (due to the hard quantization nonlinearity) and IBS spurs (due to the soft integral nonlinearity) with their simulated counterparts. When the DTC enhancement techniques are adopted, the analysis should confirm the effective spur mitigation facilitated by them. This script is executed in Line 99 of Listing 2.1.

Listing 8.2 MATLAB script to check the match between the simulated and predicted DQI and IBS spurs arising due to the DTC's nonlinearities.

```
%%%%%%%%%%%%%%%%%%%%%%%%%%%%%%%%%%%%%%%%%%%%%%%%%%%%%%%%%%%%%%
% DTC BLOCK LEVEL ANALYSIS %
%%%%%%%%%%%%%%%%%%%%%%%%%%%%%%%%%%%%%%%%%%%%%%%%%%%%%%%%%%%%%%

%% Simulated DTC residual error injection
t_original_injection = 1/F_DCO/M * q_acc;
t_residue = t_DTC - t_original_injection;
%% Calculate AC residual error PSD
npsd=2^(nextpow2(length(t_residue)/1)-1);
Fpsd=1/(npsd/F_ref);
fpsd=0:Fpsd:Fpsd*(floor(npsd/2)-1);
[PSD,~]=pwelch(t_residue-mean(t_residue),window(@hann,npsd),
    npsd/2,npsd,F_ref,'Two-sided');
PSD_t_residue=PSD(1:floor(npsd/2));
%% Prediction of DQI spurs
k = 1:ncycles;
if DTC_quant_ON == 1
    % Raw DQI waveform
    q_in_raw_DTC_predict = -(mod(-X*k,res_DTC)) * 1/F_DCO/M;
    % AC PSD calculation
    [PSD,~]=pwelch(q_in_raw_DTC_predict-mean(
        q_in_raw_DTC_predict),window(@hann,npsd),npsd/2,npsd,
        F_ref,'Two-sided');
    PSD_q_in_raw_DTC_predict=PSD(1:floor(npsd/2));
end
%% Prediction of IBS spurs (for MASH 1-1 as an example)
if (DTC_nonlin_ON) == 1 && (MASH_order == 2)
    % PNN waveform prediction
    tau = - mod(X/M*k,1);
    PNN0 = alpha0 * ( (1+tau)+ (-tau));
    PNN1 = alpha1 * ((tau*M/res_DTC).* (1+tau)+((tau+1)*M/
        res_DTC).* (-tau));
    PNN2 = alpha2 * ((tau*M/res_DTC).^2.*(1+tau)+((tau+1)*M/
        res_DTC).^2.*(-tau));
    PNN3 = alpha3 * ((tau*M/res_DTC).^3.*(1+tau)+((tau+1)*M/
        res_DTC).^3.*(-tau));
    PNN_out_INL = res_DTC/M/F_DCO * (PNN0 + PNN1 + PNN2 + PNN3)
        ;
    % AC PSD calculation
    [PSD,~]=pwelch(PNN_out_INL-mean(PNN_out_INL),window(@hann,
        npsd),npsd/2,npsd,F_ref,'Two-sided');
```

```
    PSD_PNN_out_INL=PSD(1:floor(npsd/2));
end
%% Result plotting
if (DTC_quant_ON ==1) && (DTC_nonlin_ON ~= 1)
    %% Quantization nonlinearity only
    % Time-domain analysis
    figure('Position',[0 0 900 700]+100)
    stairs(t_residue(2:5*res_DTC),'linewidth',10);
    hold on
    stairs(q_in_raw_DTC_predict(2:5*res_DTC),'linewidth',3);
    hold off
    xlim([-inf +inf])
    grid on
    legend('Sim','Pred','location','southwest')
    title('Time-Domain Analysis of Raw DQI Spurs (Quant Only)')
    ax = gca;
    ax.LineWidth=3;
    ax.FontSize=20;
    % Frequency-domain analysis
    figure('Position',[0 0 900 700]+100)
    semilogx(fpsd,10*log10(PSD_t_residue),'linewidth',3);
    hold on
    semilogx(fpsd,10*log10(PSD_q_in_raw_DTC_predict),'--','
        linewidth',3);
    hold off
    grid on
    xlim([2e2 inf]);
    yyaxis left
    ylim([-450 -250]);
    yyaxis right
    ylim(10*log10(Fpsd)+[-450 -250]);
    ylabel('Spur [dB{\cdot}s^2]')
    set(gca,'YColor','k');
    legend('Sim','Pred');
    title('Frequency-Domain Analysis of Raw DQI Spurs (Quant
        Only)')
    ax=gca;
    ax.LineWidth=3;
    ax.FontSize=20;
elseif (DTC_quant_ON ~=1) && (DTC_nonlin_ON == 1)
    %% Integral nonlinearity only
    % Frequency-domain analysis
    figure('Position',[0 0 900 700]+100)
    semilogx(fpsd,10*log10(PSD_t_residue),'linewidth',3);
    hold on
    if MASH_order == 2
    semilogx(fpsd,10*log10(PSD_PNN_out_INL),'--','linewidth',3)
        ;
    end
    hold off
    grid on
```

```
82       xlim([2e2 inf]);
83       yyaxis left
84       ylim([-450 -250]);
85       ylabel('PSD [dB{\cdot}s^2/Hz]')
86       yyaxis right
87       ylim(10*log10(Fpsd)+[-450 -250]);
88       ylabel('Spur [dB{\cdot}s^2]')
89       set(gca,'YColor','k');
90       legend('Sim','Pred');
91       title('Frequency-Domain Analysis of IBS Spurs (Nonlin Only)')
92       ax=gca;
93       ax.LineWidth=3;
94       ax.FontSize=20;
95   elseif (DTC_quant_ON ==1) && (DTC_nonlin_ON == 1)
96       %% Both nonlinearities
97       % Frequency-domain analysis
98       figure('Position',[0 0 900 700]+100)
99       semilogx(fpsd,10*log10(PSD_t_residue),'linewidth',3);
100      hold on
101      semilogx(fpsd,10*log10(PSD_q_in_raw_DTC_predict),'--','linewidth',3);
102      if MASH_order == 2
103      semilogx(fpsd,10*log10(PSD_PNN_out_INL),'--','linewidth',3);
104      end
105      hold off
106      grid on
107      xlim([2e2 inf]);
108      yyaxis left
109      ylim([-450 -250]);
110      ylabel('PSD [dB{\cdot}s^2/Hz]')
111      yyaxis right
112      ylim(10*log10(Fpsd)+[-450 -250]);
113      ylabel('Spur [dB{\cdot}s^2]')
114      set(gca,'YColor','k');
115      legend('Total Sim','DQI Pred','IBS Pred');
116      title('Frequency-Domain Analysis of DQI & IBS Spurs (Both Nonlins)')
117      ax=gca;
118      ax.LineWidth=3;
119      ax.FontSize=20;
120  end
```

For the conciseness of presentation, in this script, we perform the nonlinear analysis at the DTC block level, namely analyzing the signals as shown in Fig. 6.3. The residual error after DTC cancellation (t_ϵ in Fig. 6.3) is extracted by calculating the difference between the original time-domain injection of the MMDC's AQE (t^{acc}) and the corrective delay produced by the nonlinear DTC ($t_{\mathrm{DTC}}^{\mathrm{NI}}$); this is coded in Line 7.

Nevertheless, the quantitative analysis of these spurs at the DPLL system level, as shown in, e.g., Fig. 6.9a, can easily be performed by transferring the predicted and simulated signals to the output PN domain via the discussed NTF.

The script is multiplexed by the quantization and integral nonlinearity enablers `DTC_quant_ON` and `DTC_nonlin_ON` that are defined in the parameterization stage of the main file, i.e., Lines 28 and 29 of Listing 2.1, in order to produce the corresponding results automatically. When the quantization nonlinearity is enabled for simulation, Line 18 predicts the raw DQI spurs' waveform as per (6.10); when the polynomial INL is enabled, Lines 24–35 calculate the PNN waveform manifesting as the IBS spurs according to the method discussed in Sect. 8.2.1. Here, a MASH 1–1's PNN prediction is shown for example.

Time and frequency-domain comparisons are plotted after Line 37. When both nonlinearities are enabled simultaneously, the script compares the simulated result with both families of spurs predicted by the two prediction methods to show the coexistence and separability of the spurs.

When the methods such as ENOP MMDC and IDQ introduced in this chapter are used, the corresponding spurs should be mitigated accordingly.

As a final remark, the insights and conclusions drawn from the observations of the simulations performed using the DTC-based fractional-N DPLL described in Table 3.3 can be extended to other systems that have different parameterizations. The manifestation of the PN and spurs and the qualitative effects of mitigation by the techniques analyzed in this work do not change significantly with different loop designs. Finally, we present a survey of state-of-the-art DTC enhancement techniques and their effectiveness in Table 8.1. We conclude that one should use (i) a DCD linearity enhancement technique to linearize the DCD's transfer characteristic as much as possible, (ii) a statistical method to gain conventional-fractional-spur immunity, (iii) the DEM method to eliminate REM spurs, and (iv) the IDQ method to eliminate DQI spurs. The DTC so designed can, in principle, recover in fractional-N PLLs insofar as possible the integer-N PN performance. Furthermore, it does not suffer from any of the architecture-dependent nonlinearity-induced fractional spurs studied in this work (see Fig. 8.9c and d).

8.4.3 Optimality of Jitter Minimization for Integer-N DPLLs in DTC-Based Fractional-N Mode

As a final remark, we comment on the jitter-minimized system design yielded by the jitter-minimization-oriented design strategy for integer-N DPLLs proposed in Chap. 5, when the system operates in the DTC-enhanced fractional-N mode.

The analysis in this subsection is based on the DTC-based fractional-N jitter analysis performed in Sect. 7.3.2, which takes into account the impact of the DTC's

Table 8.1 Survey of DTC enhancement techniques mitigating the adverse PN/Spurs studied in this section

DTC enhancement		Target phenomenon			
Class	Technique	Close-In PN	IBS Spurs	DQI Spurs	REM Spurs
DCD linearity enhancement	Adaptive pre-distortion [1, 3, 4], CSC [2, 7, 14], isolated CSC [24], memory-effect suppression [8], DTC range reduction [8, 9, 12, 25], second-/third-order nonlinearity cancellation [10, 11], inverse-constant-slope DTC [13], RCVS DTC [15], pseudo-differential DTC [16], etc.	Mitigate	Mitigate	Ineffective	Ineffective
Stochastic MMDC	Successive requantizer [17–19], PMR DDSM [20], time-invariant-probability modulator [21], PDS $\Delta\Sigma$M [10–12], ENOP DDSM [22], etc.	Ineffective	Mitigate	Ineffective	Ineffective
DTC quantization mitigation	IDQ (Sect. 7.2), Δ-Σ DTC requantizer [8, 25, 39]	Ineffective (Increased but suppressible as per Sect. 7.3)	Ineffective	Eliminate by DQI Mitigate by Δ-Σ	Ineffective
DCD REM mitigation	DEM [23]	Ineffective	Ineffective	Ineffective	Eliminate

residual error. Based on the derived analytical jitter expression, we are motivated to redo jitter minimization like that done in Sect. 5.2 to discover any necessary change that needs to be made to the DPLL in order to minimize the jitter, while accommodating the DTC's noise contribution.

In the DTC-based fractional-N mode, the system jitter is derived as in (7.21). Before performing jitter minimization on this, let us compare the fractional-N expressions, (7.18), (7.20), and (7.21)—representing the input time difference, input jitter power, and RMS input jitter, respectively—with their counterparts in the integer-N mode, namely Eqs. (4.3), (4.7), and (4.13) in Chap. 4.

One can notice that, apart from those terms related to the DTC's QE in the fractional-N expressions—namely the last term of (7.18), the second line of (7.20), and the $\sigma^2_{Q_{\text{DTC}}}$ term in (7.21)—the rest of the expressions are identical to their integer-N counterparts. This means that the DTC *per se* does not change the jitter induced by the intrinsic noises. Furthermore, it is interesting to note that, in (7.18), the last term in relation to the DTC's QE has a form and polarity that are the same as the term related to the reference jitter . This results in the coefficient of the DTC's QE term in (7.20), i.e., $2/(2-X)$, being the same as that of the reference term. It means that, in Eq. (7.21) predicting the RMS system jitter, the jitter power induced by the DTC's QE, $\left(\frac{\Delta}{M}\cdot T_{\text{out}}\right)^2\cdot\sigma^2_{Q_{\text{DTC}}}$, can be lumped together with the reference jitter power, $\sigma^2_{t_{n,\text{ref}}}$.

Now, recall the jitter minimization strategy in Sect. 5.2. It first optimizes the loop dynamics by finding the best DLF proportional gain, $\hat{\beta}$, that zeros the first-order derivative of the system jitter with respect to β, i.e., $\partial\sigma_{\Delta t}/\partial\beta = 0$. Similarly to the integer-N case, it is trivial to show that, when calculating this derivative of (7.21), the real-valued solution can only be found when $\partial\eta/\partial\beta = 0$, where η in the fractional-N case is expressed by (4.14). Note that Eq. (4.14) only comprises the jitter contributed by the TDC and DCO and is the same as its counterpart in the integer-N case. This means that the jitter minimization procedure following the calculation of $\partial\eta/\partial\beta = 0$ is the same as that in Sect. 5.2 following Eq. (5.2). As discussed in that chapter, the subsequent jitter minimization by optimizing the loop dynamics and TDC design will only change the jitter induced by the TDC and DCO and will not alter that due to the reference jitter. In the case of the DTC-based fractional-N DPLL, since the DTC-induced jitter is lumped together with the reference component in (7.21), the former will not affect the following jitter minimization steps for the same reason.

Therefore, we can conclude that taking into account the jitter contributed by the DTC's residual error will not disturb the optimality of the jitter-minimized DPLL design that has been optimized in the integer-N mode, i.e., the optimal integer-N loop design is still optimal for fractional-N.

To verify the optimality of the jitter-minimized DPLL design in the fractional-N mode, we first plot in black in Fig. 8.10 the prediction for $\sigma_{\Delta t}$ as a function of β in the case of the 2-bit-TDC DPLL optimized as per Table 5.2(iii) in the integer-N mode. The constituent RMS jitter components contributed by the sources, namely the TDC QE (the first term in (4.14)), DCO jitter (the second term in (4.14)), reference jitter ($\sigma_{t_{n,\text{ref}}}$ in (7.21)), and DTC QE ($\left(\frac{\Delta}{M}T_{\text{out}}\right)\cdot\sigma_{Q_{\text{DTC}}}$ in (7.21)), are also shown in color in the figure. Clearly, the jitter components jointly determine the predicted trajectory of the overall RMS jitter. The DTC-induced jitter (magenta in Fig. 8.10), similar to the reference one (red), contributes a horizontal jitter floor that is independent of the β value. Hence, it does not affect the value of $\hat{\beta}$ that minimizes the system jitter; the optimized loop design still yields the minimum jitter in the fractional-N mode. The predicted minimum jitter is $\sigma_{\Delta t} = 407.2$ fs, which occurs at $\beta = \hat{\beta} = 25.3$.

Behavioral simulations of the DTC-based DPLL can also be performed to verify quantitatively the optimality of the jitter minimization strategy. Five cases are

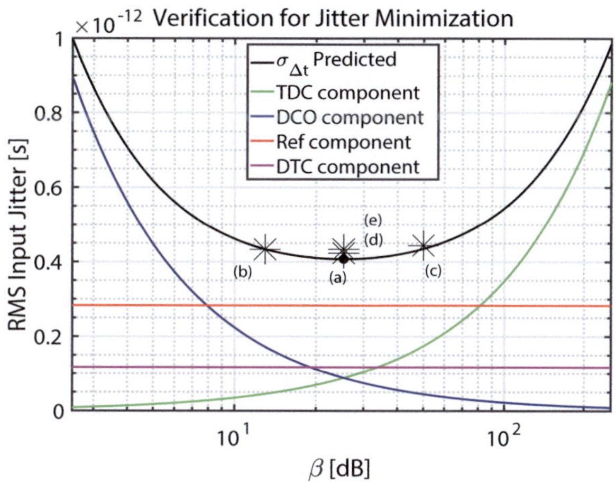

Fig. 8.10 Plot of $\sigma_{\Delta t}$ in (7.21) by sweeping the β value from 2.53 to 253

simulated, namely Case (a) sets the optimized $\hat{\beta}$ and optimized TDC design as per Sect. 5.2, the simulated $\sigma_{\Delta t} = 409.2$ fs is marked by the black dot in Fig. 8.10; Cases (b) and (c) set the optimized TDC and $\beta = 0.5 \cdot \hat{\beta}$ and $2 \cdot \hat{\beta}$, respectively, their simulated $\sigma_{\Delta t} = 433.5$ fs and 443.9 fs; Cases (d) and (e) set the optimized $\hat{\beta}$ and the TDC with resolution ±80% of the optimum, the simulated $\sigma_{\Delta t} = 423.3$ fs and 433.9 fs. Results of the latter four cases are marked by stars in Fig. 8.10. Matched predictions and simulations in Cases (a)–(c) confirm the accuracy of this analysis. The results also confirm that, taking into account the DTC's noise contribution, the jitter-minimized design proposed in Sect. 5.2 is still optimal in fractional-N mode.

Finally, one can also notice that, as long as the DTC's residual error is memoryless, i.e., resulting from memoryless DTC nonlinearities such as element mismatch, requantization, and the DCD's INL, one can update the kth-cycle residual error in the second term in (7.16) and this update will not change the analysis. This implies that the conclusions derived in this subsection (the optimality of jitter-minimized integer-N design in fractional-N mode) can be generalized to other memoryless DTC's residual error cases.

8.5 Conclusion

In this chapter, the effects, in terms of mitigating the excess PN and spurs, of state-of-the-art DCD linearity enhancement techniques, statistical methods used in the MMDC, the DEM method targeting REM spurs, and the IDQ method targeting DQI spurs are studied and compared. A comprehensive DTC enhancement strategy comprising these techniques is described; in principle, this can completely cancel the architecture-dependent spurs discussed in this work and thereby enable fractional-spur-free integer-N PN performance in fractional-N frequency synthesizers. Behavioral simulations at DTC-block and DPLL-system levels underpin and confirm our insights and design guidelines.

8.6 Summary

1. Section 8.2 addresses the polynomial INL associated with the DCD. It analyzes and compares the effectiveness of DCD linearity enhancement techniques and stochastic DDSMs in mitigating conventional IBS spurs induced within the DTC.
2. Section 8.3 addresses the REM INL of the DCD; the DEM method is shown to be able to eliminate REM spurs at source.
3. Section 8.4 proposes a comprehensive DTC enhancement strategy comprising these techniques in order to cancel completely the architecture-dependent spurs discussed in this chapter and thereby to enable fractional-spur-free integer-N level PN performance in fractional-N digital frequency synthesizers.

References

1. S. Levantino, G. Marzin, C. Samori, An adaptive pre-distortion technique to mitigate the DTC nonlinearity in digital PLLs. IEEE J. Solid-State Circuits **49**(8), 1762–1772 (2014)
2. J.Z. Ru, C. Palattella, P. Geraedts, E. Klumperink, B. Nauta, A high-linearity digital-to-time converter technique: constant-slope charging. IEEE J. Solid-State Circuits **50**(6), 1412–1423 (2015)
3. S. Levantino, L. Grimaldi, C. Samori, Analysis of adaptive pre-distortion in DTC-based digital fractional-N PLLs, in *2016 12th Conference on Ph.D. Research in Microelectronics and Electronics (PRIME)*, pp. 1–4 (2016)
4. A. Mahmoud, P. Andreani, F. Pepe, A 2.8–3.8-GHz low-spur DTC-based DPLL with a class-D DCO in 65-nm CMOS. IEEE Microwave Wirel. Compon. Lett. **27**(11), 1010–1012 (2017)
5. H. Liu, D. Tang, Z. Sun, W. Deng, H.C. Ngo, K. Okada, A. Matsuzawa, A 0.98mW fractional-N ADPLL using 10b isolated constant-slope DTC with FOM of –246dB for IoT applications in 65nm CMOS, in *2018 IEEE International Solid-State Circuits Conference—(ISSCC)*, pp. 246–248 (2018)
6. W. Wu, C.-W. Yao, K. Godbole, R. Ni, P.-Y. Chiang, Y. Han, Y. Zuo, A. Verma, I.S.-C. Lu, S.W. Son, T.B. Cho, A 28-nm 75-fs_{rms} analog fractional-N sampling PLL with a highly linear DTC incorporating background DTC gain calibration and reference clock duty cycle correction. IEEE J. Solid-State Circuits **54**(5), 1254–1265 (2019)
7. P. Chen, F. Zhang, Z. Zong, S. Hu, T. Siriburanon, R.B. Staszewski, A 31-μW, 148-fs step, 9-bit capacitor-DAC-based constant-slope digital-to-time converter in 28-nm CMOS. IEEE J. Solid-State Circuits **54**(11), 3075–3085 (2019)
8. A. Santiccioli, M. Mercandelli, L. Bertulessi, A. Parisi, D. Cherniak, A.L. Lacaita, C. Samori, S. Levantino, A 66-fs-rms jitter 12.8-to-15.2-GHz fractional-N bang–bang PLL with digital frequency-error recovery for fast locking. IEEE J. Solid-State Circuits **55**(12), 3349–3361 (2020)

9. W. Wu, C.-W. Yao, C. Guo, P.-Y. Chiang, L. Chen, P.-K. Lau, Z. Bai, S.W. Son, T.B. Cho, A 14-nm ultra-low jitter fractional-N PLL using a DTC range reduction technique and a reconfigurable dual-core VCO. IEEE J. Solid-State Circuits **56**(12), 3756–3767 (2021)
10. H. Park, C. Hwang, T. Seong, Y. Lee, J. Choi, 32.1 a 365fs_{rms}-jitter and -63dBc-fractional spur 5.3GHz-ring-DCO-based fractional-N DPLL using a DTC second/third- order nonlinearity cancelation and a probability-density-shaping $\Delta\Sigma$M, in *2021 IEEE International Solid- State Circuits Conference (ISSCC)*, vol. 64, pp. 442–444 (2021)
11. C. Hwang, H. Park, Y. Lee, T. Seong, J. Choi, A low-jitter and low-fractional-spur ring-DCO-based fractional-N digital PLL using a DTC's second-/third-order nonlinearity cancellation and a probability-density-shaping $\Delta\Sigma$M. IEEE J. Solid-State Circuits **57**(9), 2841–2855 (2022)
12. C. Hwang, H. Park, T. Seong, J. Choi, A 188fs_{rms}-jitter and -243dB-FoM_{jitter} 5.2GHz-ring-DCO-based fractional-N digital PLL with a 1/8 DTC-range-reduction technique using a quadruple-timing-margin phase selector, in *2022 IEEE International Solid- State Circuits Conference (ISSCC)*, vol. 65, pp. 378–380 (2022)
13. S.M. Dartizio, F. Tesolin, G. Castoro, F. Buccoleri, L. Lanzoni, M. Rossoni, D. Cherniak, L. Bertulessi, C. Samori, A.L. Lacaita, S. Levantino, 4.3 a 76.7fs-Integrated-jitter and –71.9dBc in-band fractional-spur bang-bang digital PLL based on an inverse-constant-slope DTC and FCW subtractive dithering, in *2023 IEEE International Solid-State Circuits Conference (ISSCC)*, pp. 3–5 (2023)
14. Y. Liu, H. Gao, H. Xu, P. Lu, N. Yan, A 10-bit 563-fs step constant-slope digital-to-time converter in 40-nm CMOS with nonlinearity cancellation and range extension techniques. IEEE Trans. Circuits Syst. I Regul. Pap. **71**(2), 526–536 (2024)
15. M. Rossoni, S.M. Dartizio, F. Tesolin, G. Castoro, R. Dell'Orto, C. Samori, A.L. Lacaita, S. Levantino, 10.1 an 8.75GHz fractional-N digital PLL with a reverse-concavity variable-slope DTC achieving 57.3fs_{rms} integrated jitter and –252.4dB FoM, in *2024 IEEE International Solid-State Circuits Conference (ISSCC)*, vol. 67, pp. 188–190 (2024)
16. D. Xu, Z. Liu, Y. Kuai, H. Huang, Y. Zhang, Z. Sun, B. Liu, W. Wang, Y. Xiong, J. Qiu, W. Madany, Y. Zhang, A.A. Fadila, A. Shirane, K. Okada, 10.3 a 7GHz digital PLL with cascaded fractional divider and pseudo-differential DTC achieving –62.1dBc fractional spur and 143.7fs integrated jitter, in *2024 IEEE International Solid-State Circuits Conference (ISSCC)*, vol. 67, pp. 192–194 (2024)
17. A. Swaminathan, A. Panigada, E. Masry, I. Galton, A digital requantizer with shaped requantization noise that remains well behaved after nonlinear distortion. IEEE Trans. Signal Process. **55**(11), 5382–5394 (2007)
18. E. Familier, C. Venerus, I. Galton, A class of quantizers with DC-free quantization noise and optimal immunity to nonlinearity-induced spurious tones. IEEE Trans. Signal Process. **61**(17), 4270–4283 (2013)
19. E. Familier, I. Galton, Second and third-order noise shaping digital quantizers for low phase noise and nonlinearity-induced spurious tones in fractional-*N* PLLs. IEEE Trans. Circuits Syst. I Regul. Pap. **63**(6), 836–847 (2016)
20. Y. Donnelly, M.P. Kennedy, J. Breslin, S. Tulisi, S. Patil, C. Curtin, S. Brookes, B. Shelly, P. Griffin, M. Keaveney, 4.48-GHz fractional- *N* frequency synthesizer with spurious-tone suppression via probability mass redistribution. IEEE Solid-State Circuits Lett. **2**(11), 264–267 (2019)
21. T. Seong, Y. Lee, C. Hwang, J. Lee, H. Park, K.J. Lee, J. Choi, A –58dBc-worst-fractional-spur and –234dB-FoM jitter, 5.5GHz ring-DCO-based fractional-N DPLL using a time-invariant-probability modulator, generating a nonlinearity-robust DTC-control word, in *2020 IEEE International Solid- State Circuits Conference—(ISSCC)*, pp. 270–272 (2020)
22. V. Mazzaro, M.P. Kennedy, A family of $\Delta\Sigma$ modulators with high spur immunity and low folded nonlinearity noise when used in fractional-*N* frequency synthesizers. IEEE Trans. Circuits Syst. I Regul. Pap. **69**(11), 4499–4509 (2022)
23. E. Helal, A.I. Eissa, I. Galton, DTC linearization via mismatch-noise cancellation for digital fractional-N PLLs. IEEE Trans. Circuits Syst. I Regul. Pap. **69**(12), 4993–5006 (2022)

24. H. Liu, D. Tang, Z. Sun, W. Deng, H.C. Ngo, K. Okada, A sub-mW fractional-*N* ADPLL with FOM of –246 dB for IoT applications. IEEE J. Solid-State Circuits **53**(12), 3540–3552 (2018)
25. D. Tasca, M. Zanuso, G. Marzin, S. Levantino, C. Samori, A.L. Lacaita, A 2.9–4.0-GHz fractional-N digital PLL with bang-bang phase detector and 560-fs_{rms} integrated jitter at 4.5-mW power. IEEE J. Solid-State Circuits **46**(12), 2745–2758 (2011)
26. A. Santiccioli, C. Samori, A. Lacaita, S. Levantino, Power-jitter trade-off analysis in digital-to-time converters. Electron. Lett. **53**(5), 306–308 (2017). [Online]. https://ietresearch.onlinelibrary.wiley.com/doi/abs/10.1049/el.2016.4577
27. B. Miller, R. Conley, A multiple modulator fractional divider. IEEE Trans. Instrum. Meas. **40**(3), 578–583 (1991)
28. M. Zanuso, S. Levantino, C. Samori, A.L. Lacaita, A wideband 3.6 GHz digital $\Delta\Sigma$ fractional-N PLL with phase interpolation divider and digital spur cancellation. IEEE J. Solid-State Circuits **46**(3), 627–638 (2011)
29. M.P. Kennedy, Y. Donnelly, J. Breslin, S. Tulisi, S. Patil, C. Curtin, S. Brookes, B. Shelly, P. Griffin, M. Keaveney, 4.48GHz 0.18μm SiGe BiCMOS exact-frequency fractional-N frequency synthesizer with spurious-tone suppression yielding a –80dBc in-band fractional spur, in *2019 IEEE International Solid- State Circuits Conference—(ISSCC)*, pp. 272–274 (2019)
30. M.P. Kennedy, V. Mazzaro, S. Tulisi, M. Scully, N. McDermott, J. Breslin, 10.4 a 45.5fs-integrated-random-jitter and –75dBc-integer-boundary-spur BiCMOS fractional-N PLL with suppression of fractional, horn, and wandering spurs, in *2024 IEEE International Solid-State Circuits Conference (ISSCC)*, vol. 67, pp. 194–196 (2024)
31. Y. Donnelly, M.P. Kennedy, Prediction of phase noise and spurs in a nonlinear fractional-*N* frequency synthesizer. IEEE Trans. Circuits Syst. I Regul. Pap. **66**(11), 4108–4121 (2019)
32. V. Mazzaro, M.P. Kennedy, Spur immunity in MASH-based fractional-N CP-PLLs with polynomial nonlinearities. IEEE Trans. Circuits Syst. I Regul. Pap. **68**(6), 2295–2306 (2021)
33. M.P. Kennedy, X. Lu, X. Wang, Insights into architectural spurs in high-performance fractional-*N* frequency synthesizers. IEEE Open J. Solid-State Circuits Soc. **4**, 238–251 (2024)
34. X. Lu, M.P. Kennedy, Reconciliation of statistical approaches to predicting nonlinearity-induced spurs in fractional-*N* frequency synthesizers. IEEE Trans. Circuits Syst. I: Regul. Pap. 1–14 (2024)
35. Y. Hu, W. Tao, R.B. Staszewski, Nonlinearity-induced spur analysis in fractional-N synthesizers with $\Delta\Sigma$ quantization cancellation. IEEE Open J. Solid-State Circuits Soc. 1–1 (2024)
36. K. Hosseini, M.P. Kennedy, Maximum sequence length MASH digital Delta-Sigma modulators. IEEE Trans. Circuits Syst. I Regul. Pap. **54**(12), 2628–2638 (2007)
37. S. Pamarti, J. Welz, I. Galton, Statistics of the quantization noise in 1-bit dithered single-quantizer digital delta–sigma modulators. IEEE Trans. Circuits Syst. I Regul. Pap. **54**(3), 492–503 (2007)
38. V. Mazzaro, M.P. Kennedy, Immunity of ENOP-based fractional-N frequency synthesizer to wandering and horn spurs, in *2022 17th Conference on Ph.D. Research in Microelectronics and Electronics (PRIME)*, pp. 1–4 (2022)
39. Z. Zong, P. Chen, R.B. Staszewski, A low-noise fractional-*N* digital frequency synthesizer with implicit frequency tripling for mm-wave applications. IEEE J. Solid-State Circuits **54**(3), 755–767 (2019)

Chapter 9
Conclusions

The key contributions of this book can be clustered into the following three groups.

1. In Chap. 3, the TDC linearized analysis framework based on MMSE estimation is formulated; this is demonstrated to provide accurate linearized modeling for coarse-resolution TDCs, including the BPD, in various operation modes. Linearized models of such TDCs taking into account the interaction between the TDC's transfer characteristic and the probability distribution of its input signal in different integer-N and fractional-N modes are derived and verified.
2. Chapters 4 and 5 are about linearized analysis and performance optimization at the DPLL-system level. Based on the linearized TDC model, Chap. 4 presents the most accurate linearized DPLL system model to date and provides comprehensive predictions for the integer-N and fractional-N system jitters and PN. Based on the analytical expressions of the predicted jitter and PN, Chap. 5 proposes a jitter-minimization-oriented design strategy that optimizes the DPLL's noise performance whilst using minimum hardware.
3. Chapters 6, 7, and 8 focus on the analysis and enhancement of DTCs that are used in fractional-N DPLLs to cancel the MMDC's QE. Chap. 6 provides DTC macromodels that take into account the DTC's quantization nonlinearity and the polynomial and REM contributions to the INL of its DCD. It reveals the phenomenon of DQI spurs induced by the resolution mismatch between the DTC's control word and the feasible output delay. The distinct PN and spurious-tone spectral signatures caused by different nonlinearities are separated and distinguished. Chap. 7 describes IDQ methods in order to eliminate DQI spurs at source whilst introducing a minimum PN penalty. A comprehensive DTC design methodology is proposed in Chap. 8 that provides immunity against DTC-related spurs. The DTC designed according to this methodology can be used to help fractional-N DPLLs to approach the fractional-spur-free integer-N PN performance limit.

© The Author(s), under exclusive license to Springer Nature Switzerland AG 2026

X. Wang and M. P. Kennedy, *Jitter and Spur Minimization in Fractional-N Digital Frequency Synthesizers*, Analog Circuits and Signal Processing,
https://doi.org/10.1007/978-3-032-27186-0_9

In terms of practical utility, the MATLAB models and scripts that facilitate accurate and efficient behavioral simulation and analysis to demonstrate all the concepts, analysis, and design methods presented in this book are included chapter by chapter, as needed.

MIX
Papier aus verantwortungsvollen Quellen
Paper from responsible sources
FSC® C105338

If you have any concerns about our products,
you can contact us on
ProductSafety@springernature.com

In case Publisher is established outside the EU,
the EU authorized representative is:
Springer Nature Customer Service Center GmbH
Europaplatz 3, 69115 Heidelberg, Germany

Printed by Libri Plureos GmbH
in Hamburg, Germany